Metallurgical Failure Analysis

Metallurgical Failure Analysis

Charlie R. Brooks
Materials Science and Engineering Department
The University of Tennessee
Knoxville, Tennessee

Ashok Choudhury
Oak Ridge National Laboratory
Metals and Ceramics Division
Oak Ridge, Tennessee

McGraw-Hill, Inc.
New York St. Louis San Francisco Auckland Bogotá
Caracas Lisbon London Madrid Mexico Milan
Montreal New Delhi Paris San Juan São Paulo
Singapore Sydney Tokyo Toronto

Library of Congress Cataloging-in-Publication Data

Brooks, Charlie R.,
  Metallurgical failure analysis / Charlie R. Brooks, Ashok
Choudhury.
      p.   cm.
  Includes bibliographical references and index.
  ISBN 0-07-008078-X
  1. Metals—Fracture. 2. Fracture mechanics. I. Choudhury, A.
(Ashok)   II. Title.
  TA460.B755   1993
  620.1'66—dc20   92-24839
                                                                 CIP

Copyright © 1993 by McGraw-Hill, Inc. All rights reserved. Printed in the United States of America. Except as permitted under the United States Copyright Act of 1976, no part of this publication may be reproduced or distributed in any form or by any means, or stored in a database or retrieval system, without the prior written permission of the publisher.

1 2 3 4 5 6 7 8 9 0   DOC/DOC   9 8 7 6 5 4 3 2

ISBN   0-07-008078-X

The sponsoring editor for this book was Robert W. Hauserman, the editing supervisor was Joseph Bertuna, and the production supervisor was Pamela A. Pelton. It was set in Century Schoolbook by McGraw-Hill's Professional Book Group composition unit.

Printed and bound by R. R. Donnelley & Sons Company.

> Information contained in this work has been obtained by McGraw-Hill, Inc., from sources believed to be reliable. However, neither McGraw-Hill nor its authors guarantees the accuracy or completeness of any information published herein and neither McGraw-Hill nor its authors shall be responsible for any errors, omissions, or damages arising out of use of this information. This work is published with the understanding that McGraw-Hill and its authors are supplying information but are not attempting to render engineering or other professional services. If such services are required, the assistance of an appropriate professional should be sought.

Contents

Preface   ix

Chapter 1. Introduction   1

   1.1.   **Objectives**   1
   1.2.   **Approach to Metallurgical Failure Analysis**   2
   1.3.   **Tools of Metallurgical Failure Analysis**   8
       1.3.1. Optical microscopy   8
       1.3.2. Transmission electron microscopy   15
       1.3.3. Scanning electron microscopy   17
       1.3.4. Comparison of OM, TEM, and SEM   22
       1.3.5. Related tools and techniques   23
   1.4.   **Sample Preparation**   31
       1.4.1. Cleaning of surfaces   31
       1.4.2. Preparation of replicas for the TEM   32
       1.4.3. Preparation of samples for the SEM   34
   References   34
   Bibliography   35
   Appendix 1A. Stereomicroscopy   36
   Appendix 1B. Care and Handling of Fractures   42
   Appendix 1C. Preparation and Preservation of Fracture Specimens   43
   Appendix 1D. Cleaning of Fracture Surfaces   57
   Appendix 1E. Examination of Cleaning Techniques for Postfailure Analysis   62
   Appendix 1F. Recommended Cleaning Solutions for Metallic Fractures   71
   Appendix 1G. Scale and Rust Removal Solution   71

Chapter 2. Mechanical Aspects and Macroscopic Fracture-Surface Orientation   73

   2.1.   **Introduction**   73
   2.2.   **Tensile Test**   74
   2.3.   **Principal Stresses**   79

2.4.	Stress Concentration	84
2.5.	Triaxial Stress and Constraint	88
2.6.	Plane Stress	91
2.7.	Plane Strain	94
2.8.	Fracture of Tensile Samples	95
2.9.	Effect of Strain Rate and Temperature	100
2.10.	Crack Propagation	101
2.11.	Meaning of Ductile and Brittle Fracture	105
2.12.	Fracture Mechanics and Failure	107
2.13.	Fatigue Loading	112
2.14.	Creep Deformation	115
References		117
Bibliography		117

Chapter 3. Fracture Mechanisms and Microfractographic Features — 119

3.1.	Introduction	119
3.2.	Slip and Cleavage	120
3.3.	Twinning	127
3.4.	Cleavage Fracture Topography	128
3.5.	Void Coalescence	139
3.6.	Mixed Mechanism and Quasicleavage Fracture	151
3.7.	Tearing Topography Surface	153
3.8.	Intergranular Separation	153
3.9.	Fatigue Fracture Topography	157
3.10.	High-Temperature Fracture Topography	177
3.11.	Environmentally Assisted Fracture	184
3.12.	Flutes	187
3.13.	Wear	189
3.14.	Stereo Examination of Fracture Surfaces	190
3.15.	Comparision of SEM and TEM Fractographs	190
3.16.	Artifacts	200
References		210
Bibliography		211

Chapter 4. Fracture Modes and Macrofractographic Features — 213

4.1.	Introduction	213
4.2.	Tensile Overload	214
4.3.	Torsion Overload	225
4.4.	Bending Overload	227
4.5.	Fatigue Fracture	232
4.6.	Correlation of Micro- and Macrofractographic Features	248
References		268
Bibliography		268

Chapter 5. Case Studies — 271

- 5.1. Introduction — 271
- 5.2. Case A: A Cracked Vacuum Bellows — 271
- 5.3. Case B: Failure of a Large Air-Conditioning Fan Blade — 280
- 5.4. Case C: A Cracked Automobile Flywheel Flex Plate — 296
- 5.5. Case D: Failed Welded Railroad Rails — 304
- 5.6. Case E: Broken Stainless-Steel Wires from an Electrostatic Precipitator — 312
- 5.7. Case F: Broken Wire Cutters — 319
- 5.8. Case G: Broken Steel Punch — 328
- 5.9. Case H: Broken Stainless-Steel Hinge for a Check Valve — 334
- References — 348
- Bibliography — 349

Appendix A. Temperature Conversions — 351

Appendix B. Metric Conversion Factors — 357

Appendix C. Converting Common Units from the English to the Metric (SI) System — 359

Appendix D. Rockwell C and B Hardness Numbers for Steel — 361

Appendix E. The Relations Between ASTM Grain Size and Average Grain "Diameter" — 365

Appendix F. Comments on Magnification Markers — 367

Glossary 369
Index 401

Preface

Despite the care taken in their design, installation, and operation, machine and structural components fail. Their failure is such a common occurrence that failure analysis remains an extremely important subject. In general, the main justification for performing a failure analysis lies in the application of the knowledge gained about the failure for the minimization of future problems, and also in the prominent role such analysis usually plays in litigation subsequent to failures.

A treatise on failure analysis can be approached by first covering the general aspects of the failure of materials, then examining the failure of specific classes of materials, such as metallic, polymeric, ceramic, or electronic. However, the majority of machine and structural components are made of metallic materials, and thus we have chosen to limit this book to the failure analysis of metals. Also, since most failures involve the breakage of components, the main thrust of the book is the analysis of fracture failures.

There are many books and special tracts which treat failure analysis of materials, and metallic materials specifically. Most of these cover the results of research related to failures, but these are introductory books. However, we felt that there was no proper balance between introductory material and examples of fracture-surface appearances. The introductory books were weak on fractographs, and handbooks, such as the *Metals Handbooks*, were designed for those already having a rather detailed understanding of metallurgical failure analysis. Thus, in our book we have placed emphasis on the appearance of fracture surfaces, and therefore have included copious fractographs to illustrate the points made in the text. The scope, however, is limited to an introductory range so as not to get bogged down in too much detail. The introductory nature of the book is designed for metallurgists and materials scientists and engineers who are novices in the subject, as well as for those working in this field who desire a refresher. In addition, it will be useful to engineers and scientists who encounter the subject as a peripheral, but important, aspect of their work. It is, furthermore, a useful reference book—for example, as a quick source of definitions and examples of fractographic terms. Also, the book will be helpful in design courses for senior undergraduate and graduate students in materials engineering and mechanical engineering.

In Chapter 1, following an introduction to the methodology of failure analysis, we present a brief treatment of the common tools used as an aid in understanding how the experimental information is obtained, and the advantages, disadvantages, and limitations of them. Included in this chapter is a treatment of sample preparation. In Chapter 2, the relation of the external loading of components to the macroscopic fracture-surface orientation is treated. This is followed, in Chapter 3, by a treatment of the microscopic fracture mechanisms and the common microfractographic features associated with known loading conditions. Chapter 4 covers the macroscopic fracture-surface appearances associated with known loading conditions. Both the macro- and the microfractographic aspects of fracture are illustrated by examples. Finally, several case studies are described in detail with reference to information and fractographs in the preceding chapters to illustrate the use of the introductory concepts.

In the text, we have usually given values of quantities in both U.S. customary and SI units. Commonly used units, such as centimeters, inches, angstroms, and micrometers have been retained. Data in figures are given in the units of the source from which they were taken. However, we have supplied convenient conversion tables in the appendixes.

Understanding the physical metallurgy of metallic materials can be central to a successful metallurgical failure analysis. However, the physical metallurgy involved is too complicated to be reviewed in this book, which is only an introduction to metallurgical failure analysis. There exist many books and articles which can be consulted for information about the metallurgy of specific metals and alloys.

Finally, a word about definitions: A glossary of terms follows Appendix F. Terms not included there can be found by consulting the index. The common terms used in metallurgical failure analysis are defined and described in the text.

Acknowledgments

C. R. Brooks thanks former students David Dellinger, Hwa-Perng Kao, Clayton Crouse, B. D. Cutler, and Brian Cruse, who made the preliminary failure analyses of some of the case studies presented in Chapter 5. Their names are listed with the appropriate case studies. We express our appreciation to the authors and publishers for allowing us to use information from their work and sources, which are cited in the text. Especially we thank ASM International for permitting the extensive use of information from various *Metals Handbooks*. We thank Dr. William T. Becker for critiquing Chapter 2 and making useful suggestions for its improvement; Sue Brooks for proofreading some of the manuscript; and especially Sue Turner for excellent typing (and retyping) of the manuscript. We also thank our wives, Sue Brooks and Kim Choudhury, for their patience.

Charlie R. Brooks
Ashok Choudhury

Metallurgical Failure Analysis

Chapter 1

Introduction

The general conclusion is this—Frost does not make either iron (cast or wrought) or steel brittle, and accidents arise from neglect to submit wheels, axles, and all other parts of the rolling stock to a practical and sufficient test before using them.

JAMES PRESCOTT JOULE
Philosophical Magazine, 1871

1.1 Objectives

The analysis of failures of metallic components is an extremely important aspect of engineering. Establishing the causes of failures provides information for improvements in design, operating procedures, and the use of components. Also, determining the cause of a failure can play a pivotal role in establishing liability in litigation. Failure analysis is often difficult and frustrating, but understanding how to approach an analysis and how to interpret observations provides a basis for assuring meaningful results.

The objective of this book is to introduce the important aspects associated with the failure analysis of metallic components. Emphasis is placed on the analysis of broken components, where observations of the fracture surface play a key role. Thus a treatment of both macroscopic and microscopic observations of fracture surfaces is given. Since loading conditions are often an important aspect of the possible causes of failures, a simplified treatment of the mechanics involved is presented. It is to be noted that some information about prior loading conditions can often be gleaned from a careful observation of the general macroscopic orientation of the fracture surface. Also included is a section which reviews the common experimental methods used in metallurgical failure analysis. Finally, some case studies of metallurgical failure analyses are introduced, and the approaches taken are related to the information presented in preceding chapters.

1.2 Approach to Metallurgical Failure Analysis

Metallurgical failure analysis deals with the determination of the causes of the failure of metallic parts or components. In the broad, and correct, sense, *failure* can be defined as the inability of a component to function properly, and this definition does not imply fracture. *Failure analysis* can be defined as the examination of a failed component and of the failure situation in order to determine the causes of the failure. The purpose of a failure analysis is to define the mechanism and causes of the failure and usually to recommend a solution to the problem.

The causes of failures can be broken down into the following categories:

1. *Misuse:* The component is placed under conditions for which it was not designed. This is a common cause of failure, and its establishment sometimes relies on determining that the assembly of the component and the design were correct, leaving misuse as a suspected cause.
2. *Assembly errors and improper maintenance:* Assembly errors involve such factors as leaving off a bolt or using incorrect lubricant. Maintenance of equipment ranges from painting surfaces to cleaning and lubrication, and its neglect may lead to failure. It is also pointed out that a failure may be caused by some other part of the system not functioning properly, thereby placing the component which failed under conditions for which it was not designed. Thus failure of a component may point to a problem elsewhere in the system.
3. *Design errors:* This is a very common cause of failure. In this category the following items are considered to be specified by the design process:
    a. Size and shape of the part. This is usually determined by stress analysis or geometric constraints.
    b. Material. This refers to the chemical composition *and* the treatment (for example, heat treatment) necessary to achieve the required properties.
    c. Properties. This is related to stress analysis, but other properties such as corrosion resistance must also be considered.

It is interesting to examine some information about the causes of failures and compare it to the preceding list.

TABLE 1.1 Frequency of Causes of Failure in Some Engineering Industry Investigations

Origin	%
Improper material selection	38
Fabrication defects	15
Faulty heat treatments	15
Mechanical design fault	11
Unforeseen operating conditions	8
Inadequate environment control	6
Improper or lack of inspection and quality control	5
Material mixup	2

SOURCE: Adapted from Davies.[1]

1. *Improper material selection:* Table 1.1 shows that improper material selection is a common problem.
2. *Improper maintenance:* The data in Table 1.2 show that improper maintenance is the main problem in failed aircraft components.
3. *Faulty design considerations:* Causes of failures due to faulty design considerations or misapplication of material include the following (adapted from Dolan[2]):
    a. Ductile failure (excess deformation, elastic or plastic; tearing or shear fracture)
    b. Brittle fracture (from flaw or stress raiser of critical size)
    c. Fatigue failure (load cycling, strain cycling, thermal cycling, corrosion fatigue, rolling contact fatigue, fretting fatigue)
    d. High-temperature failure (creep, oxidation, local melting, warping)
    e. Static delayed fractures (hydrogen embrittlement, caustic embrittlement, environmentally stimulated slow growth of flaws)
    f. Excessively severe stress raisers inherent in the design

TABLE 1.2 Frequency of Causes of Failure of Aircraft Components (Laboratory Data)

Origin	%
Improper maintenance	44
Fabrication defects	17
Design deficiencies	16
Abnormal service damage	10
Defective material	7
Undetermined cause	6

SOURCE: Adapted from Davies.[1]

g. Inadequate stress analysis, or impossibility of a rational stress calculation in a complex part
h. Mistake in designing on the basis of static tensile properties, instead of the significant material properties that measure the resistance of the material to each possible failure mode

4. *Faulty processing:* Causes of failures due to faulty processing include the following (adapted from Dolan[2]):
   a. Flaws due to faulty composition (inclusions, embrittling impurities, wrong material)
   b. Defects originating in ingot making and casting (segregation, unsoundness, porosity, pipes, nonmetallic inclusions)
   c. Defects due to working (laps, seams, shatter cracks, hot-short splits, delamination, excess local plastic deformation)
   d. Irregularities and mistakes due to machining, grinding, or stamping (gouges, burns, tearing, fins, cracks, embrittlement)
   e. Defects due to welding (porosity, undercuts, cracks, residual stress, lack of penetration, underbead cracking, heat-affected zone)
   f. Abnormalities due to heat treating (overheating, burning, quench cracking, grain growth, excessive retained austenite, decarburization, precipitation)
   g. Flaws due to case hardening (intergranular carbides, soft core, wrong heat cycles)
   h. Careless assembly (such as mismatch of mating parts, entrained dirt or abrasive, residual stress, gouges or injury to parts)
   i. Parting-line failures in forging due to poor transverse properties

5. *Deterioration in service:* Causes of failures due to deterioration during service conditions include the following (adapted from Dolan[2]):
   a. Overload or unforeseen loading conditions
   b. Wear (erosion, galling, seizing, gouging, cavitation)
   c. Corrosion (including chemical attack, stress corrosion, corrosion fatigue, dezincification, graphitization of cast iron, contamination by atmosphere)
   d. Inadequate or misdirected maintenance or improper repair (such as welding, grinding, punching holes, cold straightening)
   e. Disintegration due to chemical attack or attack by liquid metals or platings at elevated temperatures
   f. Radiation damage (sometimes must decontaminate for examination, which may destroy vital evidence of cause of failure); varies with time, temperature, environment, and dosage
   g. Accidental conditions (such as abnormal operating temperatures, severe vibration, sonic vibrations, impact or unforeseen collisions, ablation, thermal shock)

Most metallurgical failures involve fracture of the component, and thus most failure analyses involve examination of the mechanical loading situation. In this book, the term *mode* of fracture will be used to reflect the type of loading involved, such as tensile overload, fatigue, or creep, and is covered in Chapter 4. The term fracture *mechanism* will be used to define the type of microscopic process whereby the material fractured. This refers to processes such as cleavage or void coalesence, which are covered in detail in Chapter 3. The frequencies of the various types of fracture modes which have been identified are illustrated by the data in Tables 1.3 and 1.4.

The steps involved in conducting a failure analysis and their sequence depend upon the failure. One sequence to do this includes the following eight steps:

1. *Description of the failure situation:* Here the history of the failure should be documented. Any information pertaining to the failure, such as the design of the component (including the material and properties), and how the component was being used, is important to obtain. Especially useful are photographs of the part and of associated components.

2. *Visual examination:* Here the general appearance of the part should be documented. Care should be exercised in handling the part so as not to damage any of the fracture surfaces or other important features.

3. *Mechanical design analysis (stress analysis):* When the part clearly involved mechanical design as a major design component, a stress analysis should be carried out. This will help to establish whether the part was of sufficient size and of proper shape, and what mechanical properties were required. In some cases this analysis may establish the cause of failure. For example, if the load on a part can be determined and estimates of the mechanical properties made, then it may be possible to establish that the part is too small for this load.

TABLE 1.3 Frequency of Causes of Failure in Some Engineering Industry Investigations

Cause	%
Corrosion	29
Fatigue	25
Brittle fracture	16
Overload	11
High-temperature corrosion	7
Stress corrosion/corrosion fatigue/hydrogen embrittlement	6
Creep	3
Wear, abrasion, and erosion	3

SOURCE: Adapted from Davies.[1]

TABLE 1.4 Frequency of Causes of Failure of Aircraft Components

Cause	%
Fatigue	61
Overload	18
Stress corrosion	8
Excessive wear	7
Corrosion	3
High-temperature oxidation	2
Stress rupture	1

SOURCE: Adapted from Davies.[1]

4. *Chemical design analysis:* This step refers to an examination of the suitability of the material from the standpoint of corrosion resistance.
5. *Fractography:* Examination of the fracture surface with the unaided eye, with optical microscopes, and with electron microscopes should be carried out in order to establish the mechanism of fracture.
6. *Metallographic examination:* This requires sectioning and metallographic preparation. It may require agreement between all parties involved before sectioning. This step will help to establish such facts as whether the part had the correct heat treatment.
7. *Properties:* The properties pertinent to the design should be determined. This is not always possible because the test to determine a property may destroy the part. In terms of mechanical properties, hardness is especially important. Hardness will frequently correlate with many other mechanical properties (such as yield strength). It is a simple test to perform, and it usually will not damage the part.
8. *Failure simulation:* A very useful approach is to take an identical (supposedly) part and subject it to the exact condition under which it is designed to operate. This may be too expensive to carry out and is not done frequently.

An alternative procedure for metallurgical failure analysis as adapted from Ryder et al.[3] follows these steps:

1. Collection of background data and selection of samples
2. Preliminary examination of the failed part (visual examination and record keeping)
3. Nondestructive testing
4. Mechanical testing (including hardness and toughness testing)

5. Selection, identification, preservation, and/or cleaning of all specimens
6. Macroscopic examination and analysis (fracture surfaces, secondary cracks, and other surface phenomena)
7. Microscopic examination and analysis
8. Selection and preparation of metallographic sections
9. Examination and analysis of metallographic sections
10. Determination of failure mechanism
11. Chemical analyses (bulk, local, surface corrosion products, deposits or coatings, and microprobe analysis) ·
12. Analysis of fracture mechanics
13. Testing under simulated service conditions (special tests)
14. Analysis of all the evidence, formulation of conclusions, and writing the report (including recommendations)

The importance of obtaining background information cannot be overemphasized. Also, prior to the physical destruction of any broken components and the assembly of which they are a part, it is important to document, usually in pictures, their external features. In cases that may involve litigation it is recommended that all parties involved agree on any testing in which physical destruction may occur. Determination of the mechanical properties frequently plays a prominent role in failure analysis, and these properties can often be estimated from hardness measurements. This procedure is essentially nondestructive. Thus consideration of hardness measurements usually occurs as an early step.

A critical step is examination of the fracture surface. This usually is best made by sectioning the broken component for ease of handling. However, it is possible to reproduce the fracture surface topology by preparing a replica and making the observations on the replica. This can be very useful, and perhaps necessary, since it is nondestructive. Another critical step is usually microstructural analysis, which gives information about the processing and properties of the material. However, this nearly always requires sectioning.

A very useful step is failure simulation. Here a part identical to the one that broke is subjected to an opération that simulates what occurred during service. However, such a simulation may be too difficult or expensive to conduct.

Another feature of failure analysis is interaction with people. It is important to be able to obtain information from those involved in the failure, and to be able to interpret their statements correctly.

The type of background information that may be required for a fail-

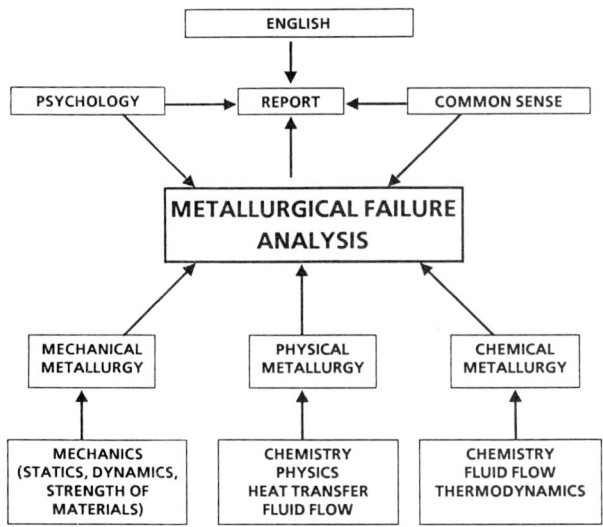

Figure 1.1 Disciplines and subjects involved in metallurgical failure analysis.

ure analysis is depicted in Fig. 1.1. The results of an analysis usually culminate in a report, which contains the findings and recommendations. Thus careful writing is very important.

1.3 Tools of Metallurgical Failure Analysis

Characterization of the microstructure and of the fracture surface topology plays a prominent role in metallurgical failure analysis. The most common tools for this are the eye and the optical microscope, the scanning electron microscope, and the transmission electron microscope. Their utilization is reviewed in this section. Less commonly used techniques are covered in the volume in the 9th edition of the *Metals Handbook* on materials characterization.[4]

1.3.1 Optical microscopy

In an optical microscope (OM) utilized in fractography (and in metallography), a light source passes through an objective lens which, if at the proper distance from the surface, will form an image of the surface. To further magnify this image, this light passes through another lens (the eyepiece) and is then focused on the retina of the eye. The general scheme for an inverted-stage metallurgical microscope is shown in Fig. 1.2. The light from the source passes through a condenser lens to collimate the beam. There is also an adjustable aper-

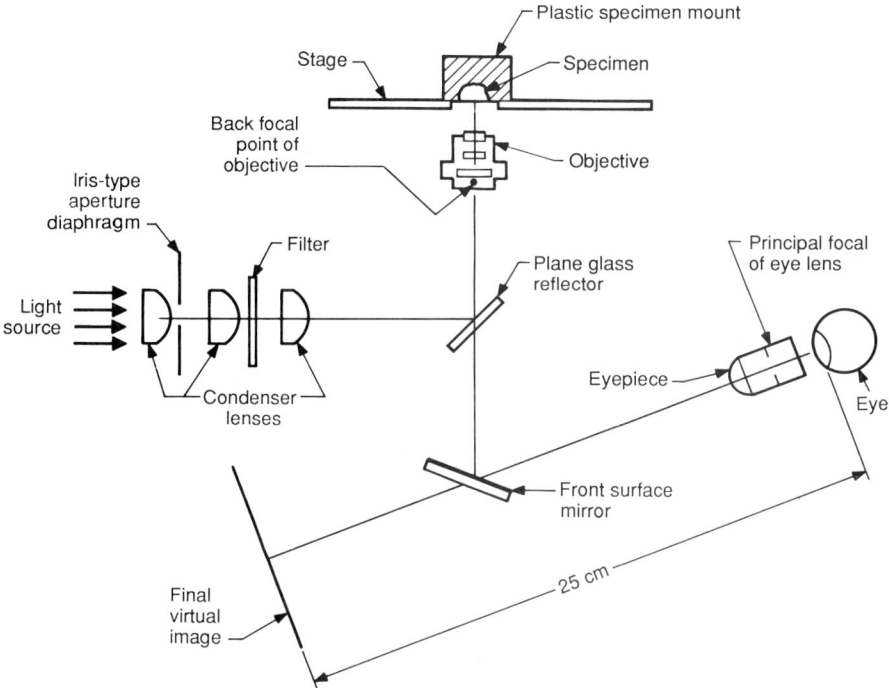

Figure 1.2 Schematic diagram of the optical system typical of that of an inverted-stage metallurgical microscope.

ture to control the intensity and to limit off-axis rays. A filter can be placed in the beam to create a monochromatic beam. The light impinges upon a plane glass reflector, is reflected up through the objective lens to the specimen surface, and is then reflected back, forming an image. These rays pass through the plane glass reflector onto a front surface mirror, to the eyepiece lens, and then to the eye.

The angle of refraction depends on the wavelength of the light. Thus two rays of different wavelengths, coming in the same direction from the same point on an object, will refract differently when striking the surface of the lens. This leads to a focusing error called *chromatic aberration*. To avoid it, a filter can be placed in the beam to generate a monochromatic beam.

Another lens error is *spherical aberration*. If a lens were ideal, two rays emitted at different angles from an object, upon passing through the lens, should image at the same point. However, the correct angle of refraction will be attained only for the correct surface geometry. Most lenses are ground to an approximately spherical surface, and this is not the correct geometry. Thus the two rays starting from the same point will not focus at the same point. This error can be minimized by placing an adjustable aperture near the objective lens to limit off-axis rays. The smaller the aperture, the less off-axis rays there are to contribute to the image, and the sharper the image. However, as the aperture becomes smaller, the light intensity decreases; hence there is a lower limit on the aperture size. More importantly, those rays from the object which strike the edge of the aperture diffract and generate a wavefront. This front, at the image plane, interacts with the waves which pass through the aperture opening without striking the edge, and this creates an interference.

If the aperture is sufficiently wide, the lack of sharpness of the image caused by the off-axis rays will mask the lack of sharpness from this diffraction effect (see Fig. 1.3). But upon decreasing the aperture size, the diffraction effect becomes more important. Thus there is an optimum aperture size to use.

This limitation imposed by the diffraction effect is quite important, since it essentially determines how fine the detail can be on a surface and yet still be discerned. That is, if on the surface there are two point protrusions too close together, they will image as one large spot instead of two distinct spots. The minimum separation distance between two points that can be allowed and yet obtain separate images is referred to as the *resolving power* of the microscope. The ability to resolve fine detail is referred to as *resolution,* and the resolving power is a measure of resolution. Note that the resolving power is a distance, and to see (or resolve) fine detail, the resolving power should be small. The resolving power (RP) depends on some optical parameters, and the mathematical relation can be stated as

$$RP \cong \frac{0.6\lambda}{2NA} \qquad (1.1)$$

where λ is the wavelength of the light being used. NA is the numerical aperture, which is given by

$$NA = \eta \sin \mu \qquad (1.2)$$

where η is the index of refraction of the medium between the object and the objective lens and μ is the light-gathering angle (Fig. 1.4).

Note that there are three methods of improving resolution (reducing resolving power). One is to use light with a smaller wavelength. This is limited by the need to use visible light and by the fact that most lenses are corrected for chromatic aberration at a single wavelength. Another method is to increase the index of refraction of the medium between the object and the objective lens. This medium is normally air, with an index of essentially 1.0. An oil-immersion lens, with an oil having an index of refraction of up to 1.4, can be used. This will

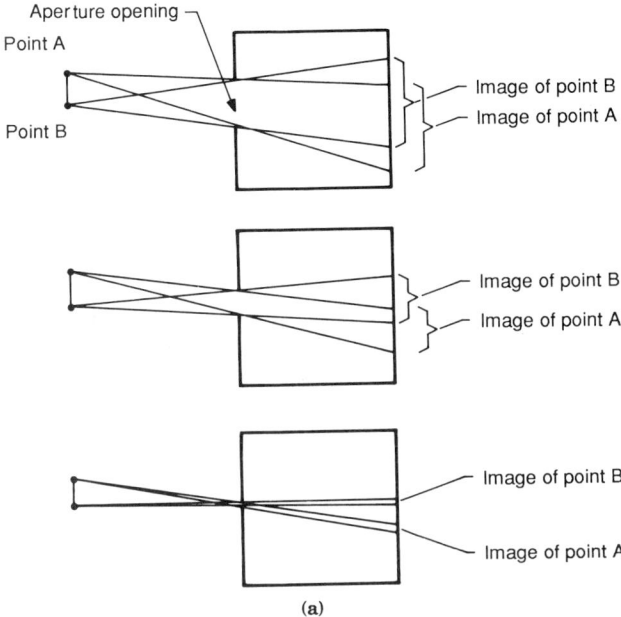

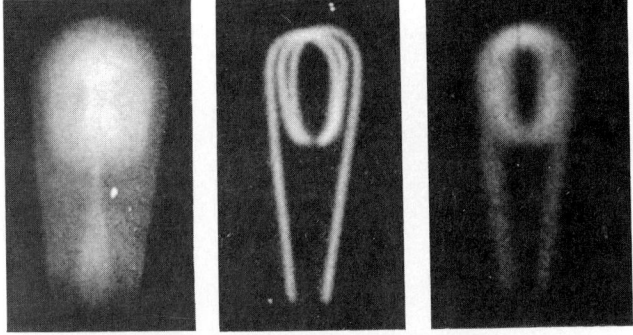

Figure 1.3 (a) Schematic illustration of the effect of reduction of aperture size on image sharpness. However, at too small an aperture opening, diffraction from the edges limits the resolution. (b) Pinhole photographs of incandescent filament, where image sharpness improves, then deteriorates. (*From Ruechardt.*[5])

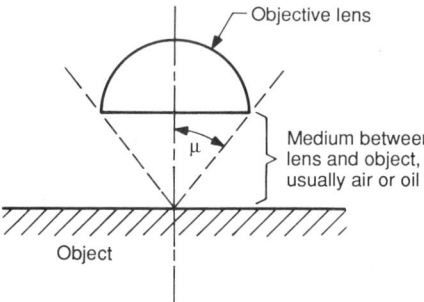

Figure 1.4 Schematic diagram defining the angle μ used in the definition of numerical aperture.

improve the resolving power by about 40 percent. Fluids with higher indices of refraction are not suitable; for one thing, the transmission of light is reduced to an unusable level. Another method is to increase the light-gathering angle. This is accomplished by the design of the lens. Note that this is limited, as the angle cannot exceed 90° (Fig. 1.4).

Figure 1.5 illustrates vividly the meaning of resolution. The two photographs are at the same magnification. [This can be calculated by measuring the bar at the bottom of the pictures, dividing by 10 micrometers (μm), and multiplying by 10^{-4} (to convert micrometers to centimeters); see Appendix F.] However, different objective lenses were used to obtain the pictures. The lens with the numerical aper-

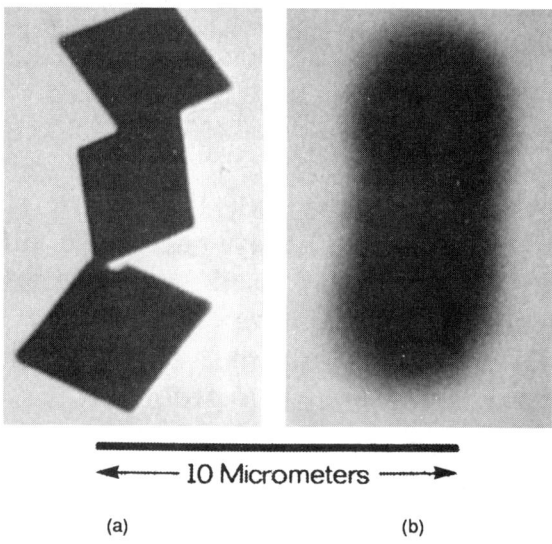

Figure 1.5 Photomicrograph of silver chromate crystals taken with objectives of different numerical apertures. Wavelength was 546 nm (green light). (a) NA = 1.4. (b) NA = 0.1. (*From Dyer.*[6])

ture of 0.1 gave the picture in Fig. 1.5b, which appears to be the image of an oblong object. However, using an objective lens with a numerical aperture of 1.4 gives the picture in Fig. 1.5a. The same object now appears to be three separate rhomboidal objects. Using Eq. (1.1), the resolving power for the picture on the right is 1.6 μm, for that on the left it is 0.1 μm.

The example just described emphasizes that the most important characteristic of microscopy is resolution. Now, how does magnification enter into the picture? The human eye has a resolution of about 0.01 cm at a distance of 25 cm, which is the comfortable reading distance. At this distance, detail closer together than 0.01 cm on an object cannot be resolved by the eye. (This is why OMs are designed so that the virtual image is at about 25 cm; see Fig. 1.2.) The eye can see finer detail than this only by magnification. For example, consider Table 1.5, which lists the eyepieces and objective lenses for a typical metallurgical microscope. The resolving power was calculated from Eq. (1.1) using the wavelength of green light [5460 angstoms (Å)]. If an object being observed has detail separated by 10,000 Å (1 μm), then using the 5× objective will not resolve this detail, no matter what the magnification. However, using the 10× objective should allow imaging the detail, as the resolving power of this lens is 5460 Å, that is, less than 10,000 Å. If a 20× eyepiece is used with this 10× objective lens, the total magnification is 200×, and the detail which is separated by 10,000 Å will appear to the eye to be separated by 10,000 × 200 = 2,000,000 Å, or 0.02 cm. This value is greater than the resolving power of the eye, and the detail will be resolved.

Thus we see that microscope lens systems must be able to magnify the resolvable detail to about 0.01 cm for the eye to see it. In optical

TABLE 1.5 Magnification and Resolving Power (Using a Wavelength of 5460 Å) of a Set of Lenses Typical of That for a Metallurgical Microscope

Eyepiece	Objective	Total magnification	Resolving power, Å
5×	5× (NA* = 0.10)	25×	16,380
10×	5×	50×	
15×	5×	75×	
20×	5×	100×	
5×	10× (NA = 0.30)	50×	5460
10×	10×	100×	
15×	10×	150×	
20×	10×	200×	
5×	40× (NA = 0.65)	200×	2520
10×	40×	400×	
15×	40×	600×	
20×	40×	800×	

*NA—numerical aperture.

microscopy, the wavelength is approximately fixed (it must remain in the visible range), the numerical aperture is limited by the angle µ, which cannot be greater than 90° (see Fig. 1.4), and the index of refraction is limited to about 1.4. Therefore there is an approximate limit on the resolving power, and there is no need to provide a lens system that is capable of greater magnification than that required to allow the eye to see the finest detail resolvable. This magnification is approximately 1500×, and that is why OMs, even with photographic bellows attachments, do not magnify above approximately 3000×.

We have seen that one of the most important characteristics in microscopy is resolution, and we have seen how magnification is related to resolution. In fractography, another important characteristic is *depth of field*. This term refers to the allowable variation of the height of the object surface topology and yet retain common focus on the image plane (that is, objects at different levels should be focused on the same plane). This range is related to the aperture opening and to the wavelength of light. The way the aperture opening enters into the picture is shown clearly in Fig. 1.6. It appears that the smallest aperture, consistent with sufficient intensity to see the image, would be the optimum aperture. However, again the diffraction effect from the edge of the aperture must be considered, and this is why the wavelength affects the depth of field. The smaller the wavelength, the less this interference effect, and the greater the depth of field.

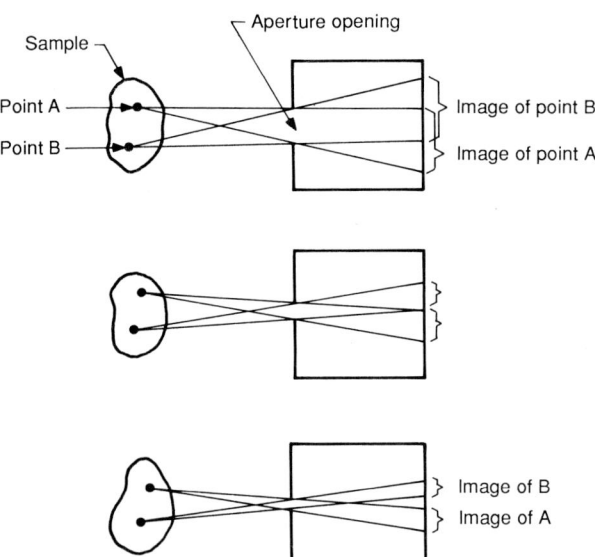

Figure 1.6 Effect of aperture opening on image sharpness from two points located at different levels in the sample.

To place the importance of depth of field in perspective, consider using an OM to observe detail at the limit of resolution of the objective lens, which we will assume to be about 2000 Å (see Table 1.5). In this case the depth of field is about 4000 Å, or 0.4 μm. Thus if this microscope were being used to observe a fracture surface, any irregularities varying in height more than 0.4 μm would not be in simultaneous focus. If the lens has a resolving power of about 16,000 Å [a lens one might use for low-magnification (such as 50×) observation], the depth of field is about 10 μm. For the unaided eye, the depth of field is about 0.01 cm.

1.3.2 Transmission electron microscopy

The operation of the transmission electron microscope (TEM) is very similar in principle to that of the OM. Figure 1.7 compares the general optical path for the two instruments. There are, of course, important practical differences between the two microscopes. By definition, the sample is viewed in transmission, so that it must be sufficiently thin for the beam to penetrate completely. This creates a problem of sample preparation (see Sec. 1.4.2). Due to the ease of scattering of electrons by air, the sample must be in a vacuum, although this is not a serious practical problem. The area that can be viewed is small, so that the examination of a large sample is difficult and time-consuming. Since the eye cannot see electrons, the image is generated by the action of the electron beam upon a phosphorescent screen, on which the observations are made.

The disadvantages of the TEM are far outweighed by its advantages, so that it has developed into a widely used tool in failure analysis. The principal advantages of the TEM compared to the OM are caused by the shorter wavelength of the electron compared to visible light. Equation (1.1), used to calculate resolving power, clearly shows that the resolution can be improved by lowering the wavelength. The wavelength of visible light is around 5000 Å, whereas that typical of electrons used in electron microscopy is about 0.04 Å. This should give an improvement in resolution by a factor of about 10^5. However, to minimize spherical aberration, small apertures are used, so that the angle μ [Eq. (1.2)] is quite small, on the order of 0.1°. Although this will degrade the resolution, the smaller wavelength of the electrons is the dominant factor, and the resolution of the TEM is considerably better than that of the optical microscope. It is this better resolution that makes the TEM so widely used.

Using Eq. (1.1), the resolution of the TEM will be found to be about 10 Å, compared to about 1500 Å for the OM. Now we can see why a TEM has a projector lens system which can magnify so greatly. To be

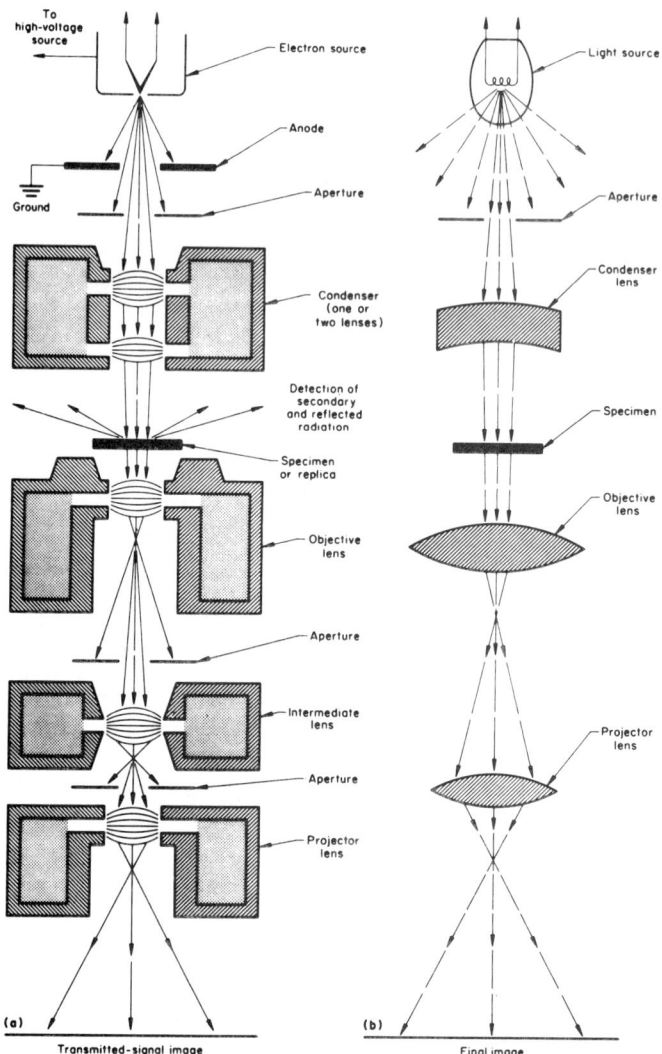

Figure 1.7 Comparison of the optical system for a transmission electron microscope and an optical microscope. (*From* Metals Handbook.[7])

able to see with the eye a detail of 10 Å, the magnification required to raise this to 0.01 cm is 100,000×. Thus the high magnification capability associated with TEM is required to magnify the fine detail so that the eye can observe it.

Also the depth of field is improved greatly by using electrons. For example, if the resolving power is 10 Å, typically the depth of field will be about 5000 Å. This is generally the maximum thickness allowable, which will transmit sufficient electrons to obtain an image. This means that the entire sample will be in focus. If the TEM is operated so that the resolving power is reduced to 1000 Å (such as by using larger apertures), then the depth of field increases to about 500,000 Å (0.005 cm).

Thus we see that, compared to the OM, the TEM offers greatly improved resolution and depth of field. The two major disadvantages are sample preparation problems and the small-size sample that must be used, making the area of observation limited. Actually, the TEM and the OM are complementary tools, both of which are used in fractography.

1.3.3 Scanning electron microscopy

The scanning electron microscope (SEM) operates somewhat differently from the TEM. The electron beam is collimated by the condenser lenses, then focused by the objective lens into a small-diameter beam. This beam strikes the surface of the sample, and the interaction of the beam with the sample generates emitted electrons whose quantity is especially sensitive to the surface topography. A scanning coil causes the beam to raster the surface of the specimen. The quantity of electrons emitted from any point on the surface controls the intensity of a synchronized cathode-ray tube (CRT) display. Hence as the electron beam rasters the surface, an image is generated on the CRT which is essentially a picture of the surface. Figure 1.8 compares the optical systems of the OM, TEM, and SEM.

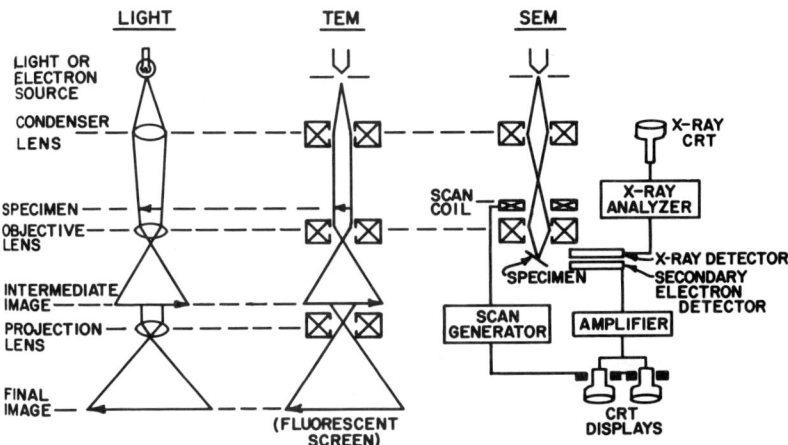

Figure 1.8 Comparison of the optical paths of the optical light microscope, the transmission electron microscope, and the scanning electron microscope. (*From Becker.*[8])

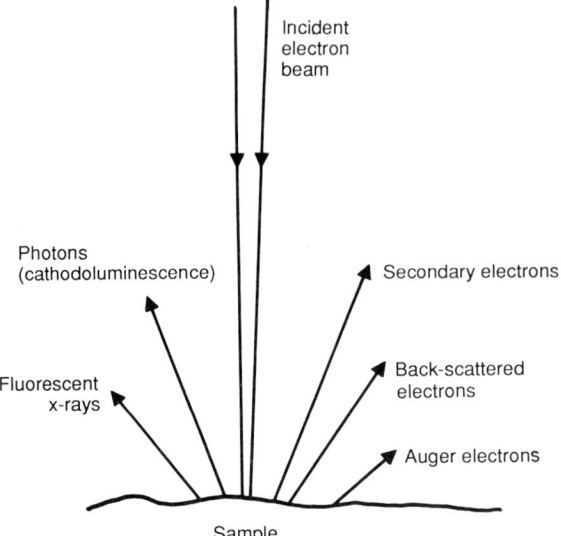

Figure 1.9 Types of signals generated by the interaction of an electron beam with a specimen. Back-scattered and secondary electrons are used to obtain an image of the surface in the scanning electron microscope. The Auger electrons are very weak and are emitted from the atoms in the first two or three layers of the surface; thus they are used for surface analysis. However, an ultrahigh vacuum system is required to prevent significant electron absorption. The fluorescent x-rays are used to obtain chemical analysis (see Sec. 1.3.5).

The interaction of the electrons with the specimen generates several types of signals (Fig. 1.9). The electrons emitted from the sample have varied energies, as shown by the typical spectrum in Fig. 1.10. Of particular interest here is the high concentration of electrons with relatively low energies and with high energies near that of the incident electron beam. When the beam penetrates the sample surface, some electrons are scattered elastically back out of the sample, retaining approximately their original energy. These are referred to as back-scattered electrons (BSE). Other electrons interact with the electrons of the atoms of the specimen and have their energy changed. Most of these lose considerable energy and are emitted with relatively low energy; these are secondary electrons (SE). The electrons penetrate the sample in a tear-drop-shaped volume (Fig. 1.11), and only those within about 1000 Å of the surface escape. The diameter of this region is larger than that of the beam striking the surface.

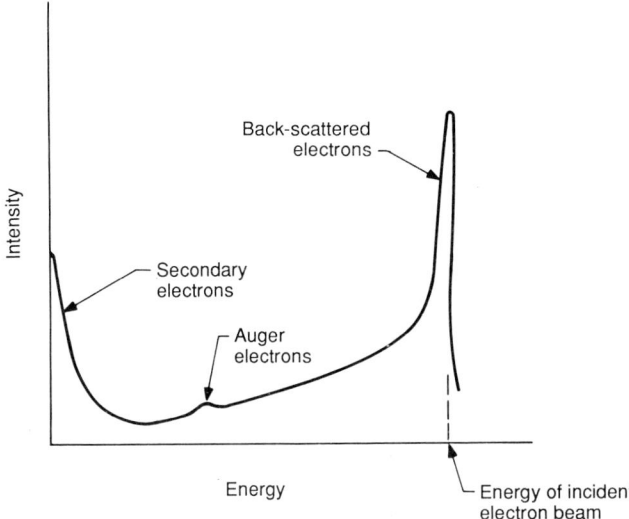

Figure 1.10 Typical electron spectrum from a sample surface in a scanning electron microscope.

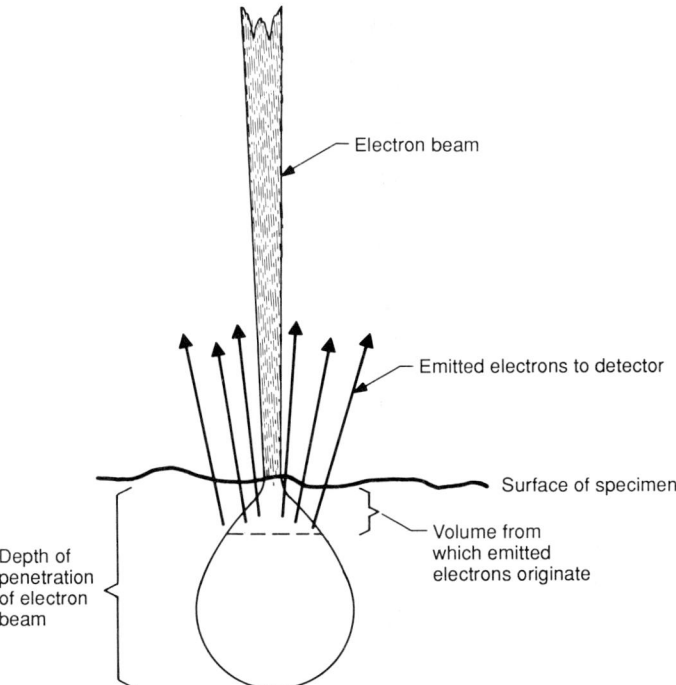

Figure 1.11 Schematic diagram to illustrate that the region from which the electrons are emitted is larger than that of the beam.

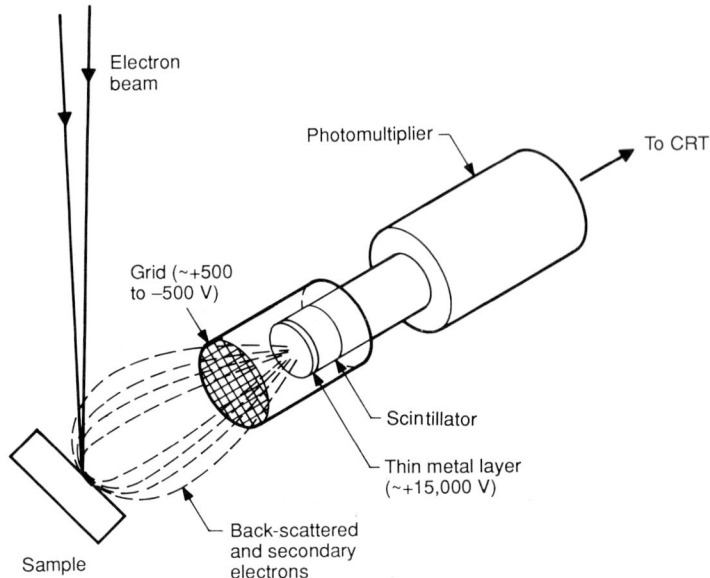

Figure 1.12 Schematic representation of the operation of the electron detector in a scanning electron microscope.

To collect all of the electrons, a rather simple device can be used (Fig. 1.12). The electrons emitted from the surface are drawn toward a scintillator, which is made of a material which gives off light when struck by the electron. To ensure that an adequate quantity of electrons reaches the scintillator, it is coated with a very thin (for example, 200 Å) layer of metal (such as aluminum), which is charged positively to about 10,000 to 15,000 V. This attracts the electrons, which penetrate the thin metal layer and reach the scintillator. To discriminate between low-energy and high-energy electrons, a grid is placed around the scintillator whose potential can be varied in the range of -500 to $+500$ V. Using a negative voltage will repel the low-energy electrons, allowing imaging with only the high-energy electrons, which sometimes is a useful mode of operation. The light from the scintillator is transmitted to a photomultiplier, where it is converted into an electrical signal, and this then modulates the intensity of the electron beam in the CRT.

The intensity of the back-scattered electrons increases with the atomic number of the elements from which they originate, and thus they can be used to generate an image which shows compositional variation. An example showing this effect is presented in Fig. 1.13.

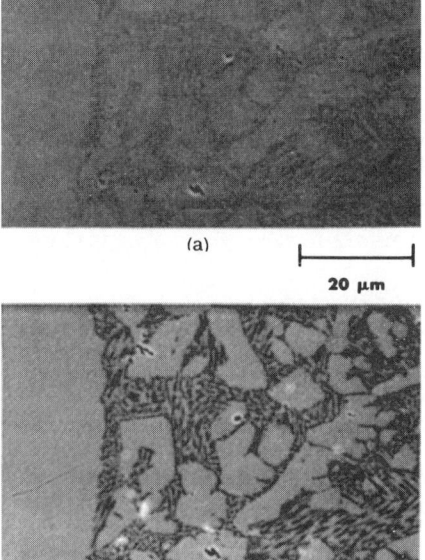

(a)

|—————|
20 μm

Figure 1.13 Micrographs of an as-polished copper-tin diffusion couple in a scanning electron microscope. (*a*) Using a secondary electron detector. (*b*) Using a back-scattered electron detector. Note the enhanced contrast using the back-scattered electron detector due to atomic number differences. (*From Verhoeven.*[9])

(b)

Figure 1.14 shows how the surface topology generates contrast in the image. For example, protrusions above the surface allow more electrons to be emitted as a greater volume of material excited by the electron beam is within the escape distance of the surface. Note that the resolution is basically fixed by the beam diameter and not the wavelength. Thus a SEM may have a resolution of 100 Å, but not the

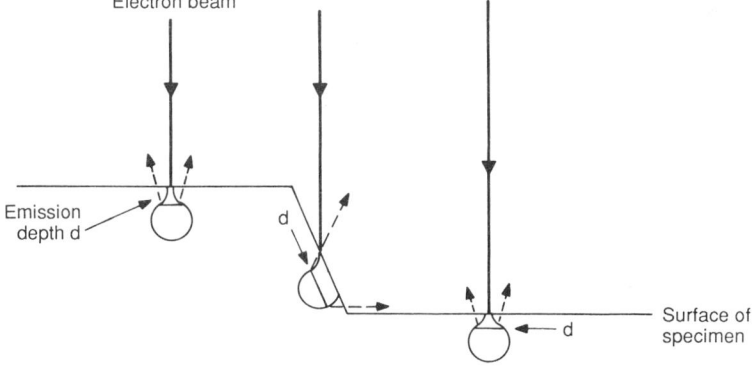

Figure 1.14 Schematic illustration of how the surface topology influences the volume of material within the emission distance d from the surface and enhances the intensity of the emitted electrons.

10 Å obtained by the TEM. However, a high depth of field is still retained. Even if it is not desired to resolve very fine detail, good depth of field makes the SEM attractive for low-magnification observations.

Unlike the TEM, in the SEM a relatively large sample can be used and rather large areas observed. Even if the sample is too large to be placed in the SEM, a replica of the surface can be prepared and placed in the SEM.

1.3.4 Comparison of OM, TEM, and SEM

The information in Table 1.6 compares the features of these microscope types. In the use of these instruments for fractography, it should

TABLE 1.6 Comparison of Characteristics of Optical Microscopes, Transmission Electron Microscopes, and Scanning Electron Microscopes

Based on a 10-cm² screen viewed at 25 cm

	Magnification	Resolution	Field	Depth of focus OM	SEM
OM	1×	0.2 mm	100 mm		
	10×	0.02 mm	10 mm	0.1 mm	10 mm
	100×	2 μm	1 mm	1 μm	1 mm
SEM	1000×	0.2 μm	0.1 mm		10 mm
	10,000×	20 nm (200 Å)	10 μm		1 μm
EM	100,000×	2 nm (20 Å)	1 μm		
	10⁶×	0.2 nm (2 Å)	0.1 μm		

(a)

	Optical	Scanning electron	Direct electron
Resolution—easy	5 μm	0.2 μm	**100 Å (10 nm)**
—skilled	0.2 μm	100 Å (10 nm)	**10 Å (1 nm)**
—special	0.1 μm	5 Å (0.5 nm)	**2 Å (0.2 nm)**
Depth of focus	*poor*	**high**	moderate
Mode—transmission	yes	yes	yes
—reflection	yes	yes	*not satisfactory*
—diffraction	yes	yes	yes
—other	some	**many**	*no*
Specimen—preparation	usually easy	**easy**	*skilled, liable to artefacts*
—range and type	versatile real or replica	versatile real or replica	*only thin, or replica*
—maximum thickness for transmission	**thick**	medium	*very thin*
—environment	**versatile**	usually vacuum but can be modified	*vacuum*
—available space	small	**large**	small
Field of view	large enough	large enough	*limited*
Signal	only as image	**available for processing**	only as image
Cost	**low**	high	high

Advantages over others are indicated in bold type; disadvantages in italics.

(b)

SOURCE: From Hearle et al.[10]

1.3.5 Related tools and techniques

There are a number of tools and techniques related to microscopy which are useful in fractography. Here three are described briefly: x-ray fluorescence, x-ray diffraction, and stereomicroscopy.

X-ray fluorescence chemical analysis. This method of elemental chemical analysis is used widely. The description here will be given in terms of its use on the SEM, but it can be used in several other ways. When an atom is sufficiently excited, an inner-shell electron may be removed. When an electron from a neighboring orbital falls into this vacancy, the energy decrease is manifested as a fluorescent x-ray (see Fig. 1.15). The energy of the radiation from transitions in the outer (valence) electron shells depends on the type of atom bonding, but that from the transitions in the inner shells does not. Hence for a given element, the energy of the radiation is characteristic of the element

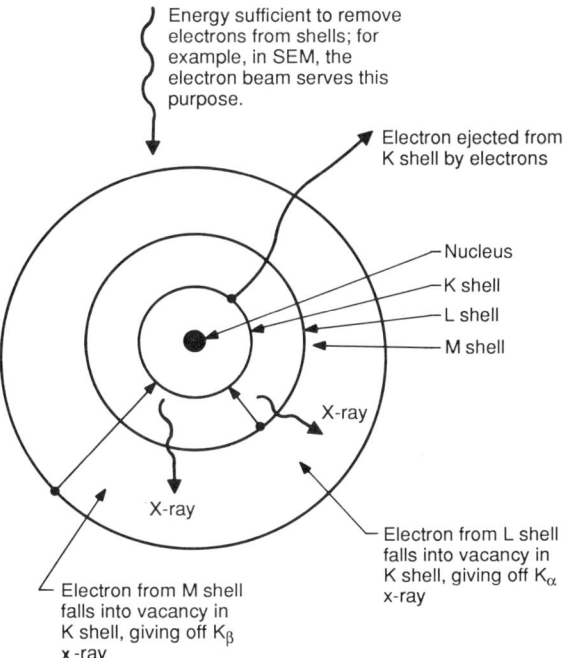

Figure 1.15 Schematic illustration of the origin of fluorescent x-rays and of the nomenclature used to describe the energy transitions.

from which it was emitted. Thus the detection of these x-rays is a method of identifying the type of atoms present. For example, the x-ray spectrum from pure iron will show strong lines or peaks at specific energies. If the spectrum of iron carbide or iron carbonate is obtained, peaks or lines in the spectrum reveal what elements are present, but not how the elements are combined.

The designation of the x-ray emissions is determined by the shell to which the electron falls and from which it comes. The scheme is shown in Fig. 1.15. For example, K_α is an x-ray photon caused by an electron from the L shell falling into a K-shell vacancy. K_β is an x-ray photon caused by an electron from the M-shell falling into a K-shell vacancy.

The energies of these electron transitions have been measured carefully, and the values can be found tabulated in several publications. Figure 1.16 shows one such listing. The energy is usually given in keV

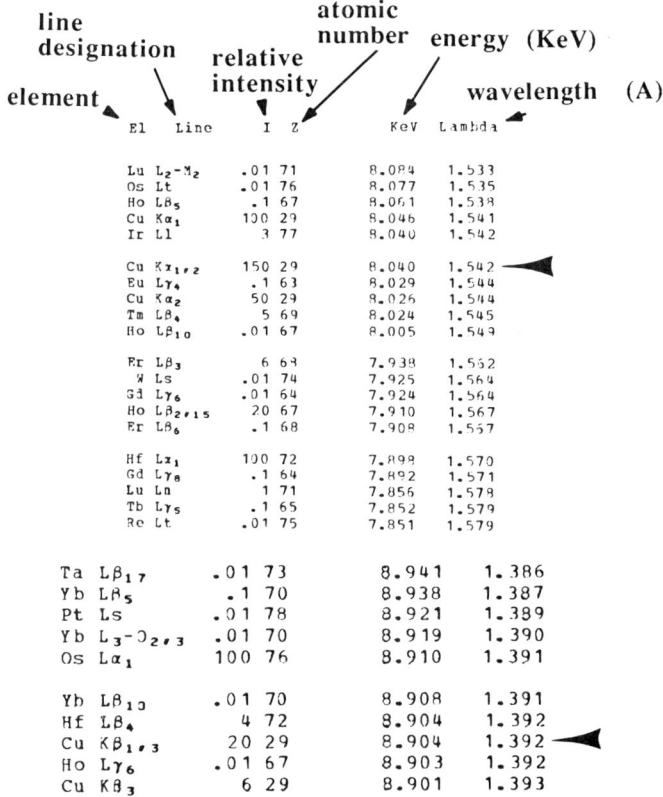

Figure 1.16 Section of a table listing the energies of x-ray emissions of the elements. CuK_α and CuK_β are noted. (*Adapted from Johnson and White.*[11])

(1000 electronvolts). Also listed are the corresponding wavelengths in angstroms. The relation connecting these two quantities is

$$E = h\nu = \frac{hc}{\lambda} \tag{1.3}$$

where E is the energy, λ is the wavelength, ν is the frequency, h is Planck's constant, and c is the speed of light. Thus as the energy increases, the wavelength decreases (see Fig. 1.16). The other listing of importance in Fig. 1.16 is I, which is the relative intensity to be expected for a transition. These values are based on experimental data and reflect the fact that all transitions do not occur with equal probability. Note that the $K_{\alpha 1-2}$ transition for copper has a relative intensity of 150 (150 percent) and that of $K_{\beta 1-3}$ a value of 20. Thus in the spectrum from copper, the peak at 8.040 keV should be only about 1/7 as strong as that at 8.904 keV. When the I value decreases to below 5, it generally will be difficult to detect the x-ray.

There are two methods used in scanning electron microscopy to measure or record the x-ray spectrum: One method uses a diffracting crystal, the other a solid-state detector. The method using the diffracting crystal is referred to as wavelength dispersive, or sometimes wavelength-dispersive x-ray analysis (WDXA) or wavelength-dispersive spectrometry (WDS). In this technique, the fluorescent x-rays are allowed to strike a single crystal. The x-rays will either pass through the crystal, will be absorbed, will be generally scattered, or will be diffracted. Those which satisfy Bragg's law

$$\lambda = 2d \sin \theta \tag{1.4}$$

will diffract, as shown in Fig. 1.17. Here d is the spacing between the planes of the crystal which are diffracting the x-rays to the detector, and its value is known for a given crystal. Thus the intensity for this specific wavelength is measured. The crystal can then be set to a different value of θ, always maintaining the θ–2θ relation shown in Fig. 1.18. Then the intensity at this wavelength is obtained. The process can be automated. It is customary to obtain a plot of intensity versus 2θ, not intensity versus wavelength, but the wavelength can be calculated from the values of 2θ by using Eq. (1.4). Figure 1.19 shows a typical spectrum. Note that for this stainless-steel sample it is quite obvious which peak is caused by which element.

There are two important disadvantages of this method of obtaining the x-ray spectrum. One is that the instrumentation may be difficult to place in the proper location for use on a SEM. Unlike the electrons, which can be focused or attracted by the high positive charge on the electron detector (see Fig. 1.12), x-rays cannot be focused so easily.

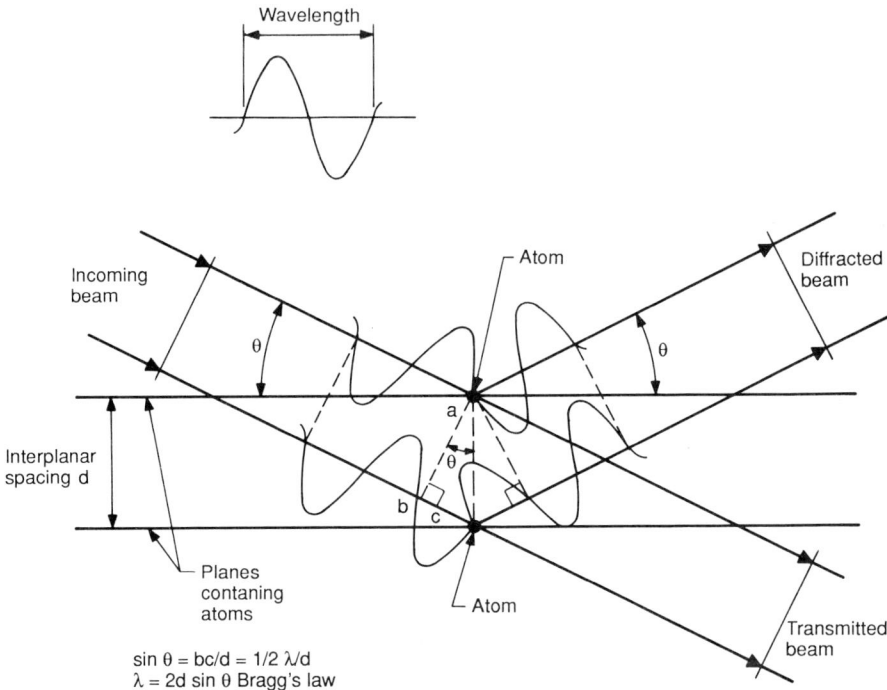

Figure 1.17 Simplified derivation of Bragg's law of x-ray or electron diffraction. If angle θ is such that the distance *bc* is ½ the wavelength λ, then Bragg's law is satisfied, and constructive interference occurs along the direction of angle θ, causing diffraction.

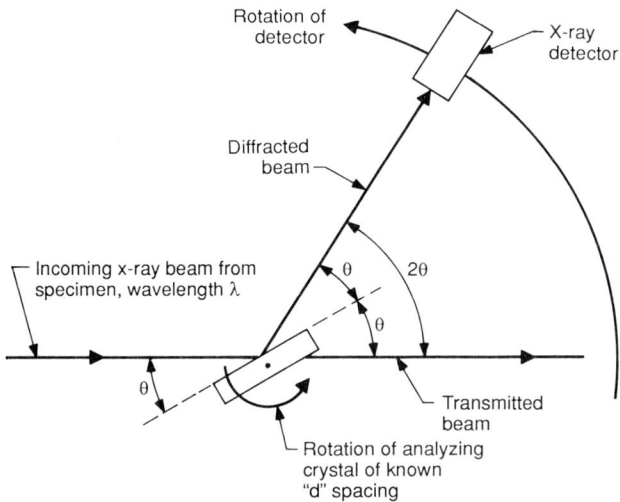

Figure 1.18 Schematic illustration of a diffractometer used to analyze the spectrum from an x-ray beam.

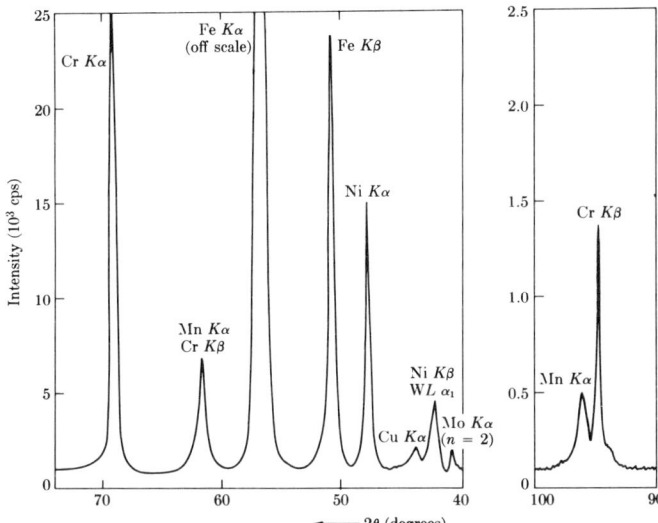

Figure 1.19 X-ray spectrum obtained by using an x-ray diffractometer (wavelength-dispersive spectrometry). The sample was stainless steel containing 19.4% Cr, 9.5% Ni, 1.5% Mo, 1.4% W, and 1.0% Mn, balance mainly Fe. A flat LiF crystal was used as the analyzer. A platinum target x-ray tube was used, operated at 50 kV and 30 mA. (*Adapted from Cullity*[12]; *courtesy of Diano Corporation.*)

Thus only those x-rays which have a direct path to the detecting crystal will be analyzed. The other limitation is that it takes from 30 to 60 min to obtain a spectrum. On the other hand, this method can detect elements as low in atomic number as carbon.

The method using a solid-state detector is referred to as the nondispersive method, as the x-rays are not dispersed by an analyzing crystal. It is commonly referred to as energy-dispersive x-ray analysis (EDXA) or energy-dispersive spectrometry (EDS). This method uses a Si (Li) counter and a FET preamplifier, both of which must be maintained at liquid nitrogen temperature. The signal from the detector is separated into a spectrum by a multichannel analyzer, so that a spectrum of intensity versus energy is displayed on a CRT. Figure 1.20 shows a typical spectrum.

This method of analysis has two important advantages in a SEM, namely, the detector can be placed close to the sample, inside the vacuum chamber, and the time to obtain a spectrum is short. Obtaining a spectrum like the one shown in Fig. 1.20 by the wavelength-dispersive spectrometry method would take perhaps 60 min. The spectrum using energy-dispersive spectrometry would require about 1 min. An important limitation, however, is that the energy-dispersive detector is not

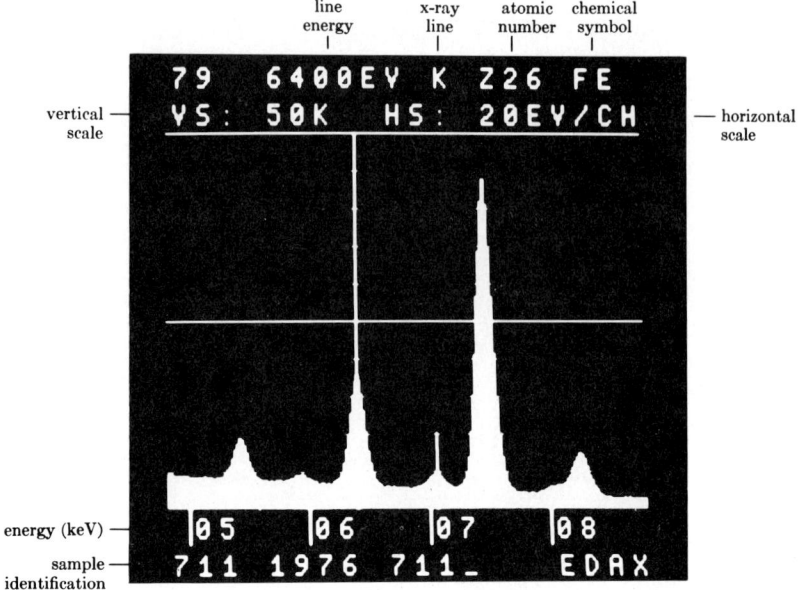

Figure 1.20 X-ray spectrum obtained by an energy-dispersive spectrometry system. (*Adapted from Cullity*[12]; *courtesy of Philips Electronic Instruments, Inc.*)

as sensitive to light elements (such as, carbon) as the wavelength-dispersive system.

In terms of the use of fluorescent x-ray analysis in the SEM, it is emphasized that usually the analysis is qualitative or semiquantitative. It is particularly useful in determining what elements are present and their relative amounts, but not in determining the exact composition.

The strength of x-ray fluorescence in fractrography lies in the microchemical analysis which can be performed with the SEM. For example, in operating the SEM, it is possible to stop the raster of the beam and move the beam spot to any location in the field of view. Thus if the chemical analysis of a particle on the surface is desired, it can be obtained by locating the beam on the particle and compiling the x-ray spectrum. This is illustrated in Fig. 1.21, which shows a fracture surface with a small particle on it. The diameter of the particle is about 4000 Å, and the electron beam was perhaps 200 Å in diameter. Recalling that this beam will blossom into a tear-dropped shape, the actual diameter of the excited region is approximately 1000 Å. The depth of penetration is approximately 1000 Å. This gives a volume of the material from which the x-rays are emitted of approximately 10^{-15} cm^3, or a mass of material of about 10^{-14} g. Hence the use of the term microchemical analysis. The spectrum from this par-

Introduction   29

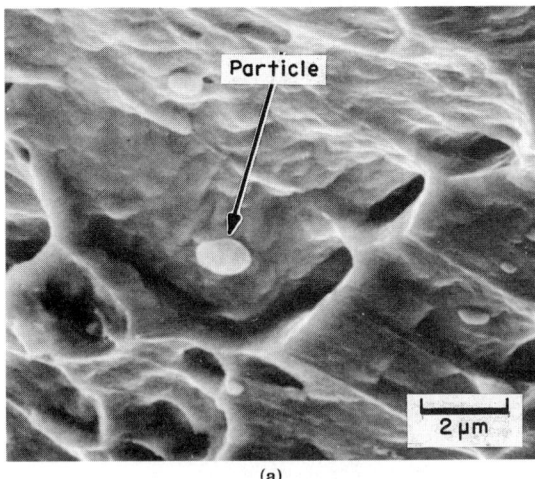

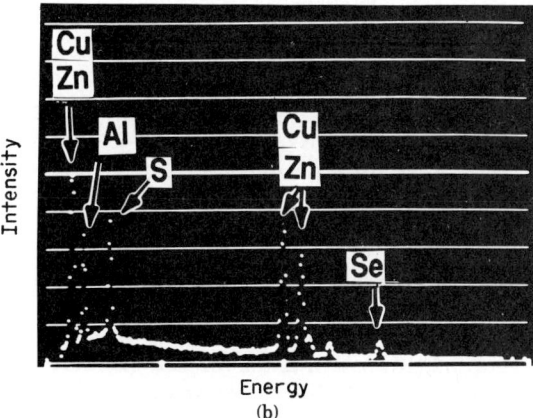

Figure 1.21 (a) Scanning electron micrograph of a fracture surface. (b) X-ray spectrum from small particle indicated by arrow.

ticle is shown in Fig. 1.21b. It was obtained in about 1 min, and it took about a minute to photograph the spectrum from the face of the CRT.

Some precautions must be taken in obtaining x-ray spectra in the SEM. It must be remembered that the electron beam is considerably smaller than the diameter of the region from which x-rays are emitted. Thus if a particle of approximately 500 Å is to be analyzed and the beam diameter is 200 Å, the x-ray spectrum may contain emissions from material outside of the particle. It is equally important to realize that secondary fluorescence can occur. This is a constant hazard in an-

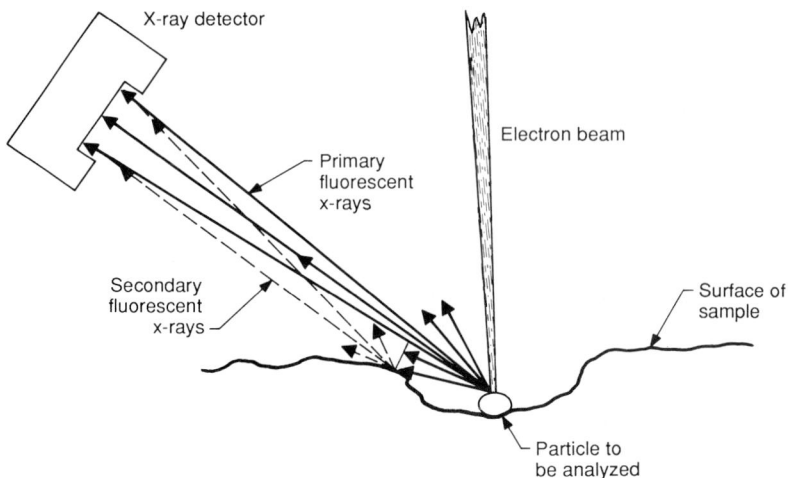

Figure 1.22 Schematic illustration of the origin of secondary fluorescence, causing a false indication of the elemental analysis of the particle due to x-rays from the edge of the matrix.

alyzing rough surfaces typical of fracture surfaces, as illustrated in Fig. 1.22. The primary x-ray emissions may be absorbed by surrounding higher regions, causing them to emit x-rays and giving a spectrum not actually characteristic of the region on which the beam is focused.

X-ray and electron diffraction. X-ray and electron diffraction is frequently used in conjunction with x-ray fluorescence to determine the nature of crystalline materials. For example, it may be important to determine the chemical and structural composition of wear debris. X-ray fluorescence can give a qualitative analysis of many elements present. However, analysis of an x-ray diffraction pattern can reveal the exact nature of the debris.

X-ray and electron diffraction relies on using a monochromatic (constant-wavelength) beam on a sample and analyzing the intensity as a function of 2θ. In this case the wavelength λ is fixed and Bragg's law [Eq. (1.4)] is used to calculate the distance d between planes in the crystals. These values can be compared to the ASTM x-ray diffraction data file to see if a match reveals what material is present. If this approach is not fruitful, then, in principle, the data can be analyzed to determine the crystal structure, although this may be quite difficult.

Electron diffraction in the TEM can be useful to determine the crystal structure of extremely fine individual particles as the diffraction pattern can be obtained by the equivalence of limiting the beam to the illumination of very small regions.

Stereomicroscopy. Stereomicroscopy refers to the observation of an object such that the entire field of view remains in focus and at the same time a three-dimensional effect is obtained. This is a particularly valuable tool in fractography. The amount of additional information obtained by viewing fracture surfaces in stereo cannot be overestimated, neither can the importance be proven in words. One has to view a surface under this condition to properly appreciate its significance.

The viewing itself is rather easy, but it may take practice. Low-magnification stereooptical microscopes are available. Where the resolution of finer detail is sought, the OM is not suitable because the high magnification required will have associated with it a poor depth of field. However, the SEM is especially appropriate for this. It is simple to obtain two pictures which, when viewed with a stereo viewer, will give the required three-dimensional effect. A photograph is taken of the area of interest, then the sample is tilted 6 to 12°, and the same area is photographed. These photographs are then properly positioned under a stereo viewer until the three-dimensional effect is observed. The same procedure can be used for samples being observed in the TEM. Volume 9 of *The Metals Handbook*,[7] has a stereo viewer in the back, and several stereo fractographs are in the book.

A proper method of obtaining stereopairs in the SEM and of determining the variation in the height of the surface topology is outlined in Appendix 1A.

1.4 Sample Preparation

1.4.1 Cleaning of surfaces

A common problem is that the fracture surface is dirty and contaminated. There are several methods of cleaning fracture surfaces. In general, the surface is washed carefully, rinsed in alcohol or acetone, then dried. The washing is best done in an ultrasonic cleaner, which will remove most loose debris. Further removal of debris and surface reaction products such as rust can be accomplished by replicating the surface several times. In this procedure the surface is wet with acetone, and one side of a piece of cellulose acetate tape (about 0.01 cm thick) is moistened with acetone. Then, with the fracture surface still wet, the wet side of the tape is pressed carefully but firmly onto the fracture surface. It is rubbed to force the sticky tape to be embedded into the surface. After several minutes of drying, the tape is peeled off, giving a replica of the surface and removing debris. (Such a replica can be observed in the SEM and the surface debris can be analyzed.) This process can be repeated several times to assist in cleaning the surface.

A major problem in surface preparation is the removal of rust on ferrous samples. How to treat the surface to remove the rust without destroying the underlying surface which may reflect the true fracture surface topology before rusting occurred is a touchy decision. Most methods of cleaning involve the use of chemicals which are designed to dissolve the oxides without attacking the underlying metal. Appendixes 1B through 1G give information on methods of cleaning surfaces. Additional sources of information can be traced from the references in the sources cited there.

1.4.2 Preparation of replicas for the TEM

In using the TEM in fractography, the actual fracture surface usually cannot be observed, of course, as it is extremely difficult to prepare the required small and thin specimen. Thus the surface topology is examined indirectly by preparing a suitable replica of the surface. One method is the one just described for cleaning the surface, namely, to use cellulose acetate tape to make the replica. However, this tape is too thick for use directly in the TEM, and in addition such polymeric materials are subject to heating by the electron beam and, hence, damage. Instead it is common to take the tape replica and put a thin (200 Å) layer of carbon on it in a vacuum coater. To enhance contrast of this carbon replica, a thin layer (200 Å) of a heavy metal, such as chromium or platinum, is deposited, or shadowed, onto the surface. Then the acetate tape is dissolved in a suitable solvent (such as acetone), freeing the thin and fragile replica into the solvent. This replica is removed onto a screen or grid, which is of suitable size (about 0.3 cm in diameter) to fit into the TEM. The replica is viewed through the openings in the grid. The process of making such a replica is depicted in Fig. 1.23.

It is possible to prepare a replica of the surface by depositing the carbon directly and not using the tape. The carbon then is removed by adhesive tape, which is dissolved to free the replica. This gives a replica of greatest fidelity, having a resolution of about 20 Å. The two-stage tape-carbon replica has a resolution of about 100 Å. The artifacts which may be present in replicas have been quite well documented (see, Sec. 3.16).

If it is desired to analyze particles on the surface, after the carbon layer is deposited, the underlying metal (original specimen) can be dissolved chemically, freeing the replica with the particles embedded in it. This is an extraction replica. It allows analysis of the particles in the TEM using electron diffraction and x-ray fluorescence. However,

Introduction   33

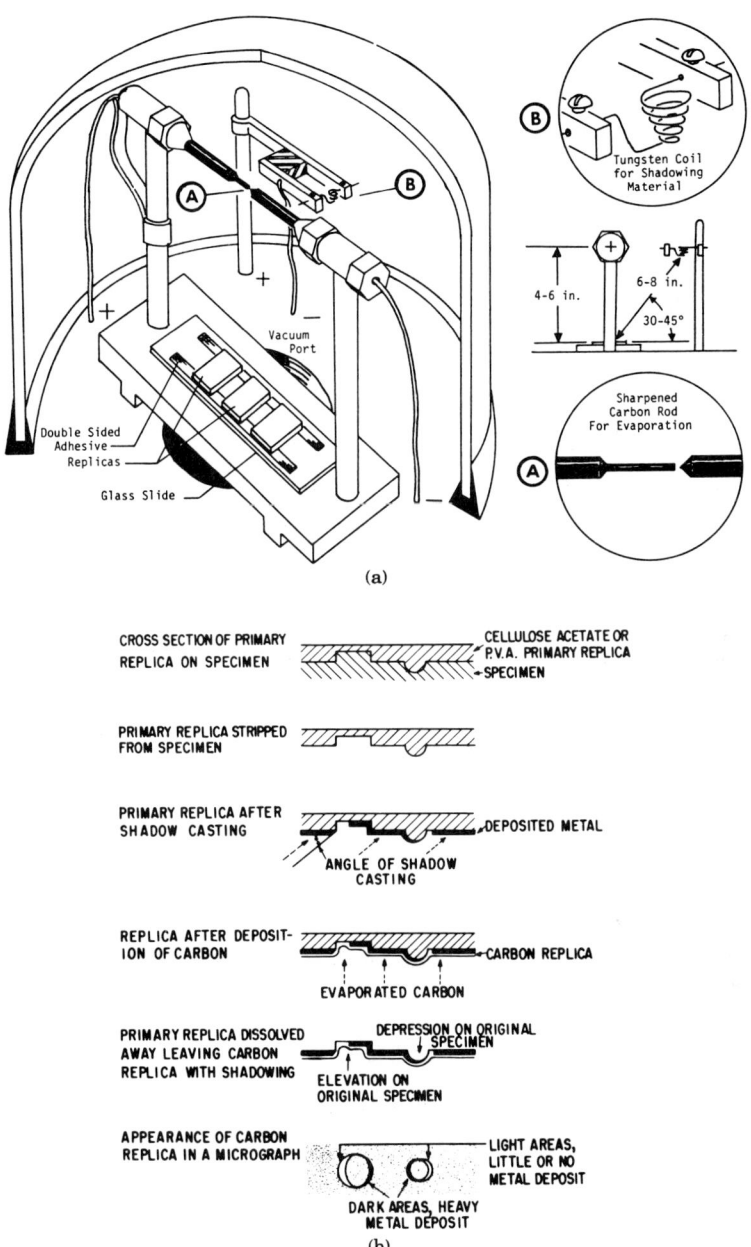

Figure 1.23 Schematic representation of a method of preparing replicas for transmission electron microscopy. (*a*) Vacuum deposition technique. (*From Nail*[13], *copyright ASTM; reprinted with permission.*) (*b*) Steps in preparing an indirect carbon replica using a cellulose acetate or polyvinyl alcohol (PVA) primary replica. (*Prepared by E. F. Koch; from Phillips*[14]; *courtesy of General Electric Company.*)

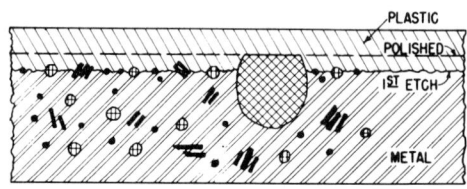

First Etch

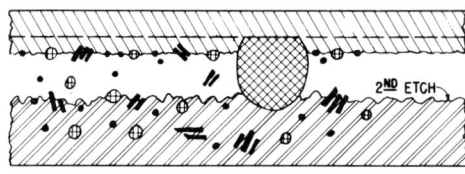

Second Etch

Replica

Figure 1.24 Schematic illustration of the preparation of an extraction replica. (*From Nail.*[13])

this process does destroy the fracture surface. The preparation of an extraction replica is illustrated in Fig. 1.24.

1.4.3 Preparation of samples for the SEM

For direct observation of metallic samples no preparation is required other than cleaning. However, the samples do have to be electrically conducting, so if there is excessive rust or debris, and the surface must be observed with this on it, then it may be necessary to coat the surface with a thin layer (200 Å) of metal (such as gold) or carbon. If x-ray fluorescence is to be carried out on the surface, some thought must be given to the excitation of x-rays from this coating and their effect on the interpretation of the x-ray spectrum obtained.

With the availability of the SEM it is normally no longer necessary to do extensive replica TEM work. Even if the sample will not fit into the SEM, the surface can be replicated with tape, the replica coated with metal, and then this replica observed in the SEM (and even optically).

References

1. G. J. Davies, "Performance in Service," in E. J. Bradbury (ed.), *Essential Metallurgy for Engineers,* Van Nostrand Reinhold (UK), London, 1985, p. 126.
2. T. J. Dolan, "Analyzing Failures of Metal Components," *Metals Eng. Quart.,* vol. 12(4), p. 32, 1972.

3. D. A. Ryder, T. J. Davies, I. Brough, and F. R. Hutchings, "General Practice in Failure Analysis," in *Metals Handbook,* 8th ed., vol. 10: *Failure Analysis and Prevention,* American Society for Metals, Metals Park, Ohio, 1975, p. 10.
4. *Metals Handbook,* 9th ed., vol. 10: *Materials Characterization,* American Society for Metals, Metals Park, Ohio, 1986.
5. E. Ruechardt, *Light,* University of Michigan Press, Ann Arbor, 1958.
6. D. L. Dyer, "Optical Limits in TV Microscopy," *Research/Develop.,* Sept. 1973.
7. *Metals Handbook,* 8th ed., vol. 9: *Fractography and Atlas of Fractographs,* American Society for Metals, Metals Park, Ohio, 1974.
8. H. C. Becker, "Scanning Electron Microscopy," *Lubrication,* vol. 61, p. 37, 1975.
9. J. D. Verhoeven, in *Metals Handbook,* 9th ed., vol. 10: *Materials Characterization,* American Society for Metals, Metals Park, Ohio, 1986.
10. J. W. S. Hearle, J. T. Sparrow, and P. M. Cross, *The Use of the Scanning Electron Microscope,* Pergamon, New York, 1972.
11. G. G. Johnson and E. W. White, "X-Ray Emission Wavelength and KeV Tables for Nondiffractive Analysis," ASTM Data Series DS 46, American Society for Testing and Materials, Philadelphia, Pa., 1970.
12. B. D. Cullity, *Elements of X-Ray Diffraction,* 2d ed., Addison Wesley, Reading, Mass., 1978.
13. D. A. Nail, "Procedures for Standard Replication Techniques for Electron Microscopy," in *Manual on Electron Metallography Techniques,* STP 547, American Society for Testing and Materials, Philadelphia, Pa., 1973, Chap. 1.
14. V. A. Phillips, *Modern Metallographic Techniques and Their Applications,* Wiley, New York, 1971.

Bibliography

The list of written material dealing, directly and indirectly, with metallurgical failures is extensive, and no attempt is made in this book to compile a bibliography. Instead, the references cited in this book can be used to trace additional information. Of special mention, though, are the different volumes of the *Metals Handbook* listed below and the article by Vander Voort which lists 531 citations.

Some of the books listed below deal specifically with failures in metallic components, and some with the more general aspects of component failures. Of special interest to illustrate the importance of failure analyses is the book by Petroski.

Barer, R. D., and B. F. Peters: *Why Metals Fail,* Gordon and Breach, New York, 1970.
Burke, J. J., and J. Weiss (eds.): *Risk and Failure Analysis for Improved Performance Reliability,* Plenum, New York, 1980.
Carper, K. L. (ed.): *Forensic Engineering,* Elsevier, New York, 1989.
Collins, J. A.: *Failure of Materials in Mechanical Design,* Wiley, New York, 1981.
Goel, V. S. (ed.): *Analyzing Failures: The Problems and the Solutions,* American Society for Metals, Metals Park, Ohio, 1986.
Lange, G. A. (ed.): *Systematic Analysis of Technical Failures,* DGM Informations gesellschaft Verlag, Braunschweig, Germany, 1986.
Metals Handbook, 8th ed., vol. 10: *Failure Analysis and Prevention,* American Society for Metals, Metals Park, Ohio, 1975.
Metals Handbook, 9th ed., vol. 11: *Failure Analysis and Prevention,* American Society for Metals, Metals Park, Ohio, 1986.
Petroski, H.: *To Engineer Is Human, The Roles of Failure in Successful Design,* St. Martin's Press, New York, 1985.
Polushkin, E. P.: *Defects and Failures of Metals,* Elsevier, New York, 1956.

Smith, A. L.: *Reliability of Engineering Materials,* Butterworths, New York, 1983.
Sourcebook in Failure Analysis, American Society for Metals, Metals Park, Ohio, 1974.
Vander Voort, G. F.: in *Metals Handbook,* 9th ed., vol. 12: *Fractography,* American Society for Metals, Metals Park, Ohio, 1987.
Whyte, R. R. (ed.): *Engineering Progress through Trouble,* Institution of Mechanical Engineers, London, 1975.

Appendix 1A Stereomicroscopy*

Although SEM images appear three-dimensional, their very format reduces them to two-dimensional representations. Their multidimensional appearance is due to high depth of focus, but perspective distortions introduced by the geometry of the beam/specimen/detector invalidate spatial measurements (both height and lateral dimensions). Our subjective impressions are based on the direction of illumination, which cannot be adequately described in a single micrograph produced by a complex geometry. These problems are aggravated when rough-surfaced specimens are examined, because the exact angle of a field of view is both unknown and unmeasurable. This in turn implies that magnification varies within a field of view.

The phenomenon of perspective distortion may be observed by recording an image having two prominent features at 20°, recording a second micrograph of the same field at 45°, and then measuring the distances between the two features on each micrograph: the measurements will differ and neither is valid. *Stereo imaging* involves recording a given field of view twice at slightly different orientations, and simultaneously viewing the stereo pair such that a three-dimensional image is perceived. Perspective is restored, and valid spatial judgments or measurements replace subjective impressions.

Recording and viewing stereo images

The four methods used to record stereo pairs are (1) the tilt method, where an angle is applied between the two micrographs; (2) the lateral-shift method, where there is a horizontal displacement between the two micrographs; (3) the rotation method, where a specimen is rotated between exposures; and (4) electromagnetic deflection of the electron beam between images (Boyde, 1975; Chatfield, 1978; Wergin and Pawley, 1980). Methods 1 and 2 are readily applied in any SEM, method 3 is difficult, and method 4 requires special accessories for the SEM. Discussed below are the tilt and lateral-shift methods of stereo image recording.

---

*From B. L. Gabriel, *SEM: A User's Manual for Materials Science,* American Society for Metals, Metals Park, Ohio, 1985.

Introduction 37

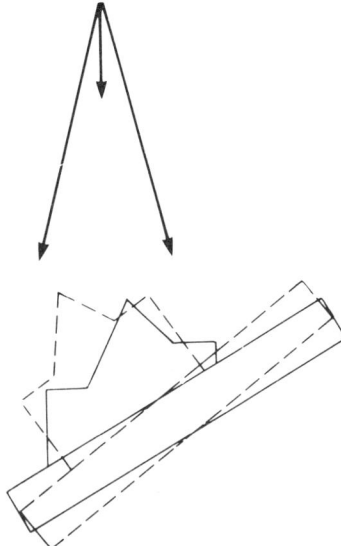

Figure 1A.1 Effect of a 7° stereo angle. Continuous line—specimen at 30°; dashed line—specimen tilted to 37°.

The *tilt method* of stereo recording is a versatile technique that may be used in any type of SEM. It is desirable but not required to have a eucentric-tilt specimen stage (the tilt axis passes through the center of the specimen, not through the center of the stage). The tilt method is as follows. Select and record the desired field of view, noting the tilt value of the specimen stage. With a wax pencil, mark the location of a prominent surface feature on the observation screen. Tilt the specimen approximately 7° (the stereo angle; see Figs. 1A.1 and 1A.2) while manipulating the stage X and Y axes to maintain the same field of view. Align the prominent surface feature beneath the wax pencil mark. Refocus the image using the Z-axis control; do not refocus with the objective lens controls. Adjust the contrast and brightness levels to match the first micrograph, and record the image.

The choice of the *stereo angle,* or the tilt difference between each micrograph of the stereo pair, is a function of the topography of the specimen. In general, smooth specimens require a stereo angle of 7 to 15°, while rough specimens require a stereo angle of 3 to 7°. The stereo angle determines *parallax* (synonym: horizontal displacement), which is a measure of the vertical position above or below a particular datum plane. With too large a stereo angle there is excessive displacement between features (i.e., excessive parallax), and the stereo image has "too much depth." With too small a stereo angle there is insufficient displacement, and the stereo image is not a true three-dimensional representation. If the microscopist is unsure of the optimal stereo angle for a given field of view, simply record several micrographs while changing the angle until the desired stereo image is obtained.

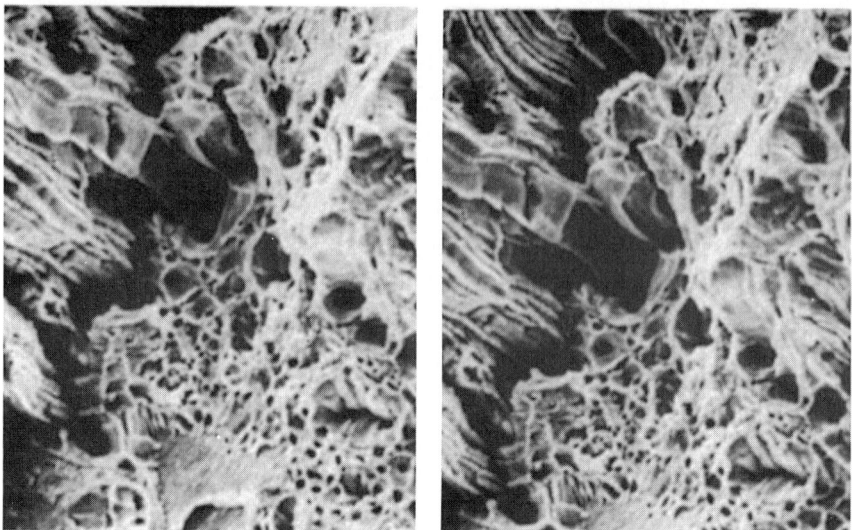

Figure 1A.2 Stereo pair of a ductile fracture prepared using the tilt method and a stereo angle of 7°. (300×.)

Another method useful for recording stereo pairs below 50× magnification is the *lateral-shift method* (synonyms: linear displacement or shift; translational shift). In this method, a micrograph is recorded, and the image is then moved horizontally while keeping the desired feature in the field of view. The second member of the stereo pair is then recorded. The distance shifted determines the depth of the stereo image; a very large lateral shift produces more depth than a small shift. Operating conditions must be the same when recording each half of the stereo pair; again, refocus (if necessary) by manipulating the Z axis, adjust contrast and brightness to match the first micrograph, and maintain the same tilt angle for both recordings. This method is appropriate only for low-magnification images, because at moderate or high levels lateral shift may displace the desired field of view before the optimal stereo image is visible (Fig. 1A.3).

Stereo pairs are viewed using simple pocket viewers, double-prism viewers, or a mirror stereoscope (Boyde, 1979). Pocket viewers are adequate for simple viewing of stereo pairs, but the more sophisticated prism and mirror stereoscopes offer significant advantages when stereo analysis is routinely conducted. The stereo effect (stereopsis) is perceived by positioning both micrographs within the viewer such that the tilt axis is vertical, i.e., rotate both micrographs 90°. The micrograph with the lower tilt value is placed to the left of the second half of the pair, and the distance between then is adjusted until the stereo image is prominent.

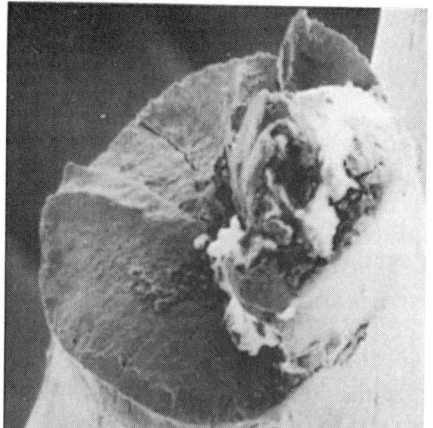 

Figure 1A.3 Stereo pair of a fractured wire prepared with the lateral-shift method. (75×.)

Because stereo viewers are designed for individual use, different methods have been developed to simultaneously project stereo pairs. The two most common projection techniques are the polarized and the anaglyph methods. The *polarized method* of stereo projection involves the simultaneous projection of each member of a stereo pair through adjacent slide projectors onto a lenticular silver screen. The projectors are equipped with filters that polarize the image from one projector at 45° to the vertical and from the other at an angle perpendicular to this. The audience must wear similarly polarized lenses to perceive the stereo effect. Wergin and Pawley (1980) thoroughly discuss the methods and equipment of the polarized method.

The *anaglyph method* of stereo projection applies a different color (usually red or green) to each half of the stereo pair. The simultaneous projection of the color-coded images produces a stereo image that is perceived by individuals wearing red-green lenses. This was the method used to project the 1950's 3-D horror movies. This method is not as popular as the polarized method, because color breakthrough and other problems may degrade the stereo effect (Barber and Brett, 1982). Nemanic (1974) and Barber and Emerson (1980) review the preparation and presentation of anaglyphs.

Quantitative stereoscopy

In addition to restoring perspective, stereo images may be used for spatial measurements. Referred to as *photogrammetry* or *quantitative stereoscopy,* valid measurements derived from stereo pairs are used in the construction of three-dimensional models. Boyde and his colleagues have contributed extensively to SEM stereoscopy (Howell and

Boyde, 1972; Boyde, 1974, 1981; Howell, 1975; and other references cited throughout this chapter). Summarized below are the more basic principles of quantitative stereoscopy.

The first stage of photogrammetry is establishment of the location of the *principal point,* defined as the point where the central ray (principal projector) of the scanning raster intersects the photographic plane. The location of the principal point does not necessarily correspond to the center of the micrograph; its location may be defined by switching off the SEM scan coils. The position of the stationary beam is visible as a bright spot. Alternatively, the position of the principal point can be established on a specimen by intentionally carbon-contaminating the surface: increase magnification roughly ten times above the desired level, permit a brief dwell time, and reduce the magnification to the desired level. A darkened spot (the site of contamination) is visible on beam-sensitive specimens, and corresponds to the location of the principal point.

Next, a *datum plane* which can be used as a baseline for measurements is identified with the tilt method. The common conventions use a datum plane parallel to one or the other halves of the stereo pair, or the midplane between the two micrographs. Using the parallel datum plane of one micrograph (Fig. 1A.4), Howell and Boyde (1972) defined the following spatial measurements, where M is magnification:

$$MZ_L = \frac{(MD)^2[X_L \cos \alpha - X_R] - MDX_LX_R \sin \alpha}{[(MD)^2 - X_LX_R]\sin \alpha + MD(X_L - X_R)\cos \alpha}$$

$$MX_L = \frac{X_L(MD - Z_L)}{MD}$$

$$MY_L = \frac{Y_L(MD - Z_L)}{MD}$$

Figure 1A.4 also shows the convention where an imaginary datum midplane between the stereo pair is located, and the following coordinates are shown:

$$MZ_C = \frac{(MD)^2(X_L - X_R)\cos(\alpha/2) + 2MDX_LX_R \sin(\alpha/2)}{[(MD)^2 + X_LX_R]\sin \alpha + MD(X_L - X_R)\cos \alpha}$$

$$MX_C = \frac{(MD)^2(X_L + X_R)\sin(\alpha/2)}{[(MD)^2 + X_LX_R]\sin \alpha + MD(X_L - X_R)\cos \alpha}$$

$$MY_C = \frac{Y_L}{MD}\left(MD + X_C \sin \frac{\alpha}{2} - Z_C \cos \frac{\alpha}{2}\right)$$

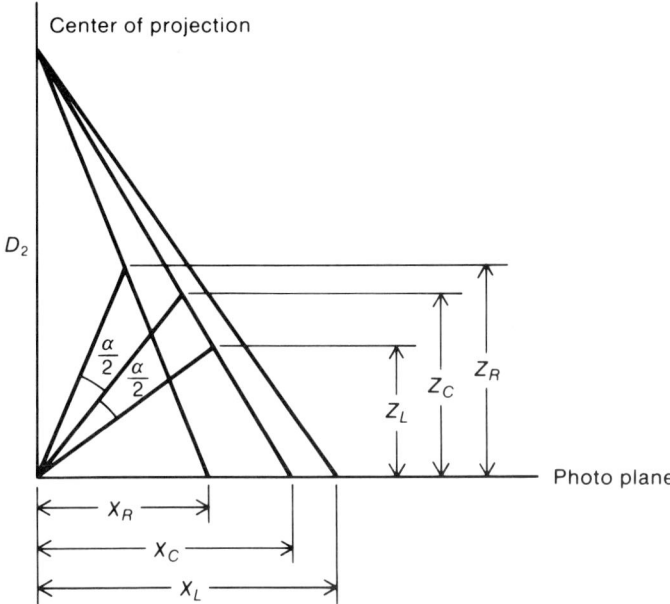

Figure 1A.4 Trigonometric relationship of the Howell and Boyde method. (*Courtesy of Howell and Boyde, 1972.*)

Individuals who regularly perform quantitative SEM measurements will be pleased to learn that modestly priced microcomputer programs are available for this purpose. Various systems are described by Howell and Boyde (1980), Roberts and Page (1980), Howell (1981), Boyde (1981), and Russ and Stewart (1983). Future developments in stereo SEM include real-time stereo imaging and image recording, computer control of the specimen stage, and continued acceptance and utilization by microscopists.

References to Appendix 1A

Barber, V. C., and D. A. L. Brett (1982): "'Colour Bombardment'—A Human Visual Problem that Interferes with the Viewing of Anaglyph Stereo Materials," *SEM, Inc.*, vol. 2, p. 495.

Barber, V. C., and C. J. Emerson (1980): "Preparation of SEM Anaglyph Stereo Material for Use in Teaching and Research," *Scanning*, vol. 3, p. 202.

Boyde, A. (1974): "Photogrammetry of Stereopair SEM Images Using Separate Images from the Two Images," *IITRI/SEM*, p. 101.

Boyde, A. (1975): "Measurement of Specimen Height Difference and Beam Tilt Angle in Anaglyph Real Time Stereo TV SEM System," *IITRI/SEM*, p. 189.

Boyde, A. (1979): "The Perception and Measurement of Depth in the SEM," *SEM, Inc.*, vol. 2, p. 67.

Boyde, A. (1981): "Recent Developments in Stereo SEM (1981 Update)," *SEM, Inc.*, vol. 1, p. 91.

Cannon, T. M., and B. R. Hunt (1981): "Image Processing by Computer," *Sci. Amer.*, vol. 245, no. 4, p. 214.

Chatfield, E. J. (1978): "Introduction to Stereo Scanning Electron Microscopy," in M. A. Hayat (ed.), *Principles and Techniques of Scanning Electron Microscopy*, vol. 6, Van Nostrand Reinhold, New York, p. 47.

Howell, P. G. T. (1975): "Taking, Presenting, and Treating Stereo Data from the SEM," *IITRI/SEM*, p. 697.

Howell, P. G. T. (1981): "Semi-Automatic Profiling from SEM Stereopairs," *Scanning*, vol. 4, p. 40.

Howell, P. G. T., and A. Boyde (1972): "Comparison of various methods for reducing measurements from stereo-pair scanning electron micrographs to 'real 3-D data,'" *IITRI/SEM*, p. 233.

Howell, P. G. T., and A. Boyde (1980): "The Use of an XY Digitiser in SEM Photogrammetry," *Scanning*, vol. 3, p. 218.

Nemanic, M. K. (1974): "Preparation of Stereo Slides from Electron Micrograph Stereopairs," in M. A. Hayat (ed.), *Principles and Techniques of Scanning Electron Microscopy*, vol. 1, Van Nostrand Reinhold, New York, p. 135.

Roberts, S. G., and T. F. Page (1980): "A Microcomputer-Based System for Stereogrammetric Analysis," *Proc. R. Micros. Soc.*, Micro 80 suppl., vol. 15, no. 5, p. 8.

Russ, J. C., and W. D. Stewart (1983): "Quantitative Image Measurement Using a Microcomputer System," *Am. Lab.*, vol. 15, no. 12, p. 70.

Wergin, W. P., and J. B. Pawley (1980): "Recording and Projecting Stereo Pairs of Scanning Electron Micrographs," *SEM, Inc.*, vol. 1, p. 239.

Appendix 1B   Care and Handling of Fractures*

When a fracture requires laboratory examination, both mating fractures should be preserved by the application of a protective coating[†] and sealed in a plastic bag containing a desiccant to prevent any accumulation of undue moisture until the examination can be made. Coatings used should be soluble in organic solvents so that they can be completely removed prior to replication. Handling the fracture faces with fingers, rubbing, or mating the fractures together can cause serious damage. Picking at the fracture with a sharp instrument should be avoided. Rough treatment or the formation of corrosion products on the fracture will obscure vital information (Fig. 1B.1). Education in the proper handling of specimens prior to any fractographic examination is strongly recommended for anyone dealing in fractures either in the field or in the laboratory.

---

*From A. Phillips, V. Kerlins, and B. V. Whiteson, *Electron Fractography Handbook*, Tech. Rep. ML-TDR-64-416, Air Force Materials Laboratory, Wright-Patterson Air Force Base, Ohio, 1965.

†For example, Krylon Crystal Clear spray Coating no. 1302, Krylon, Inc., Norristown, Pa. Solvent: trichloroethylene.

Introduction   43

HANDLE FRACTURES WITH CARE

Figure 1B.1   Care and handling of fractures.

Appendix 1C   Preparation and Preservation of Fracture Specimens*

Fracture surfaces are fragile and subject to mechanical and environmental damage that can destroy microstructural features. Consequently, fracture specimens must be carefully handled during all stages of analysis. This appendix discusses the importance of care and handling of fractures and what to look for during the preliminary visual examination, fracture-cleaning techniques, procedures for sectioning a fracture and opening secondary cracks, and the effect of nondestructive inspection on subsequent evaluation.

Care and handling of fractures[1]

Fracture interpretation is a function of the fracture surface condition. Because the fracture surface contains a wealth of information, it is important to understand the types of damage that can obscure or obliterate fracture features and obstruct interpretation. These types of damage are usually classified as chemical and mechanical damage. Chemical or mechanical damage of the fracture surface can occur dur-

---

*From R. D. Zipp and E. P. Dahlberg, in *Metals Handbook,* 9th edition, vol. 12; *Fractography,* ASM International, Metals Park, Ohio, 1987, pp. 72–77.

ing or after the fracture event. If damage occurs during the fracture event, very little can usually be done to minimize it. However, proper handling and care of fractures can minimize damage that can occur after the fracture.[2-4]

Chemical damage of the fracture surface that occurs during the fracture event is the result of environmental conditions. If the environment adjacent to an advancing crack front is corrosive to the base metal, the resultant fracture surface in contact with the environment will be chemically damaged. Cracking due to such phenomena as stress-corrosion cracking (SCC), liquid-metal embrittlement (LME), and corrosion fatigue produces corroded fracture surfaces because of the nature of the cracking process.

Mechanical damage of the fracture surface that occurs during the fracture event usually results from loading conditions. If the loading condition is such that the mating fracture surfaces contact each other, the surfaces will be mechanically damaged. Crack closure during fatigue cracking is an example of a condition that creates mechanical damage during the fracture event.

Chemical damage of the fracture surface that occurs after the fracture event is the result of environmental conditions present after the fracture. Any environment that is aggressive to the base metal will cause the fracture surface to be chemically damaged. Humid air is considered to be aggressive to most iron-base alloys and will cause oxidation to occur on steel fracture surfaces in a brief period of time. Touching a fracture surface with the fingers will introduce moisture and salts that may chemically attack the fracture surface.

Mechanical damage of the fracture surface that occurs after the fracture event usually results from handling or transporting of the fracture. It is easy to damage a fracture surface while opening primary cracks, sectioning the fracture from the total part, and transporting the fracture. Other common ways of introducing mechanical damage include fitting the two fracture halves together or picking at the fracture with a sharp instrument. Careful handling and transporting of the fracture are necessary to keep damage to a minimum.

Once mechanical damage occurs on the fracture surface, nothing can be done to remove its obliterating effect on the original fracture morphology. Corrosive attack, such as high-temperature oxidation, often precludes successful surface restoration. However, if chemical damage occurs and if it is not too severe, cleaning techniques can be implemented that will remove the oxidized or corroded surface layer and will restore the fracture surface to a state representative of its original condition.

Preliminary visual examination

The entire fracture surface should be visually inspected to identify the location of the fracture-initiating site or sites and to isolate the areas in the region of crack initiation that will be most fruitful for further microanalysis. The origin often contains the clue to the cause of fracture, and both low- and high-magnification analyses are critical to accurate failure analysis. Where the size of the failed part permits, visual examination should be conducted with a low-magnification widefield stereomicroscope having an oblique source of illumination.

In addition to locating the failure origin, visual analysis is necessary to reveal stress concentrations, material imperfections, the presence of surface coatings, case-hardened regions, welds, and other structural details that contribute to cracking. The general level of stress, the relative ductility of the material, and the type of loading (torsion, shear, bending, and so on) can often be determined from visual analysis.

Finally, a careful macroexamination is necessary to characterize the condition of the fracture surface so that the subsequent microexamination strategy can be determined. Macroexamination can be used to identify areas of heavy burnishing in which opposite halves of the fracture have rubbed together and to identify regions covered with corrosion products. The regions least affected by this kind of damage should be selected for microanalysis. When stable crack growth has continued for an extended period, the region nearest the fast fracture is often the least damaged because it is the newest crack area. Corrodents often do not penetrate to the crack tip, and this region remains relatively clean.

The visual macroanalysis will often reveal secondary cracks that have propagated only partially through a cracked member. These part-through cracks can be opened in the laboratory and are often in much better condition than the main fracture. Areas for sectioning can be identified for subsequent metallography, chemical analysis, and mechanical-property determinations.

Preservation techniques[1]

Unless a fracture is evaluated immediately after it is produced, it should be preserved as soon as possible to prevent attack from the environment. The best way to preserve a fracture is to dry it with a gentle stream of dry compressed air, then store it in a desiccator, a vacuum storage vessel, or a sealed plastic bag containing a desiccant. However, such isolation of the fracture is often not practical. Therefore, corrosion-preventive surface coatings must be used to inhibit oxidation and corrosion of the fracture surface. The primary disadvantage of using these surface coatings is that fracture surface debris, which often provides clues to the cause of fracture, may be displaced during removal of the coating. However, it is still possible to recover the surface debris from the solvent used to remove these surface coatings by filtering the spent solvent and capturing the residue.

The main requirements for a surface coating are as follows:

- It should not react chemically with the base metal.
- It should prevent chemical attack of the fracture from the environment.
- It must be completely and easily removable without damaging the fracture.

Fractures in the field may be coated with fresh oil or axle grease if the coating does not contain substances that might attack the base metal. Clear acrylic lacquers or plastic coatings are sometimes sprayed on the fracture surfaces. These clear sprays are transparent to the fracture surface and can be removed with organic solvents. However, on rough fracture surfaces, it can be difficult to achieve complete coverage and to remove the coating completely.

Another type of plastic coating that has been successfully used to protect most fracture surfaces is cellulose acetate replicating tape. The tape is softened in acetone and applied to the fracture surface with finger pressure. As the tape dries, it adheres tightly to the fracture surface. The main advantage of using replicating tape is that it is available in various thicknesses. Rough fracture surfaces can be coated with relatively thick replicating tape to ensure complete coverage. The principal limitation of using replicating tape is that on rough fracture surfaces it is difficult to remove the tape completely.

Solvent-cutback petroleum-base compounds have been used by Boardman et al. to protect fracture surfaces and can be easily removed with organic solvents.[5] In this study, seven rust-inhibiting compounds were selected for screening as fracture surface coating materials. These inhibitor compounds were applied to fresh steel fracture surfaces and exposed to 100% relatively humidity at 38°C (100°F) for 14 days. The coatings were removed by ultrasonic cleaning with the appropriate solvent, and the fracture surfaces were visually evaluated. Only the Tectyl 506 compound protected the fractures from rusting during the screening tests. Therefore, further studies were conducted with a scanning electron microscope to ensure that the Tectyl 506 compound would inhibit oxidation of the fracture surface and could be completely removed on the microscopic level without damaging the fracture surface.

Initially, steel Charpy samples and nodular iron samples were fractured in the laboratory by single-impact overload and fatigue, respectively. Representative fracture areas were photographed in the scanning electron microscope at various magnifications in the as-fractured condition. The fracture surfaces were then coated with Tectyl 506, exposed to 100% relative humidity at 38°C (100°F) for 14 days, and cleaned before scanning electron microscopy (SEM) evaluation by ultrasonically removing the coating in a naphtha solution. Figure 1C.1 shows a comparison of identical fracture areas in the steel at increasing magnifications in the as-fractured condition and after coating, exposing, and cleaning. These fractographs show that the solvent-cutback petroleum-base compound prevented chemical attack of the fracture surface from the environment and that the compound was

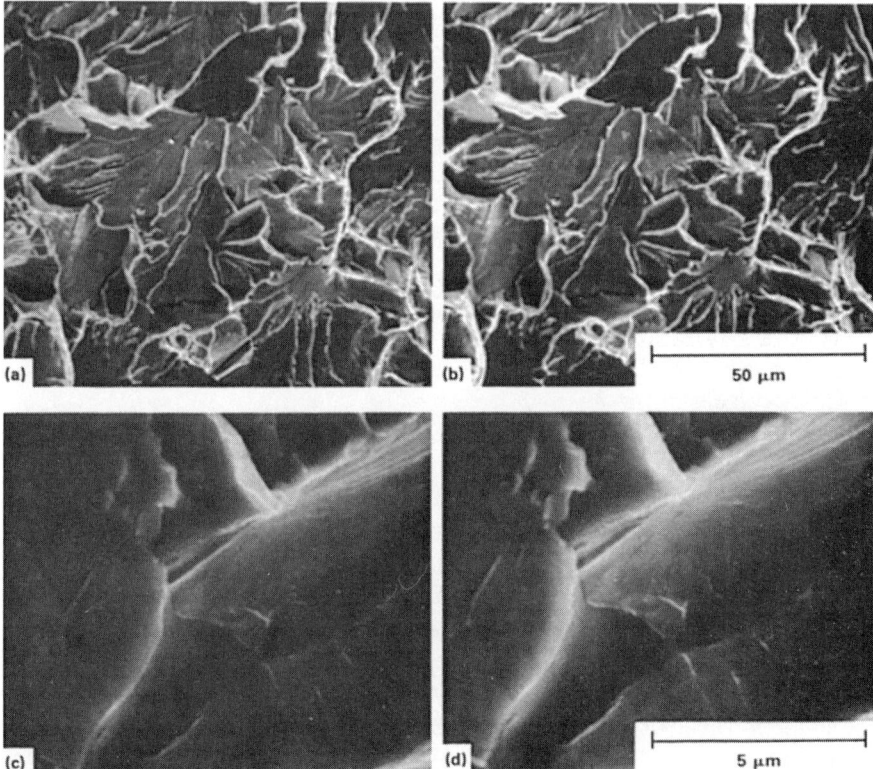

Figure 1C.1 Comparison of identical fracture areas of steel Charpy specimens at increasing magnifications. (a), (c) As-fractured surface. (b), (d) Same fracture surface after coating with Tectyl 506, exposing to 100% relative humidity for 14 days, and cleaning with naphtha.

completely removed in the appropriate solvent. It is interesting to note that Tectyl 506 is a rust-inhibiting compound that is commonly used to rustproof automobiles.

Fracture-cleaning techniques[1]

Fracture surfaces exposed to various environments generally contain unwanted surface debris, corrosion or oxidation products, and accumulated artifacts that must be removed before meaningful fractography can be performed. Before any cleaning procedures begin, the fracture surface should be surveyed with a low-power stereo binocular microscope, and the results should be documented with appropriate sketches or photographs. Low-power microscope viewing will also establish the severity of the cleaning problem and should also be used to monitor the effectiveness of each subsequent cleaning step. It is important to emphasize that the debris and deposits on the fracture surface can contain information that is vital to understanding the cause of fracture. Examples are fractures that initiate from such phenomena as SCC, LME, and corrosion fatigue. Often, knowing the nature of the surface debris and deposits, even when not essential to the fracture analysis, will be useful in determining the optimum cleaning technique. The most common techniques for cleaning fracture surfaces, in order of increasing aggressiveness, are:

- Dry air blast or soft organic-fiber brush cleaning
- Replica stripping
- Organic-solvent cleaning
- Water-base detergent cleaning
- Cathodic cleaning
- Chemical-etch cleaning

The mildest, least aggressive cleaning procedure should be tried first, and as previously mentioned, the results should be monitored with a stereo binocular microscope. If residue is still left on the fracture surface, more aggressive cleaning procedures should be implemented in order of increasing aggressiveness.

Air blast or brush cleaning. Loosely adhering particles and debris can be removed from the fracture surface with either a dry air blast or a soft organic-fiber brush. The dry air blast also dries the fracture surface. Only a soft organic-fiber brush, such as an artist's brush, should be used on the fracture surface because a hard-fiber brush or a metal wire brush will mechanically damage the fine details.

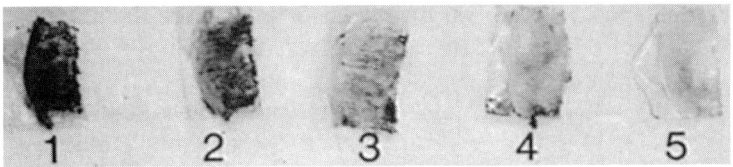

Figure 1C.2 Successive replicas (numbered 1 to 5) stripped from a rusted steel fracture surface. Note that the first replica stripped contains the most surface contaminants, while the last replica stripped is the cleanest. (Approximate size.)

Replica-stripping cleaning. This technique is very similar to that described in the section "Preservation Techniques." However, instead of leaving the replica on the fracture surface to protect it from the environment, it is stripped off of the fracture surface, removing debris and deposits. Successive replicas are stripped until all the surface contaminants are removed. Figure 1C.2 shows successive replicas stripped from a rusted steel fracture surface and demonstrates that the first replicas stripped from the fracture surface contain the most contaminants and that the last replicas stripped contain the least. Capturing these contaminants on the plastic replicas, relative to their position on the fracture surface, can be a distinct advantage. The replicas can be retained, and the embedded contaminants can be chemically analyzed if the nature of these deposits is deemed important.

The one disadvantage of using plastic replicas to clean a fracture surface is that on rough surfaces it is very difficult to remove the replicating material completely. However, if the fracture surface is ultrasonically cleaned in acetone after each successive replica is stripped from the fracture surface, removal of the residual replicating material is possible. Ultrasonic cleaning in acetone or the appropriate solvent should be mandatory when using the replica-stripping cleaning technique.

Organic solvents. Organic solvents, such as xylene, naphtha, toluene, freon TF, ketones, and alcohols, are primarily used to remove grease, oil, protective surface coatings, and crack-detecting fluids from the fracture surface. It is important to avoid use of the chlorinated organic solvents, such as trichloroethylene and carbon tetrachloride, because most of them have carcinogenic properties. The sample to be cleaned is usually soaked in the appropriate organic solvent for an extended period of time, immersed in a solvent bath where jets from a pump introduce fresh solvent to the fracture surface, or placed in a beaker containing the solvent and ultrasonically cleaned for a few minutes.

The ultrasonic cleaning method is probably the most popular of the three methods mentioned above, and the ultrasonic agitation will also remove any particles that adhere lightly to the fracture surface. However, if some of these particles are inclusions that are significant for fracture interpretation, the location of these inclusions relative to the fracture surface and the chemical composition of these inclusions should be investigated before their removal by ultrasonic cleaning.

Water-base detergent cleaning. This technique, assisted by ultrasonic agitation, is effective in removing debris and deposits from the fracture surface and, if proper solution concentrations and times are used, does not damage the surface. A particular detergent that has proved effective in cleaning ferrous and aluminum materials is Alconox. The cleaning solution is prepared by dissolving 15 g of Alconox powder in a beaker containing 350 mL of water. The beaker is placed in an ultrasonic cleaner preheated to about 95°C (205°F). The fracture is then immersed in the solution for about 30 min, rinsed in water then alcohol, and air-dried.

Figure 1C.3a shows the condition of a laboratory-tested fracture toughness sample (AISI 1085 heat-treated steel) after it was intentionally corroded in a 5% salt steam spray chamber for 6 h. Figure 1C.3b shows the condition of this sample after cleaning in a heated Alconox solution for 30 min. The fatigue precrack region is the smoother fracture segment located to the right of the rougher single overload region. Figure 1C.4a and b shows identical views of an area in the fatigue precrack region before and after ultrasonic cleaning in a heated Alconox solution. Only corrosion products are visible, and the underlying fracture morphology is completely obscured in Fig. 1C.4a. Figure 1C.4b shows that the water-base detergent cleaning has removed the corrosion products on the fracture surface. The sharp edges on the fracture features indicate that cleaning has not damaged the surface, as evidenced by the fine and shallow fatigue striations clearly visible in Fig. 1C.4b.

The effect of prolonged ultrasonic cleaning in the Alconox solution is demonstrated in Fig. 1C.5a and b, which shows identical views of an area in the fatigue precrack region after cleaning for 30 min and 3.5 h, respectively. Figure 1C.5b reveals that the prolonged exposure has not only chemically etched the fracture surface but has also dislodged

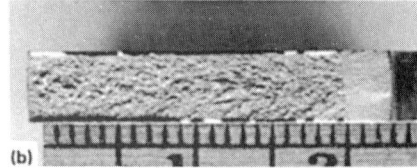

Figure 1C.3 Fracture toughness specimen that has been intentionally corroded in a 5% salt steam chamber for 6 h. (a) Before ultrasonic cleaning in a heated Alconox solution for 30 min. (b) After ultrasonic cleaning.

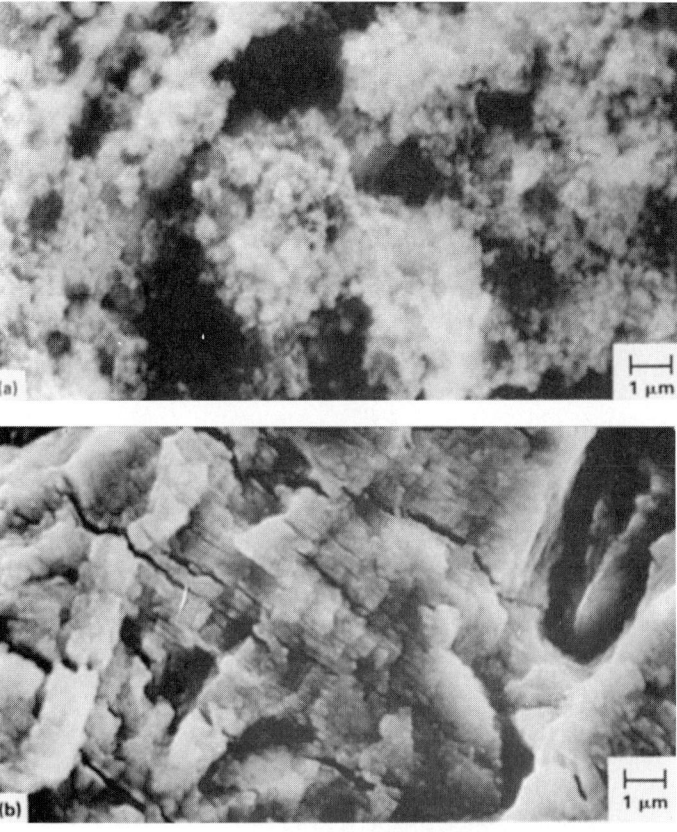

Figure 1C.4 Fatigue precrack region shown in Fig. 1C.3. (*a*) Before ultrasonic cleaning in a heated Alconox solution for 30 min. (*b*) After ultrasonic cleaning.

the originally embedded inclusions. Any surface corrosion products not completely removed within the first 30 min of water-base detergent cleaning are difficult to remove by further cleaning; therefore, prolonged cleaning provides no additional benefits.

Cathodic cleaning. This is an electrolytic process in which the sample to be cleaned is made the cathode, and hydrogen bubbles generated at the sample cause primarily mechanical removal of surface debris and deposits. An inert anode, such as carbon or platinum, is normally used to avoid contamination by plating upon the cathode. During cathodic cleaning, it is common practice to vibrate the electrolyte ultrasonically or to rotate the specimen (cathode) with a small motor. The electrolytes commonly used to clean ferrous fractures are sodium cyanide,[6,7] sodium carbonate, sodium hydroxide solutions, and inhibited sulfuric acid.[8] Because ca-

Figure 1C.5 Effect of increasing the ultrasonic cleaning time in a heated Alconox solution. (a) 30 min. (b) 3.5 h. Note the dislodging of the inclusion (left side of fractograph) and chemical etching of the fracture surface.

thodic cleaning occurs primarily by the mechanical removal of deposits due to hydrogen liberation, the fracture surface should not be chemically damaged after elimination of the deposits.

The use of cathodic cleaning to remove rust from steel fracture surfaces has been successfully demonstrated.[9] In this study, AISI 1085 heat-treated steel and EX16 carburized steel fractures were exposed to a 100% humidity environment at 65°C (150°F) for 3 days. A commercially available sodium cyanide electrolyte, ultrasonically agitated, was used in conjunction with a platinum anode for cleaning. A 1-min cathodic cleaning cycle was applied to the rusted fractures, and the effectiveness of the cleaning technique without altering the fracture morphology was demonstrated.

Figure 1C.6 shows a comparison of an as-fractured surface with a corroded and cathodically stable ductile cracking region in a quenched-and-tempered 1085 carbon steel. The relatively low magnification (1000×) shows that the dimpled topography characteristic of ductile tearing was unchanged as a result of the corrosion and cathodic cleaning. High magnification (5000×) shows that the perimeters of the small interconnecting dimples were corroded away.

Chemical etching. If the above techniques are attempted and prove ineffective, the chemical-etch cleaning technique, which involves treating the surface with mild acids or alkaline solutions, should be implemented. This technique should be used only as a last resort because it involves possible chemical attack of the fracture surface. In chemical-etch cleaning, the specimen is placed in a beaker containing the cleaning solution and is vibrated ultrasonically. It is sometimes necessary to heat the cleaning solution. Acetic acid, phosphoric acid, sodium hydroxide, ammonium citrate, ammonium oxalate solutions, and commercial solutions have been used to clean ferrous alloys.[8] Titanium alloys are best cleaned with nitric acid.[4] Oxide coatings can be removed from aluminum alloys by using a warmed solution containing 70 mL of orthophosphoric acid (85%), 32 g of chromic acid, and 130 mL of water.[10] However, it has also been recommended that fracture surfaces of aluminum alloys be cleaned only with organic solvents.[4]

Especially effective for chemical-etch cleaning are acids combined with organic corrosion inhibitors.[11,12] These inhibited acid solutions limit the chemical attack to the surface contaminants while protecting the base metal. For ferrous fractures, immersion of the samples for a few minutes in a 6N hydrochloric acid solution containing 2 g/L of hexamethylene tetramine has been recommended.[6] Ferrous and nonferrous service fractures have been successfully cleaned by using the following inhibited acid solution: 3 mL of hydrochloric acid (1.19 specific gravity), 4 mL of 2-butyne-1,4-diol (35% aqueous solution), and 50 mL of deionized water.[13] This study demonstrated the effectiveness of the cleaning solution in removing contaminants from the fracture surfaces of a low-carbon steel pipe and a Monel Alloy 400 expansion joint without damaging the underlying metal. Various fracture morphologies were not affected by the inhibited acid treatment when the cleaning time was appropriate to remove contaminants from these service fractures.

Sectioning a fracture

It is often necessary to remove the portion containing a fracture from the total part, because the total part is to be repaired, or to reduce the specimen to a convenient size. Many of the examination tools—for ex-

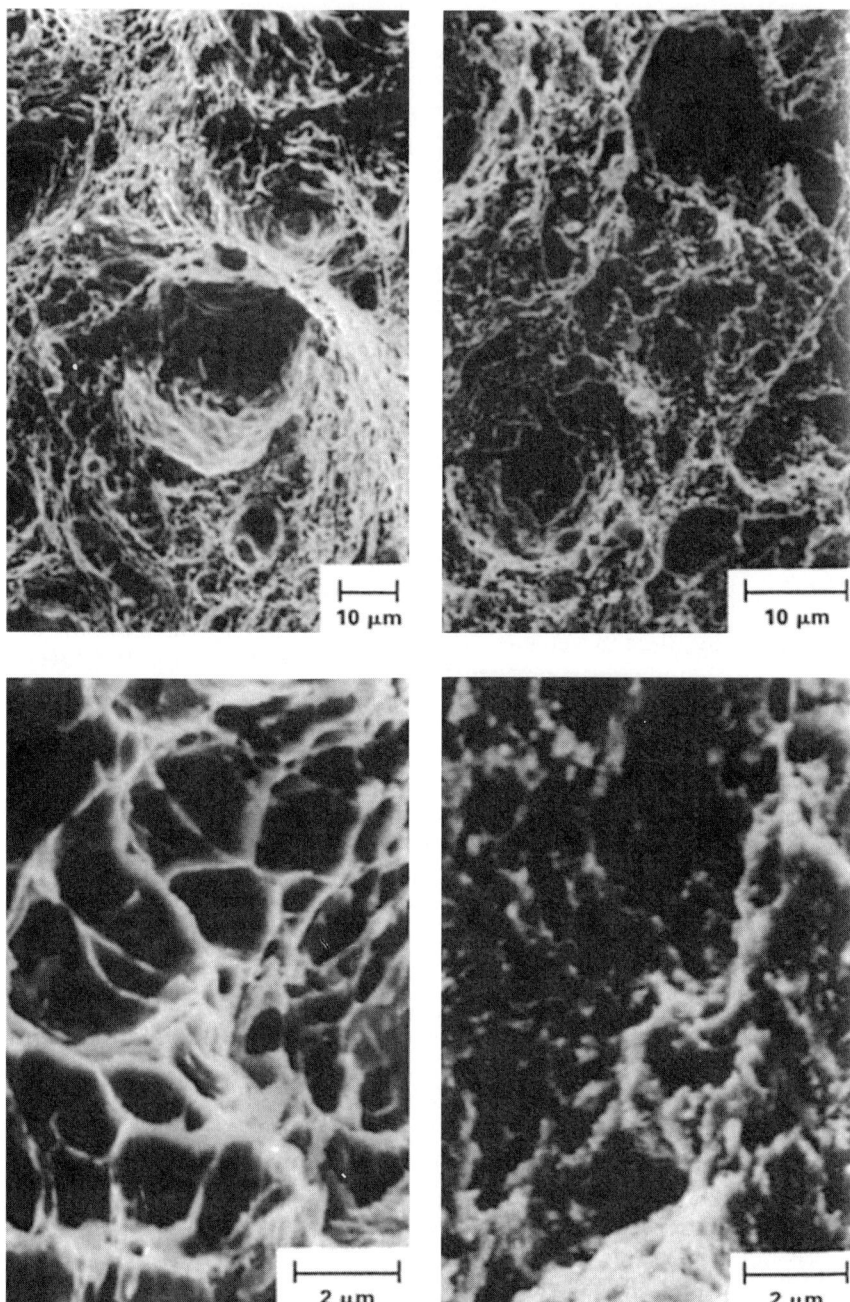

Figure 1C.6 Comparison of stable ductile crack growth areas from quenched-and-tempered 1085 carbon steel at increasing magnifications. The fractographs on the left show the as-fractured surface; those on the right show the fracture surface after corrosion exposure and cathodic cleaning.

ample, the scanning electron microscope and the electron microprobe analyzer—have specimen chambers that limit specimen size. Records, either drawings or photographs, should be maintained to show the locations of the cuts made during sectioning.

All cutting should be done such that fracture faces and their adjacent areas are not damaged or altered in any way; this includes keeping the fracture surface dry whenever possible. For large parts, the common method of specimen removal is flame cutting. Cutting must be done at a sufficient distance from the fracture so that the microstructure of the metal underlying the fracture surface is not altered by the heat of the flame and so that none of the molten metal from flame cutting is deposited on the fracture surface.

Saw cutting and abrasive cutoff wheel cutting can be used for a wide range of part sizes. Dry cutting is preferable because coolants may corrode the fracture or may wash away foreign matter from the fracture. A coolant may be required, however, if a dry cut cannot be made at a sufficient distance from the fracture to avoid heat damage to the fracture region. In such cases, the fracture surface should be solvent cleaned and dried immediately after cutting.

Some of the coating procedures mentioned above may be useful during cutting and sectioning. For example, the fracture can be protected during flame cutting by taping a cloth over it and can be protected during sawing by spraying or coating it with a lacquer or a rust-preventive compound.

Opening secondary cracks

When the primary fracture has been damaged or corroded to a degree that obscures information, it is desirable to open any secondary cracks to expose their fracture surfaces for examination and study. These cracks may provide more information than the primary fracture. If rather tightly closed, they may have been protected from corrosive conditions, and if they have existed for less time than the primary fracture, they may have corroded less. Also, primary cracks that have not propagated to total fracture may have to be opened.

In opening these types of cracks for examination, care must be exercised to prevent damage, primarily mechanical, to the fracture surface. This can usually be accomplished if opening is done such that the two faces of the fracture are moved in opposite directions, normal to the fracture plane. A saw cut can usually be made from the back of the fractured part to a point near the tip of the crack, using extreme care to avoid actually reaching the crack tip. This saw cut will reduce the amount of solid metal that must be broken. Final breaking of the specimen can be done by:

- Clamping the two sides of the fractured part in a tensile-testing machine, if the shape permits, and pulling
- Placing the specimen in a vise and bending one half away from the other half by striking it with a hammer in a way that will avoid damaging the crack surfaces
- Gripping the halves of the fracture in pliers or vise grips and bending or pulling them apart

It is desirable to be able to distinguish between a fracture surface produced during opening of a primary or secondary crack. This can be accomplished by ensuring that a different fracture mechanism is active in making the new break; for example, the opening can be performed at a very low temperature. During low-temperature fracture, care should be taken to avoid condensation of water, because this could corrode the fracture surface.

Crack separations and crack lengths should be measured before opening. The amount of strain that occurred in a specimen can often be determined by measuring the separation between the adjacent halves of a fracture. This should be done before preparation for opening a secondary crack has begun. The lengths of cracks may also be important for analyses of fatigue fractures or for fracture mechanics considerations.

Effect of nondestructive inspection

Many of the so-called nondestructive inspection methods are not entirely nondestructive. The liquid penetrants used for crack detection may corrode fractures in some metals, and they will deposit foreign compounds on the fracture surfaces; corrosion and the depositing of foreign compounds could lead to misinterpretation of the nature of the fracture. The surface of a part that contains, or is suspected to contain, a crack is often cleaned for more critical examination, and rather strong acids that can find their way into a tight crack are frequently used. Many detections of chlorine on a fracture surface of steel, for example, which were presumed to prove that the fracture mechanism was SCC, have later been found to have been derived from the hydrochloric acid used to clean the part.

Even magnetic-particle inspection, which is often used to locate cracks in ferrous parts may affect subsequent examination. For example, the arcing that may occur across tight cracks can affect fracture surfaces. Magnetized parts that are to be examined by SEM will require demagnetization if scanning is to be done at magnifications above about $500\times$.

References to Appendix 1C

1. R. D. Zipp, "Preservation and Cleaning of Fractures for Fractography," *Scan. Elec. Microsc.*, no. 1, pp. 355–362, 1979.
2. A. Phillips et al., *Electron Fractography Handbook*, MCIC-HB-08, Metals and Ceramics Information Center, Battelle Columbus Laboratories, June 1976, pp. 4–5.
3. W. R. Warke et al., "Techniques for Electron Microscope Fractography," in *Electron Fractography*, STP 436, American Society for Testing and Materials, Philadelphia, Pa., 1968, pp. 212–230.
4. J. A. Fellows et al., in *Metals Handbook*, 8th ed., vol. 9: *Fractography and Atlas of Fractographs*. American Society for Metals, Metals Park, Ohio, 1974, pp. 9–10.
5. B. E. Boardman et al., "A Coating for the Preservation of Fracture Surfaces," Paper 750967, presented at SAE Automobile Engineering Meeting, Detroit, Mich., Society of Automotive Engineers, Oct. 13–17, 1975.
6. H. DeLeiris et al., "Techniques of De-Rusting Fractures of Steel Parts in Preparation for Electronic Micro-Fractography," *Mem. Sci. Rev. de Met.*, vol. 63, pp. 463–472, May 1966.
7. P. M. Yuzawich and C. W. Hughes, "An Improved Technique for Removal of Oxide Scale from Fractured Surfaces of Ferrous Materials," *Pract. Metallog.*, vol. 15, pp. 184–195, 1978.
8. B. B. Knapp, "Preparation & Cleaning of Specimens," in *The Corrosion Handbook*, Wiley, New York, 1948, pp. 1077–1083.
9. E. P. Dahlberg and R. D. Zipp, "Preservation and Cleaning of Fractures for Fractography—Update," *Scan. Elec. Microsc.*, no. 1, pp. 423–429, 1981.
10. G. F. Pittinato et al., *SEM/TEM Fractography Handbook*, MCIC-HB-06, Metals and Ceramics Information Center, Battelle Columbus Laboratories, pp. 4–5, Dec. 1975.
11. C. R. Brooks and C. D. Lundin, "Rust Removal from Steel Fractures—Effect on Fractographic Evaluation," *Microstruc. Sci.*, vol. 3, pp. 21–33, 1975.
12. G. G. Elibredge and J. C. Warner, "Inhibitors and Passivators," in *The Corrosion Handbook*, Wiley, New York, 1948, pp. 905–916.
13. E. P. Dahlberg, "Techniques for Cleaning Service Failures in Preparation for Scanning Electron Microscope and Microprobe Analysis," *Scan. Elec. Microsc.*, pp. 911–918, 1974.

Appendix 1D Cleaning of Fracture Surfaces*

A clean fracture surface is a prerequisite for the definition of the mode of failure. As with any other type of specimen, fracture surfaces must be clean for successful SEM imaging. Because fracture surfaces are fragile, their cleaning must be approached with caution and common sense. The cleaning methods are classified in Table 1D.1. As a rule, the least aggressive method must be attempted before proceeding to more aggressive techniques, because the latter are capable of damaging the fracture surface. Dahlberg (1974, 1976), Zipp (1979), and Dahlberg and Zipp (1981) comprehensively review and compare each method, and their views are summarized below.

---

*From B. L. Gabriel, *SEM: A User's Manual for Materials Science*, American Society for Metals, Metals Park, Ohio, 1985.

TABLE 1D.1 Classification of Methods for Cleaning Fracture Surfaces According to Their Degree of Aggressiveness

Method	For removal of	Degree of aggressiveness
Soft fiber brush and dry air	Loosely adhering debris and dust	Least aggressive
Organic solvents and ultrasonic bath		↓
Toluene or xylene	Oil and grease	↓
Ketones	Varnish and gum	↓
Alcohol	Dyes and fatty acids	↓
Replica stripping	Insoluble debris and oxides	↓
Detergents (e.g., Alconox)	Corrosion products and oxides	↓
Cathodic cleaning	Deposits and oxides	↓
Corrosion-inhibited acids	Sulfides and oxides	↓
Acid etches	Oxides	Most aggressive

The least aggressive cleaning method is capable of removing loosely adhering dust or debris from the fracture surface. Short-haired *soft brushes* (e.g., a trimmed artist's paintbrush) or bursts of *compressed gas* are useful for removing dust. This method alone is rarely sufficient to clean a surface; more often, organic films (oil, grease, etc.) also obscure the surface. An ultrasonic treatment with an *organic solvent* followed by blowing with compressed gas removes both organic films and debris. Zipp (1979) recommends the various solvents listed in Table 1D.1. For heavily contaminated surfaces, it may be necessary to pass through several changes or a series of solvents.

If the debris obscuring the surface is of interest, solvent cleaning should not be used immediately. For example, if the surface is corroded, it may be desirable to first analyze the composition of the corrosion products with energy-dispersive spectroscopy, then clean the specimen and examine the native fracture surface. Alternatively, the surface may be simultaneously cleaned and the reaction products preserved by preparing *extraction replicas* (synonymous with cleaning replicas). When a strip of cellulose acetate softened with acetone is placed over a fracture and firmly pressed into position (with one's thumb), the gel will encapsulate any material that is not part of the base metal. After the acetone has evaporated, the cellulose acetate retains the entrapped reaction products. The replica is then removed from the surface, and both the particles and their location relative to the fracture surface are preserved.

Heavily contaminated/oxidized surfaces may require sequential stripping of several replicas before the native surface is exposed. Those replicas applied first will contain heavier deposits than those applied subsequently, i.e., the first replica removes the most material

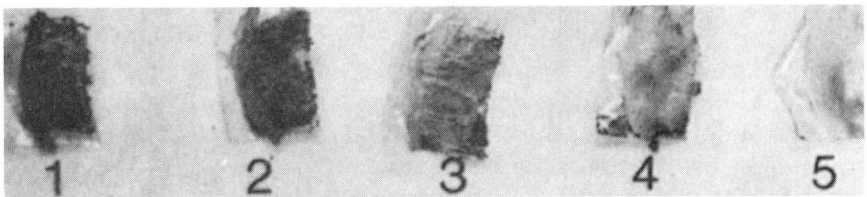

Figure 1D.1 Replicas stripped sequentially from an oxidized fracture surface. (Approximate size.) (*Courtesy of Richard Zipp.*)

(Fig. 1D.1). The fractured component is then cleaned in an ultrasonic bath with acetone followed by blowing with compressed air, and the effectiveness of cleaning is evaluated with a stereo microscope.

Although extraction replicas are very effective for cleaning oxidized surfaces, their main disadvantage is that, despite solvent cleaning, fragments of cellulose acetate adhere to very rough surfaces. Because this material is nonconductive, it will charge during SEM irradiation and degrade image quality. This problem is aggravated if the replica is stripped before it has completely dried; the cellulose acetate must be dry. If particle preservation is not an issue, this problem is minimized by ultrasonic cleaning of the specimen with acetone between replicas. The final acetone rinse should be repeated two or three times with fresh acetone to ensure that all residual cellulose acetate has been removed.

Extraction replicas are effective for removing most oxidation products, but severely oxidized surfaces may require more rigorous cleaning. It must be understood that although cleaning will expose the metal surface beneath the oxide layer, oxidation itself has consumed some of the base metal, destroying the outermost layer of the native surface. Consequently the removal of oxide scale does not restore the fracture surface to its condition at the moment of failure. Under very harsh conditions, such as the environmental exposure of a failed component for months or years, the native fracture surface may be completely destroyed. Under less severe conditions, enough of the surface usually survives for definition of the fracture mode. Further, the very aggressive cleaning methods may themselves attack the base metal and erase all fine structural features. With these factors in mind, the analyst must be cautious and prepared to interrupt any of the following cleaning processes. One can readily resume the cleaning method, but the fracture surface cannot be restored once it has been obliterated by inappropriate cleaning methods.

Moderately aggressive cleaning with *water-based detergents* removes adherent oxides and corrosion products. Alconox is a popular detergent available from many laboratory suppliers. Zipp (1979) recommends a solution of 15 g Alconox/350 mL water; the solution is heated to 90°C and fractures are cleaned for 30 min. Simultaneous ultrasonic treatment is desirable. The specimen is then thoroughly rinsed with water followed by acetone and dried. Prolonging this treatment or increasing the concentration of the Alconox is ineffective and may cause attack of the base metal.

Cathodic cleaning is another moderately aggressive cleaning method for removal of oxides or heavy surface deposits. The specimen is made the cathode, an inert metal or graphite the anode, and both are submerged within an electrolytic bath of sodium cyanide, sodium carbonate, or sodium hydroxide (DeLeiris et al., 1966; Yuzawich and Hughes, 1978). Commercially available Endox 214 is another popular electrolyte (Dahlberg and Zipp, 1981). During the electrolytic reaction, the specimen is mechanically cleaned by the scrubbing action of hydrogen bubbles generated by the specimen. The specimen is rinsed in water and then acetone, dried, and examined. The cleaning should be periodically interrupted and its effectiveness evaluated with a stereo microscope; if required, cathodic cleaning may be repeated. High-strength steels or other alloys susceptible to hydrogen-induced cracking may be adversely affected by cathodic cleaning.

Sulfides and oxides are removed from fracture surfaces using *corrosion-inhibited acids*. The acid attacks and displaces the reaction products while the inhibitor protects the base metal from attack. However, the base metal will be attacked (etched) if the progression of cleaning is not carefully monitored. Ferrous alloys have been cleaned with $6N$ HCl containing 2 g/L of hexamethylene tetramine (DeLeiris et al., 1966; Lane and Ellis, 1971; Dahlberg, 1974). Kayafas (1980) used 1,3-di-n-butyl-2-thiourea to remove iron sulfide films. Both ferrous and nonferrous alloys may be cleaned with 2-butyne-1,4-diol inhibited HCl (Nathan, 1965; Farrar, 1974; Dahlberg, 1976). This corrosion-inhibited acid is prepared as follows:

HCl (1.190 specific gravity)	3 mL
2-Butyne-1,4-diol (35% aqueous)	4 mL
Distilled water	50 mL

The specimen is cleaned by immersion in an ultrasonic bath for 30 s, followed by rinsing with water and then acetone, and drying. Do not exceed a 30-s exposure to these solutions; a prolonged treatment increases the probability of base-metal attack.

If the corrosion-inhibited acid method fails, as a last resort the specimen may be cleaned in a *weak acid* or *base*. This extremely aggressive method will attack the base metal unless constantly monitored.

Knapp (1948) recommends weak acetic acid, phosphoric acid, or sodium hydroxide for cleaning ferrous alloys. Titanium alloys may be cleaned with nitric acid (Zipp, 1979). Aluminum alloys are cleaned with a mixture of orthophosphoric acid (70 mL of 85% aqueous solution), chromic acid (32 g), and distilled water (130 mL), as described by Pittinato et al. (1975). Following the brief submersion in an acid or base, the specimen is rinsed in water and then acetone, and dried. Water washing must be thorough to stop the reaction; residual acid or base will consume the base metal, and vapors will attack stereo microscope objective lenses.

To summarize, the analyst should always choose the least aggressive cleaning method that effectively exposes the fracture surface. The objectives of the study should be known before the specimen is cleaned; if debris removed from the surface requires preservation, extraction replicas should be prepared, because with other methods the surface materials are lost. One should also evaluate the effectiveness of the treatment while it is in progress; cleaning will not be adversely affected if stopped and restarted. Always evaluate cleaning with a low-power stereo microscope, and when the cleaning appears adequate, examine the specimen in the SEM. By following these precautions, damage of the surface of interest is avoided.

References to Appendix 1D

Dahlberg, E. P. (1974): "Techniques for Cleaning Service Failures in Preparation for Scanning Electron Microscopy and Microprobe Analysis," *IITRI/SEM*, p. 911.

Dahlberg, E. P. (1976): "Failure Analysis by Examination of Fracture Surfaces. Analytical Procedures and Cleaning Techniques for Field Failures," *IITRI/SEM*, vol. 1, p. 715.

Dahlberg, E. P., and R. D. Zipp (1981): "Preservation and Cleaning of Fractures for Fractography—Update," *SEM, Inc.*, vol. 1, p. 423.

DeLeiris, H., et al. (1966): "Techniques of De-Rusting Fractures of Steel Parts in Preparation for Electronic Micro-Fractography," *Mem. Sci. Rev. de Met.*, vol. 63, p. 463.

Farrar, J. C. M. (1974): "The Role of the SEM in the Failure Analysis of Welded Structures," *IITRI/SEM*, p. 859.

Kayafas, I. (1980): "Corrosion Product Removal from Steel Fracture Surfaces for Metallographic Examination," *Corrosion*, vol. 36, no. 8, p. 443.

Knapp, B. B. (1948): "Preparation and Cleaning of Specimens," *The Corrosion Handbook*, Wiley, New York, p. 1077.

Lane, G. S., and J. Ellis (1971): "The Examination of Corroded Fracture Surfaces in the Scanning Electron Microscope," *Corr. Sci.*, vol. 11, p. 661.

Nathan, C. C. (1965): "Corrosion Inhibitors," in *Encyclopedia of Chemical Technology*, vol. 6, Wiley, New York, p. 317.

Pittinato, G. F., et al. (1975): *SEM/TEM Fractography Handbook*. Metals and Ceramics Information Center, Battelle Columbus Laboratories, Columbus, Ohio.

Yuzawich, P. M., and C. W. Hughes (1978): "An Improved Technique for Removal of Oxide Scale from Fractured Surfaces of Ferrous Material," *Pract. Metallog.*, vol. 15, p. 184.

Zipp, R. D. (1979): "Preservation and Cleaning of Fractures for Fractography," *SEM, Inc.*, vol. 1, p. 355.

Appendix 1E Examination of Cleaning Techniques for Postfailure Analysis*

The metallurgist has found fractographic examination of fracture surfaces to be a valuable tool in failure analysis. A meaningful examination, however, may not be possible if the fracture surfaces become contaminated with grease, oil, debris, oxidation, and corrosion products. Typical cleaning practices include the use of a dry air blast, organic solvents (such as acetone, toluene, and alcohol), and repeated stripping of cellulose-acetate replicas. These procedures are useful in the removal of loosely adhering particles, grease, and oils, whereas more tightly bonded oxidation and corrosion products require more aggressive removal methods. Several previous studies[1,2] have demonstrated the effectiveness of acid-based corrosion removal techniques. For example, Kayafas[2] has shown that cleaning with an inhibited hydrochloric acid pickling solution yields adequate fractographic results on the internal surface of hydrogen blisters in pipe line steel. Lohberg et al.[3] have reported success in removing oxidation products from ferritic materials using a deoxidizing agent known as Endox. In addition, Zipp[4] has recommended that if the benign cleaning methods (air blast, organic solvents, and repeated replica stripping) are unsuccessful in the removal of tightly bonded oxidation and corrosion products, then the fracture surface could be cleaned in the following manner: submerge the sample in a water-based detergent (15 g Alconox powder† + 350 cm^3 water) heated to 90°C and agitated in an ultrasonic cleaner for 30 min. *194 °F*

The object of this study is to evaluate the effectiveness of the cleaning procedure recommended by Zipp. To this end, a comparison is made of fracture markings from surfaces of as-fractured samples and corroded samples which were subsequently cleaned with the benign techniques and with the Alconox solution.

Experimental procedures

The material chosen for this study was ASTM A36 bridge steel because of its wide commercial use in structures and its tendency for fracture surface deterioration in the presence of aggressive environ-

---

*From R. S. Vecchio and R. W. Herzberg, in J. J. Mecholsky, Jr., and S. R. Powell, Jr. (eds.), *Fractography of Ceramic and Metal Failures,* STP 827, American Society for Testing and Materials, Philadelphia, Pa., 1982. (Copyright ASTM; reprinted with permission.)

†Alconox powder is a detergent made by Alconox Inc., New York, N.Y. 10003; ingredients are not available.

ments. Three different fracture mechanisms were generated and examined in an ETEC Autoscan scanning electron microscope (SEM) at an accelerating voltage of 20 keV. None of the surfaces was coated before viewing. Cleavage fracture was generated in simple tension at liquid nitrogen temperature. A fatigue fracture surface was generated by load cycling a compact tension specimen at an R ratio of 0.7 over a stress intensity factor range of 13 to 28 MPa · $m^{1/2}$. Unstable crack growth in the remaining ligament of the fatigue sample led to the formation of microvoid coalescence.

The as-fractured samples were exposed to a corrosive environment of 5% salt water at 100°C for 2 h and then left exposed in room air (60% humidity) for 48 h. This resulted in the formation of corrosion products on the fracture surfaces. An energy dispersive x-ray analysis using a Tracor Northern 1710 system was performed to examine the nature of corrosion products resulting from the salt water exposure and reaction products which might have resulted from the Alconox cleaning treatment. In addition to the laboratory-generated fracture surfaces a polished metallographic section was treated in the Alconox solution for different lengths of time (15 and 30 min) and examined in the SEM.

Results and discussion

The typical appearance of the as-fractured samples is shown in Fig. 1E.1a to c, which reveals examples of microvoid coalescence, cleavage, and fatigue striations, respectively, as developed in this material. Isolated regions of fracture through pearlite packets, as evidenced by the observed carbide and ferrite lamellae, were also noted on all fracture surfaces. A photomicrograph typical of all fracture regimes in the corroded condition is shown in Fig. 1E.1d. This photograph clearly illustrates how difficult it is to examine a corroded fracture surface for evidence of specific micromechanisms of failure. After use of the benign cleaning techniques some oxidation and corrosion product remained on the surface, though the microvoid coalescence and cleavage regions (Fig. 1E.2a and b) could be identified. The fatigue-damaged region did not lend itself to easy analysis owing to the persistent corrosion products found on the fracture surface (Fig. 1E.2c and d).

Figure 1E.3 shows the appearance of the fracture surfaces after being cleaned with Alconox solution; much of the corrosion debris remaining after the standard cleaning procedure was removed, though there is evidence that the fracture surfaces were etched by the Alconox. Note the considerable presence of pearlite colonies on the fracture surfaces (see especially Fig. 1E.3b and c). As a result, the in-

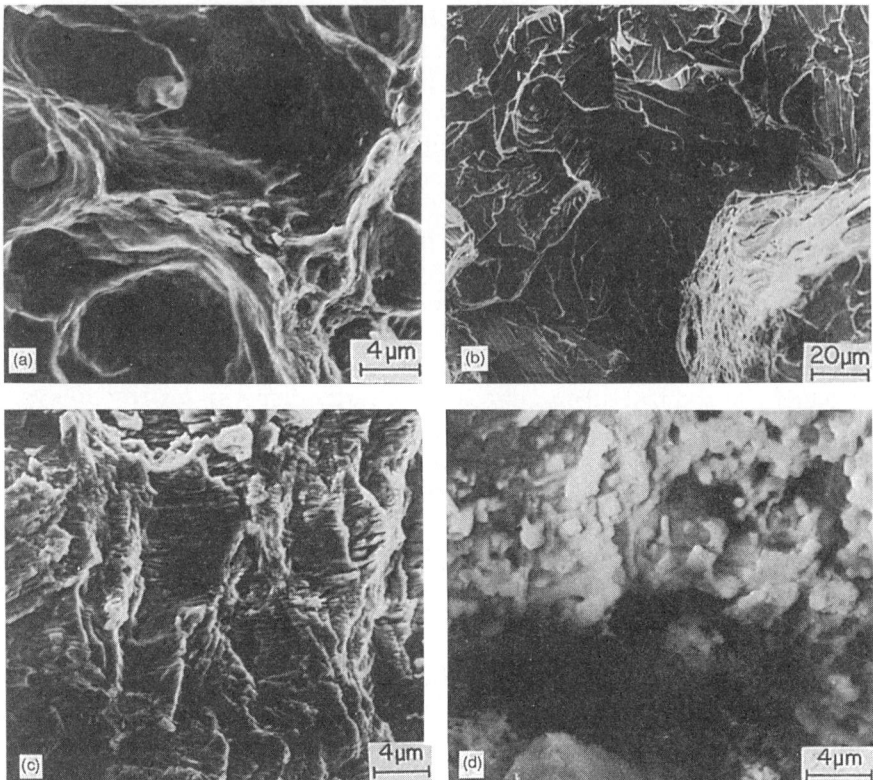

Figure 1E.1 Photomicrographs of as-fractured and as-corroded surfaces. (*a*) Microvoid coalescence. (*b*) Cleavage. (*c*) Fatigue striations. (*d*) Typical appearance of fracture surfaces after salt water exposure.

terpretation of fatigue fracture markings is greatly complicated by the simultaneous presence of parallel lamellae within the etched pearlite regions and parallel fatigue striation markings. Some parallel fracture surface markings were verified as fatigue striations by matching their spacings with the striation spacings predicted from the prevailing stress intensity factor conditions associated with the fatigue test.[5] Fatigue striations were also differentiated from most pearlite lamellae, since the striations were generally parallel to the advancing crack front whereas pearlite lamellae assumed a random orientation relative to the crack front. Overall, however, it was difficult to make a reliable determination of striations on the fatigue fracture surfaces; while much of the corrosion products were removed, the etching introduced complicating surface artifacts.

In general, it was found that the fatigue surfaces exhibited the poor-

Introduction 65

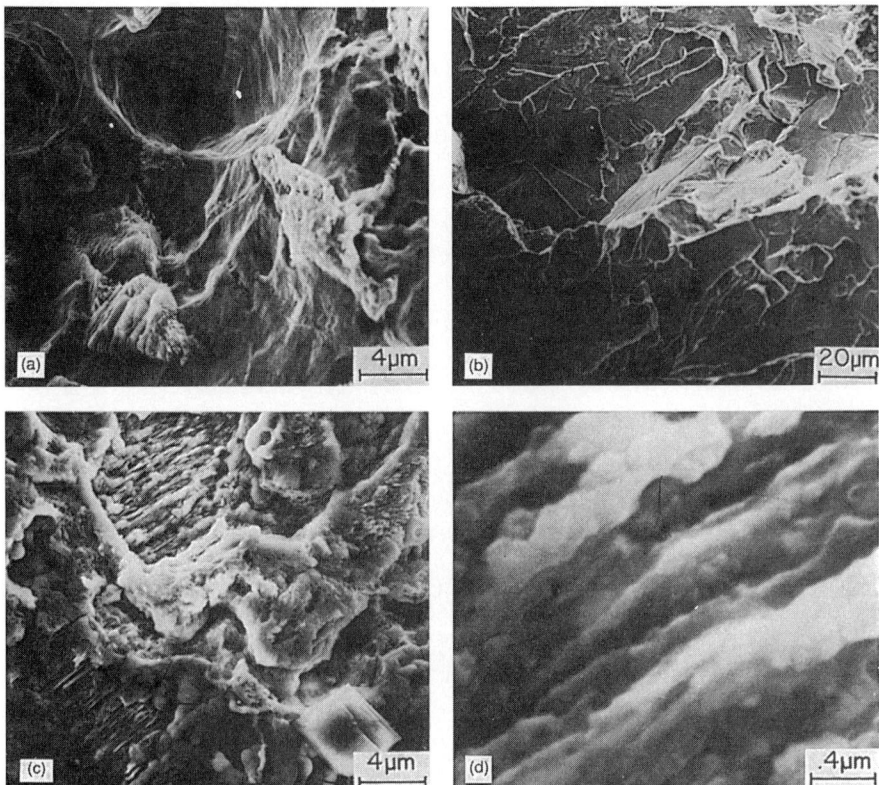

Figure 1E.2 Fracture surface appearance after cleaning by benign techniques. (*a*) Microvoid coalescence. (*b*) Cleavage. (*c*), (*d*) Fatigue striations.

est response to corrosion removal techniques. Other experiences in our laboratory have shown similar results with failed turbine disks.* Here, intergranular failure was easily identified, while evidence for fatigue damage remained obscured after both fracture zones were cleaned with the benign techniques. It appears that these cleaning techniques are most effective in removing corrosion products from morphological features that exhibit long-range flatness (that is, 20 to 40 μm). The (100) cleavage facets in body-centered-cubic materials, the side walls of the larger microvoids, and the intergranular facets all exhibit such long-range flatness and appear to clean more easily than the fatigue fracture surfaces. By comparison, the sizes of fatigue striations are on the average one to two orders of magnitude smaller

---

*Unpublished research, Lehigh University, Bethlehem, Pa. 18015.

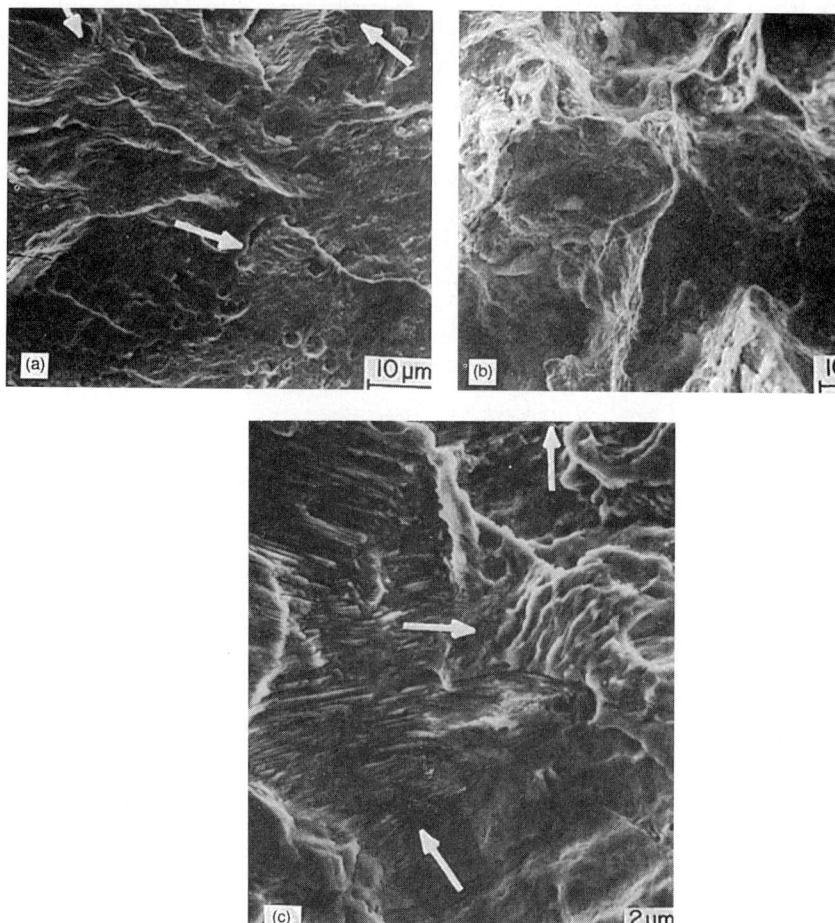

Figure 1E.3 Typical fracture surface appearance following cleaning in Alconox solution. (*a*) Microvoid coalescence. (*b*) Cleavage. (*c*) Fatigue fracture region. Note considerable cleavage and evidence of pearlite on fatigue fracture surfaces (see arrows).

than these other fracture markings, and thus are more likely to be obscured by corrosion debris. If the benign cleaning techniques leave numerous corrosion-free regions on cleavage, microvoid coalescence, and intergranular fracture surfaces, then the absence of large corrosion-free regions may serve as an indication of the existence of fatigue damage. If this is the case, then these observations may aid in failure analysis. More studies are needed to clarify this point.

In many failure investigations an x-ray analysis is essential in determining the nature of fracture surface contaminants. Figure 1E.4*b*

Introduction 67

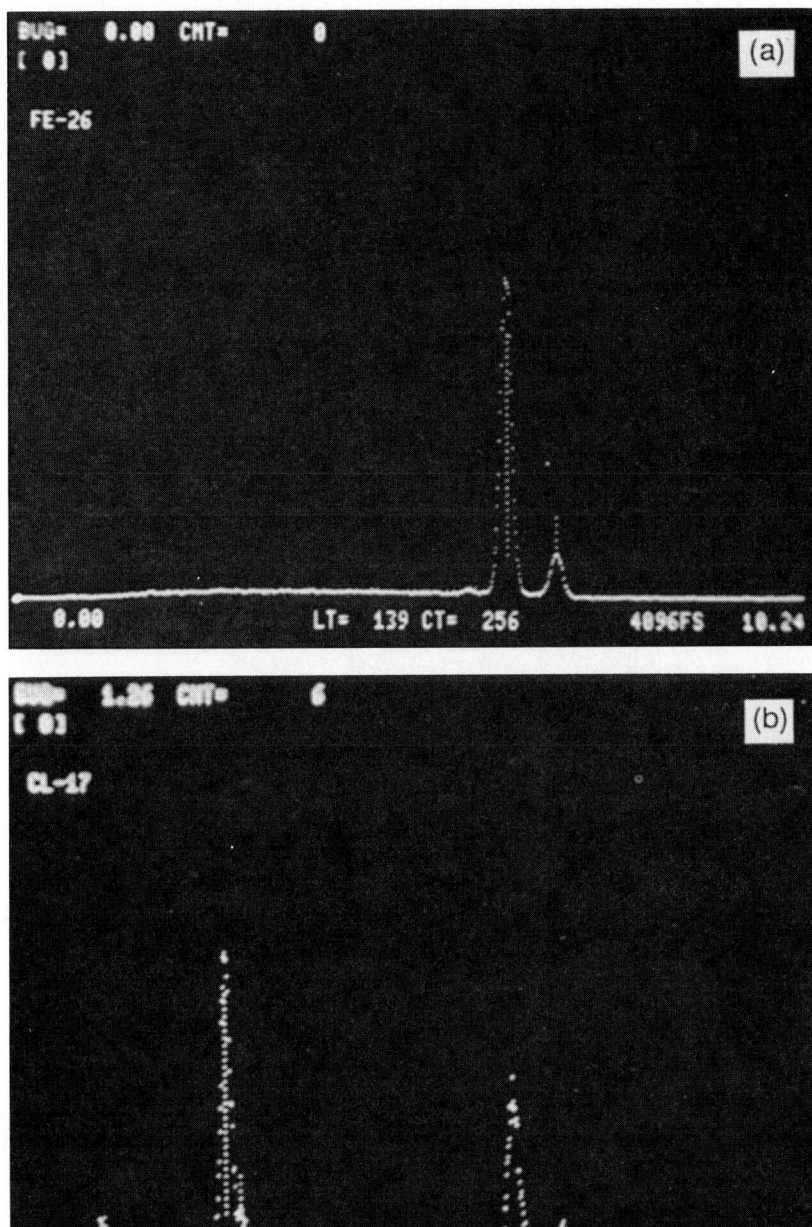

Figure 1E.4 Photospectragraphs of fracture surfaces. (*a*) As-fractured. (*b*) As-corroded.

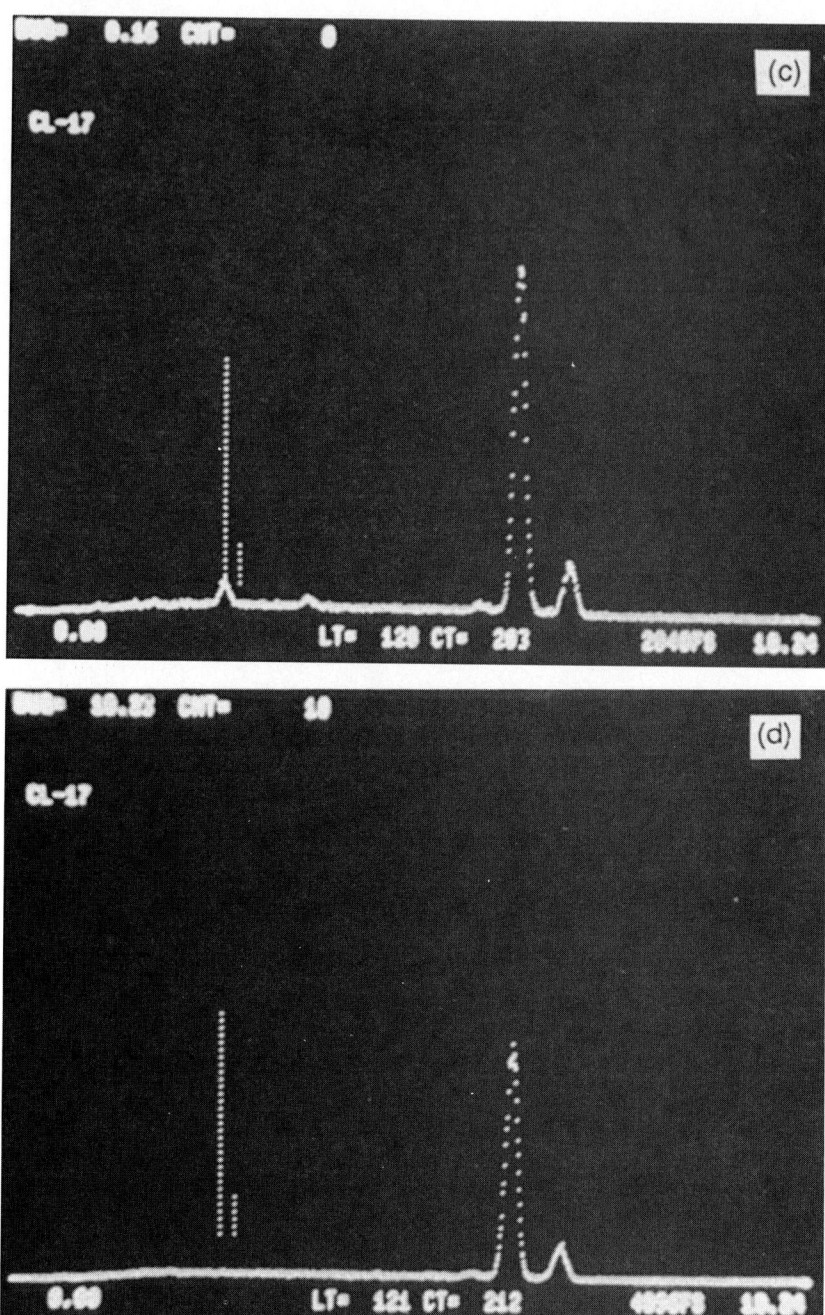

Figure 1E.4 (*Continued*) (*c*) After being cleaned with benign techniques. (*d*) After being cleaned in Alconox solution. Note the chlorine peak after cleaning by benign procedures and the absence of the peak following the cleaning in Alconox.

indicates the presence of large amounts of chlorine on the fracture surfaces of the as-corroded samples. If this were an actual service failure, the chlorine peak would serve as a clue in determining the nature and possible source of the corrosive medium.[6] The presence of the corrosive environment may have contributed to the failure (that is, stress corrosion cracking or corrosion-assisted fatigue). After cleaning the surface by the benign techniques, small amounts of chlorine were still detectable (Fig. 1E.4c). After use of the Alconox detergent, however, no indications remained of surface contaminants. Therefore it is suggested that if an x-ray analysis is indicated with failure examination, then use of the Alconox detergent should be avoided or deferred until after the analysis is completed.

To further study the nature of cleaning-induced damage, a metallographic section was examined after a 15-min detergent exposure. No pearlite colonies were revealed (Fig. 1E.5b), though the surface exhibited a moderate degree of degradation. After a 30-min exposure in Alconox, pearlite colonies were clearly evident. Figure 1E.5c shows dramatically how the surface of the metallographic section has been etched. We conclude that if cleaning in this detergent is necessary for the purpose of a fractographic examination, the exposure time should be limited to 15 min.

Conclusions

Based on the results of this investigation, the following conclusions have been reached:

1. Benign cleaning techniques are effective in the removal of loosely adhering particles, grease, and oils.

2. Fatigue-damaged surfaces exhibited the poorest response to the benign cleaning techniques. Based on this observation, identification of mechanisms responsible for fatigue damage may be difficult and unreliable when the fracture surface is obscured by corrosion debris.

3. Cleaning in Alconox detergent removes most tightly adhering oxidation and corrosion debris, though there is evidence that this procedure results in etching the surface. This degradation complicates the interpretation of fatigue-damaged regions.

4. If cleaning in Alconox detergent is necessary for the purpose of a fractographic examination the exposure should be limited to 15 min.

5. Insomuch as Alconox detergent removes most surface contaminants, dispersive x-ray analysis should be completed before such cleaning.

70   Chapter One

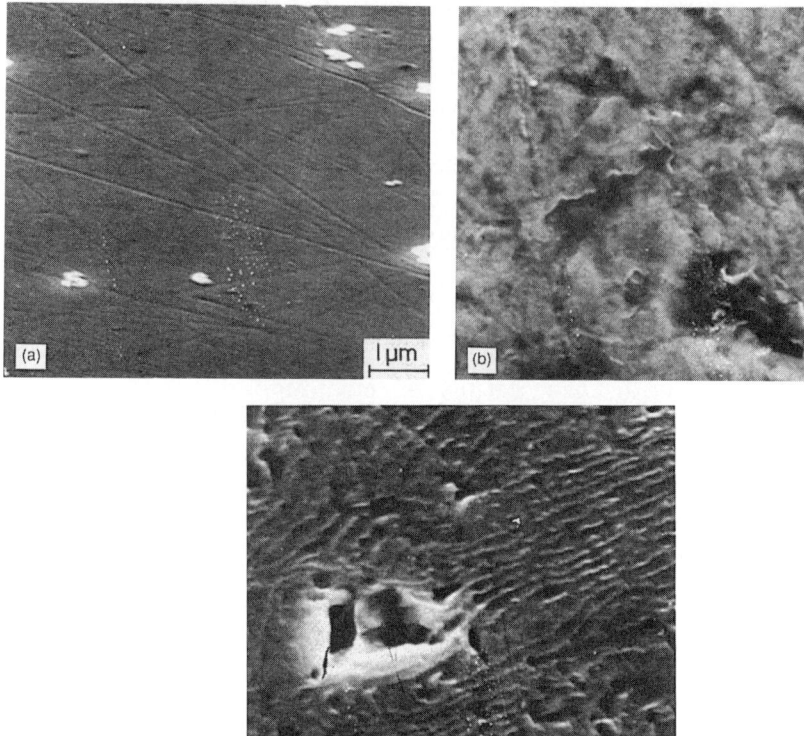

Figure 1E.5 Photomicrographs of metallographically prepared section exposed to Alconox solution for different lengths of time. (*a*) As-polished. (*b*) After 15-min exposure. (*c*) After 30-min exposure. Note the extensive amount of pearlite revealed after the 30-min exposure.

Acknowledgments

This work was supported in part by the Materials Research Center and the Department of Metallurgy and Materials Engineering, Lehigh University, Bethlehem, Pa., and Swiss Aluminum Ltd. Co., Zurich, Switzerland. The authors wish to acknowledge the help of Steve Paterson and Raymond Stofanak for their valuable discussions and assistance.

References to Appendix 1E

1. H. DeLeiris, et al., *Mem. Sci. Rev. de Met.*, vol. 63, pp. 463–472, May 1966.
2. I. Kayafas, *Corrosion NACE*, vol. 36, pp. 443–445, Aug. 1980.
3. R. Lohberg, et al., in *Microstructural Science*, vol. 9, Elsevier-North Holland, New York, 1981, pp. 421–427.
4. R. D. Zipp, in *Scanning Electron Microscopy I*, SEM Inc., AMF O'Hare, Ill., 1979, pp. 355–362.

5. R. C. Bates and W. G. Clark, Jr., *Trans. Am. Soc. Met.*, vol. 62, p. 380, 1969.
6. D. B. Ballard and H. Yakowitz, in *Scanning Electron Microscopy 1970*, IITRI, Chicago, Ill., April 1970, p. 32.

Appendix 1F Recommended Cleaning Solutions for Metallic Fractures*

Solution	Purpose
Organic solvents	To remove oil, grease, or plastic coatings from fractured surfaces
Acetic acid Phosphoric acid Sodium hydroxide	Used either cold or warm to clean Fe-base alloy fractures
Ammonium citrate Ammonium oxalate Sulfamic acid Immersion dip for 1 to 15 min in 6 N HCl containing 2 g per liter of hexamethylenetetramine	For cleaning rust, scale, or oxidation products from Fe-base alloy fractures
Nitric acid	Titanium alloy fractures
Orthophosphoric acid (70 mL) + chromic acid (32 g) + water (130 mL) or organic solvents	Used either cold or warm to clean aluminum alloy fractures

Appendix 1G A Scale and Rust Removal Solution†

For scale and rust removal without attack on iron, use either solution A at 150°F *or* solution B at room temperature.

Solution A	Solution B
Water 80 mL	Water 49 mL
HCl (conc.) 20 mL	HCl (conc.) 49 mL
Reilly Inhibitor #22 4 drops	Rodine #50 2 mL
Use at 150°F	Use at room temperature

- *Immerse* the iron parts until the rust and scale are removed.
- *Wash* the parts with water and with alcohol.
- *Coat* or *paint* the clean dry surfaces with Rustarest D14-50-S.

---

*From S. Battacharyya, V. E. Johnson, S. Agarwal, and M. A. H. Howes (eds.), *Failure Analysis of Metallic Materials by Scanning Electron Microscopy*, IIT Research Institute, Chicago, Ill., 1979.

†Origin unknown.

Note: Reilly Inhibitor #22 and probably Rodine #50 give off gas on standing. Leave the closure loose so that the gas can escape without building up an excessive pressure.

- Reilly Inhibitor #22 from Reilly Tar & Chemical Co., 20310 Chagrin Blvd., Cleveland 22, Ohio.
- Rodine #50 from Amchem Products, Inc., P.O. Box 33, Ambler, Pa.
- Rustarest D14-50-S from International Rustproof Co., 1061 East 260th St., Cleveland 32, Ohio.

Chapter 2

Mechanical Aspects and Macroscopic Fracture-Surface Orientation

As is usual in a severe frost, we have recently heard of many severe accidents consequent upon the fracture of the tires of the wheel of railway-carriages.

JAMES PRESCOTT JOULE
Philosophical Magazine, 1871

2.1 Introduction

The overall macroscopic orientation of the fracture surface of a broken component is generally related to the loading conditions. The relationship between this orientation and the load may be complex and difficult to deduce, but in spite of this, in many cases considerable information may be gleaned from the fracture-surface orientation. For example, if the fracture clearly did not involve any plastic deformation (as deduced from the absence of significant geometric or dimensional change) and hence the fracture was brittle, and the macroscopic fracture surface was relatively flat, then fracture probably occurred by a stress normal to the fracture plane.

The purpose of this chapter is to review mechanics aspects which are related to the cause of the fracture-plane orientation. The principles are outlined only in sufficient detail to illustrate their application, and examples of macroscopic fracture surfaces are given to demonstrate the ideas. It is emphasized that it is the macroscopic fracture-surface orientation that is the subject of this chapter. The detailed fracture-surface topography is the subject of Chaps. 3 and 4.

2.2 Tensile Test

To introduce the concepts of loading and fracture, it is useful to examine the simple tensile test. This provides a definition of the common tensile mechanical properties and also allows distinguishing between elastic and plastic deformation.

Consider a cylinder loaded along its axis, and the tensile machine operated so that the cylinder elongates at an approximately constant rate. Let the load be increased so that the cylinder increases in length, and then let the load be reduced to zero. If the length is now the same as before the load, the material is said to have been deformed elastically. Thus *elastic* deformation can be defined as deformation such that any changes in dimensions or shape are recovered when the external load is removed. However, if the axial load is increased beyond a certain value and then reduced to zero, the length will be greater than that before loading, and thus the cylinder has deformed plastically. *Plastic* deformation can be defined as deformation in which there is permanent change in dimensions or shape after the removal of the external load. The load beyond which plastic deformation occurs is called the *elastic limit*. This behavior is shown schematically in Fig. 2.1.

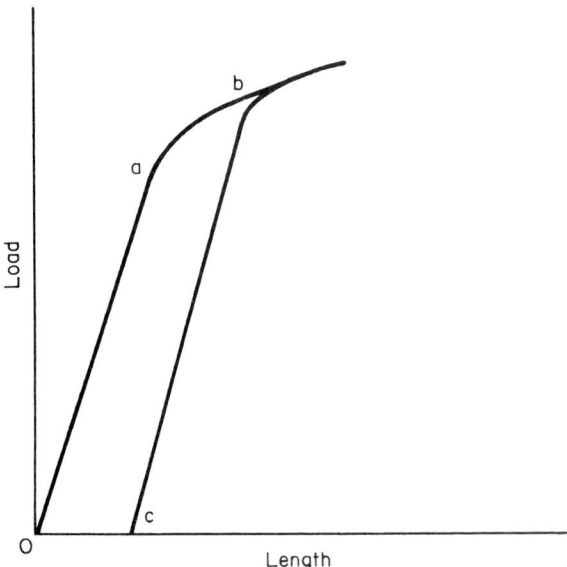

Figure 2.1 Illustration of the meaning of elastic and plastic deformation. If the cylinder is loaded from zero load to *a* then unloaded, the length will return to the original value, with an elongation of zero. Thus only elastic elongation has occurred. If then the cylinder is loaded from zero to *b* then unloaded, the length will be greater than the original length by the elongation 0c. Thus plastic deformation has occurred.

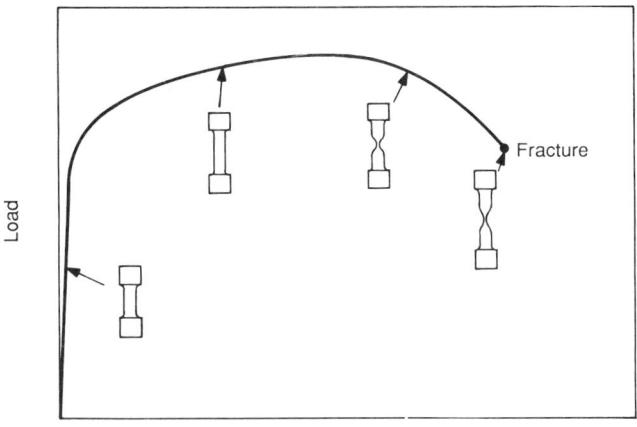

Figure 2.2 Typical load-elongation curve, showing the change in shape of the cylinder at various test stages.

For most metallic materials, to cause the cylinder to continue to elongate requires a continually increasing load. This is because the material becomes stronger as it is plastically deformed, and the material is said to *work* or *strain harden*. However, beyond a certain load and elongation, it is observed that subsequent plastic deformation occurs in a very localized region, which exhibits a decrease in cross-sectional area. This is called *necking*. Since the supporting cross-sectional area is now reduced, less load is required to continue to elongate the cylinder. Therefore the load-elongation curve passes through a maximum, and the load decreases until fracture occurs. A complete load-elongation curve is presented schematically in Fig. 2.2, which also shows the change in the profile of the cylinder. Figure 2.3 shows a tensile sample in the necking stage.

In a common standard tensile test, a specimen of specified dimensions (such as a gage length of 2.000 in and a diameter of 0.505 in) is used. The increase in the gage length is recorded as a function of load until fracture occurs. To make the results comparable to cases where the sample might have a different cross-sectional area, the load is divided by the original cross-sectional area to obtain the *nominal* or *en-*

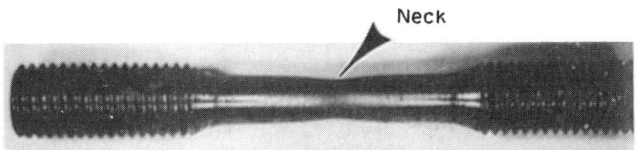

Figure 2.3 Tensile sample showing the necked region. (*Adapted from Smith.*[1])

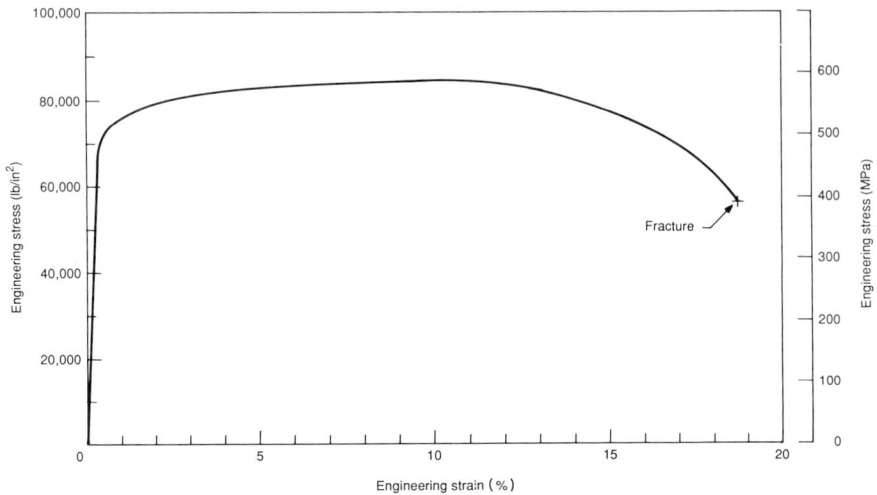

Figure 2.4 Typical tensile-test engineering-stress–engineering-strain diagram for cold-rolled 0.18% C steel.

gineering stress. Also the elongation (change in length) is divided by the original length to obtain the *nominal* or *engineering strain.* It is customary to plot the engineering stress versus the engineering strain to obtain a stress-strain curve. A typical result is shown in Fig. 2.4. On the scale plotted, the linear elastic range is not well resolved. Figure 2.5 shows a stress-strain curve with the low-strain region expanded so that the elastic region is revealed.

Because the stress at which plastic deformation first occurs is difficult to locate, an approximation to this value is used. It is called the *yield strength* and is based on the stress that corresponds to a small but specified plastic strain. Usually the value of strain of 0.002 is used, which is 0.2 percent. The procedure for obtaining such a yield strength is illustrated in Fig. 2.5. A line parallel to the linear elastic line is drawn from a strain of 0.002 (0.2 percent), and its intersection with the stress-strain curve defines the 0.2 percent yield strength. In some materials (for example, normalized low and medium carbon steels), yielding is very prominent in the stress-strain curve. This is illustrated in Fig. 2.6. In this case, the *yield point* or *yield-point stress* is used as a measure of the stress to induce yielding, and an effective yield strength is not quoted.

A prominent feature of the engineering-stress–engineering-strain curve is the maximum, and the stress at the maximum is called the *ultimate strength, ultimate tensile strength,* or more commonly *tensile strength.* Note that it is an artifact of the way the test is made, associated with necking.

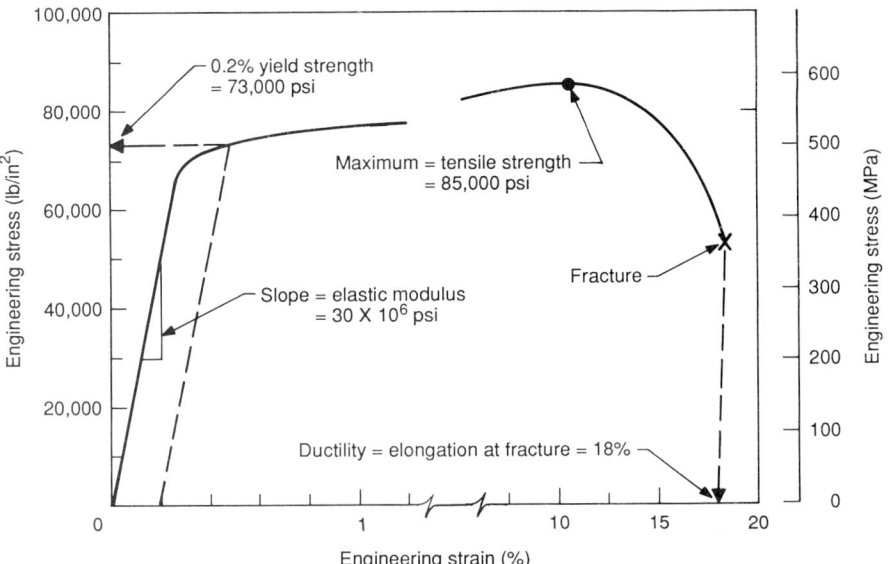

Figure 2.5 Engineering-stress–engineering-strain diagram for cold-rolled 0.18% C steel, showing how the 0.2 percent yield strength and other tensile mechanical properties are determined.

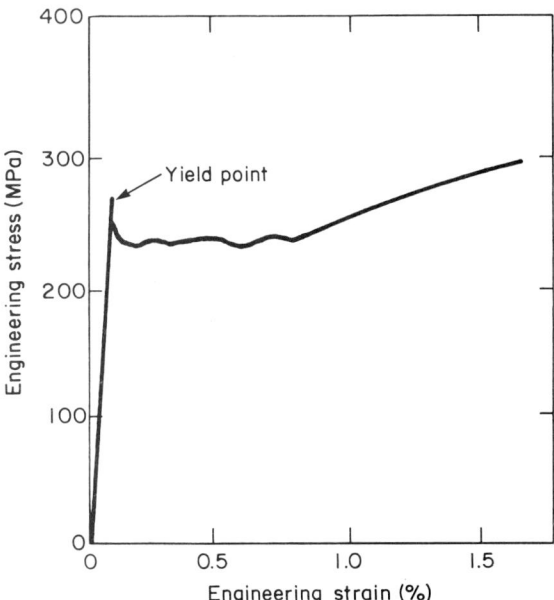

Figure 2.6 Engineering-stress–engineering-strain diagram for annealed 0.18% C steel, showing yield-point behavior.

The slope of the linear part of the curve in the elastic region gives the *elastic* or *Young's modulus*. The strain at fracture × 100 is defined as the *elongation,* more properly called the elongation at fracture. It is taken as a measure of the ductility of the material.

Note that the yield strength and the tensile strength are both measures of the strength of a material. However, the yield strength is often more applicable to design since it approximates the stress required to initiate plastic deformation, and since the part will cease to function well before fracture.

If, during the tensile test, the diameter at the minimum cross section is monitored (for example, at the neck during this stage) and the corresponding area is divided into the load, instead of the original (nominal) area, then the *true stress* is obtained. Also, if the infinitesimal change in length dl is divided by the instantaneous length l, then integrated from the initial length to the actual length, the *true strain* ln l/l_0 is obtained. A plot of these data gives the true-stress–true-strain curve. Figure 2.7 compares the two curves.

It is important to realize that the stress-strain curve, and hence the mechanical properties derived from it, may be sensitive to test variables such as test temperature and strain rate (see Sec. 2.9). Thus this behavior has to be taken into account in using tensile mechanical properties.

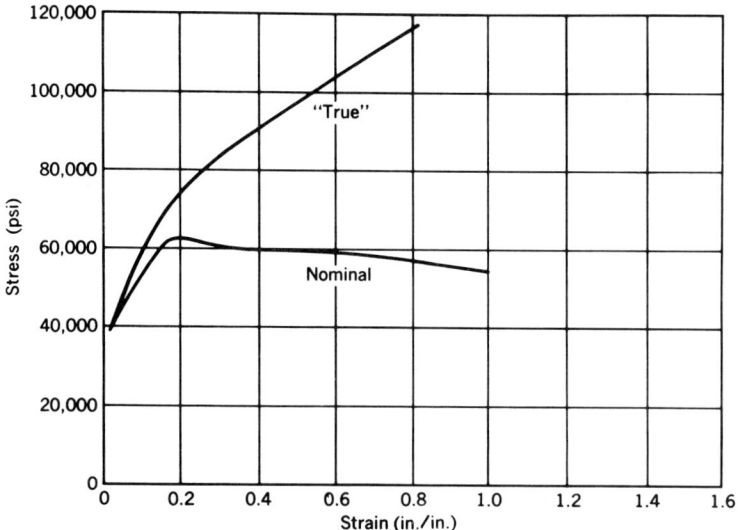

Figure 2.7 Comparison of the engineering-stress–engineering-strain and the true-stress–true-strain diagram for a semikilled steel. (*From Smith.[2]*)

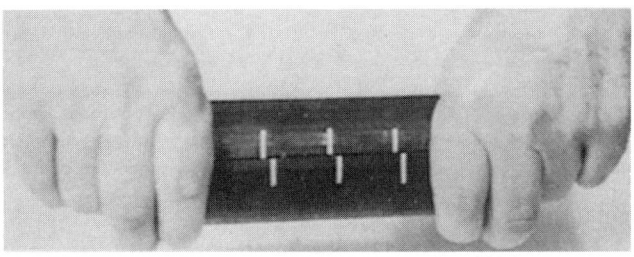

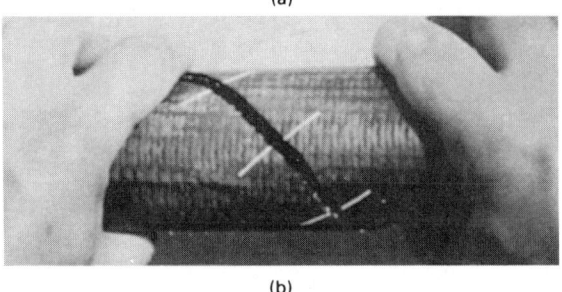

Figure 2.8 Illustration of the concept of normal and shear stresses. (*From Wulpi.*[3])

2.3 Principal Stresses

The loading on most components is usually neither unidirectional nor uniform. Thus the three-dimensional nature of the loading must be considered. The loads are expressed in terms of stress, which is the load at any point divided by an area of specified orientation relative to the direction of the force, with the area reduced to an infinitesimal size. Two types of stresses are commonly used: a *normal stress* [which can be positive (tension) or negative (compression)], based on an area normal to the direction of the force, and a *shear stress*, based on an area parallel to the force. The concept of normal and shear stresses is illustrated in Fig. 2.8. Upon twisting the hose, the vertical markers across the longitudinal slit are offset (Fig. 2.8a), showing that a shear stress would be generated if the tube were not cut. If the tube is slit at 45°, then upon twisting, the tube separates normal to this slit, showing that a normal stress would exist perpendicular to the line at 45° (Fig. 2.8b).

The external load can be resolved into three orthogonal components to produce three orthogonal stresses. This is easier to visualize in two dimensions, as illustrated in Fig. 2.9. The normal stress at the location shown is that of the resolved forces acting on an infinitesimal plane normal to the plane of the page and perpendicular to the force.

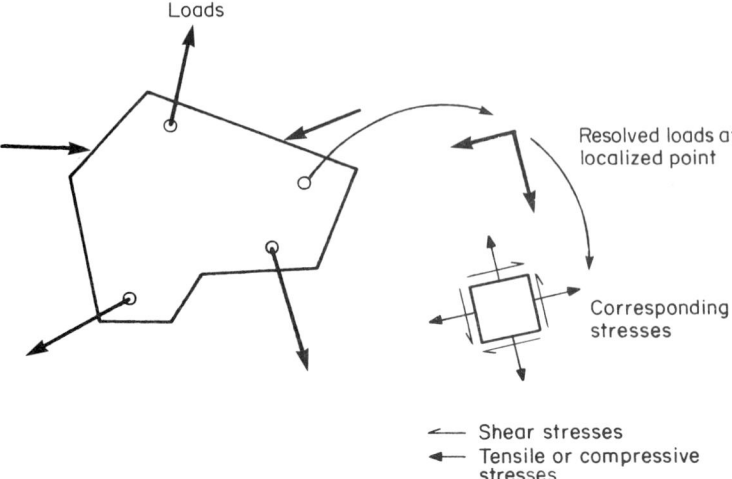

Figure 2.9 Resolution of external loads on a component into two orthogonal loads, and the corresponding normal and shear stresses, at a point.

Since the small region is constrained from rotating, the overall loading causes shear loads to act on adjacent faces to prevent this. The corresponding shear stresses are associated with the same areas (normal to the plane of the page) as the normal stresses.

A fundamental result of the application of mechanics to such a configuration is that there are orientations for which the normal stresses are a maximum and a minimum, and on these planes the shear stress is zero. This is illustrated in Fig. 2.10 for a simple two-dimensional loading case. The resolution of the forces onto different planes clearly gives a maximum and a minimum normal stress for the orientation in Fig. 2.10a. The maximum and minimum normal stresses are called the *principal normal stresses,* and the planes on which these extreme values of normal stress are obtained are called the *principal planes*. There also exists a plane on which the shear stress is a maximum and one on which it is a minimum, and these stresses are called *principal shear stresses*. (The normal stresses are not necessarily zero on these planes.) It is necessary to determine these principal stresses at all locations within the component in order to know where failure is likely to commence. These principal stresses are used in almost all modeling, stress analysis, and failure criteria.

In some loading configurations the orientation of the principal stresses is obvious, but in most cases an analytical treatment is required to determine this. The relation of the orientation of the external loading to that of the principal stresses is illustrated by a few ex-

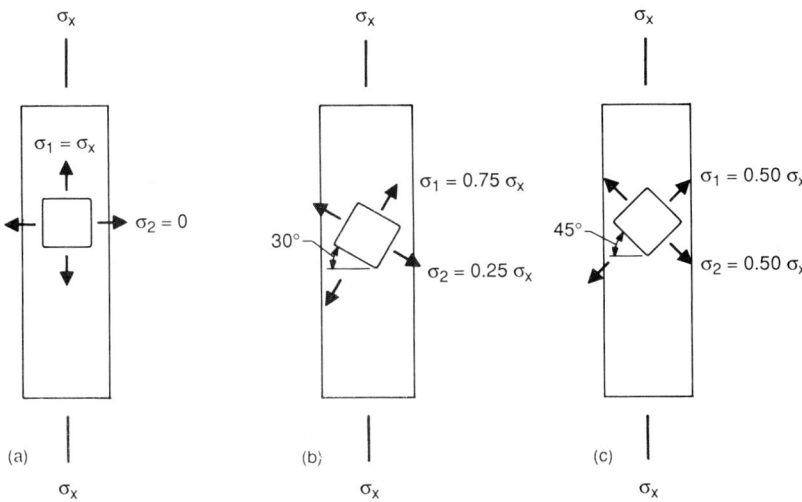

Figure 2.10 Illustration of the concept of maximum and minimum normal stress. The sample is a cylinder loaded axially with stress σ_x. (a) On the element shown, the maximum normal stress is σ_1 which is equal to σ_x. The minimum normal stress is zero. (b), (c) As the element is rotated, the maximum normal stress never equals the value of $\sigma_1 = \sigma_x$ in a.

amples, which were chosen for their simplicity and also because they involve common types of loading. Consider a round cylinder loaded axially, and the load uniformly distributed across the area normal to the loading direction (Fig. 2.11a); there is no other external load. The maximum stress σ_1 is the axial tensile stress. The other two normal stresses σ_2 and σ_3 are both zero, and hence are the minimum normal stresses. In this case the maximum shear stress is at 45° to the cylinder axis. In the case of simple bending of a beam of circular cross section (Fig. 2.11b), the upper surface is in tension in the longitudinal direction, and the lower surface is in compression (negative stress). The maximum normal stress is at the top surface, and the minimum normal stress is at the bottom (and negative). The maximum shear stress is on two orthogonal planes: perpendicular and parallel to the cylinder axis. It is a maximum at the axis. For the case of simple torsion of a cylinder (Fig. 2.11c), the maximum and minimum normal stresses are at 45° to the cylinder axis (the third normal stress is zero), and the maximum shear stress is parallel to the axis. This is illustrated in Fig. 2.8.

The effect of the principal stresses on the macroscopic fracture-surface plane can be visualized in the case of a material which fractures on a plane normal to the principal stress. Such a material would be considered brittle (but see Sec. 2.11). A ductile material tends to

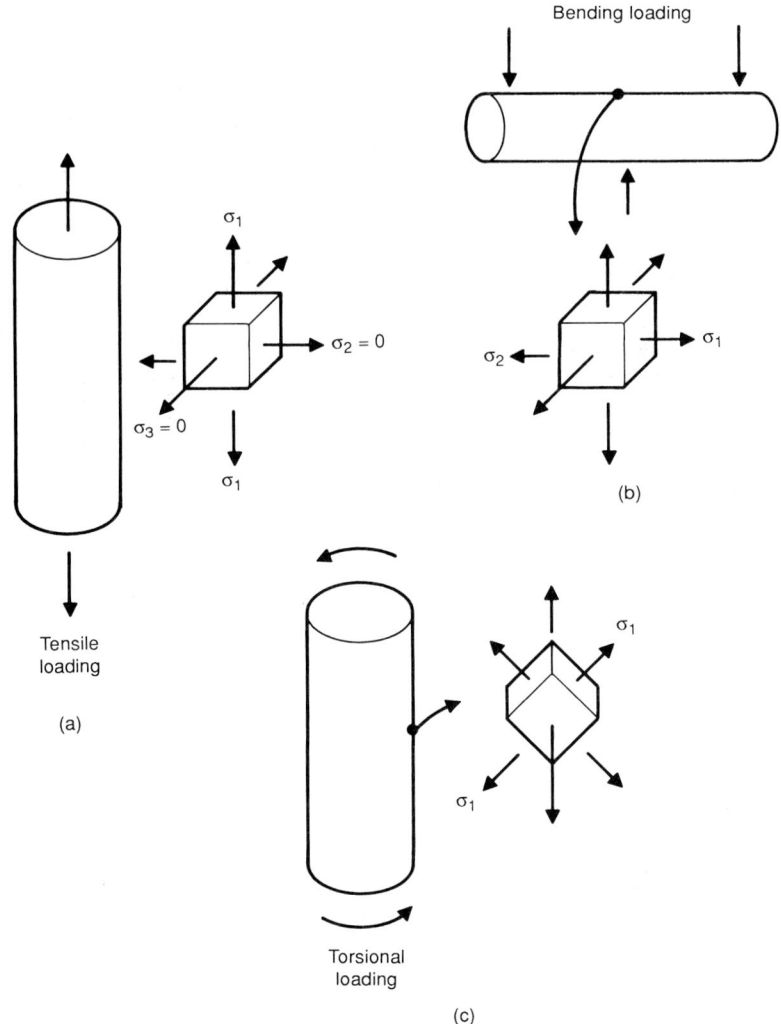

Figure 2.11 Illustration of the principal normal stresses and their orientation for cylinders loaded in tension, bending, and torsion. (The principal shear stresses are not shown.)

flow along the planes of maximum shear stress. Thus for the case of simple tensile loading, the maximum normal stress is parallel to the cylinder axis, and if the material is brittle, fracture occurs across the plane normal to the axis (Fig. 2.12). However, if the material is ductile, so that extensive plastic deformation can occur, then the fracture-plane orientation is more difficult to visualize, but should involve flow on planes of maximum shear stress; that is, on planes at 45° to the load. For simple torsional loading, the maximum normal stresses are at 45° to the cylinder axis, and the fracture-surface path is spiral and

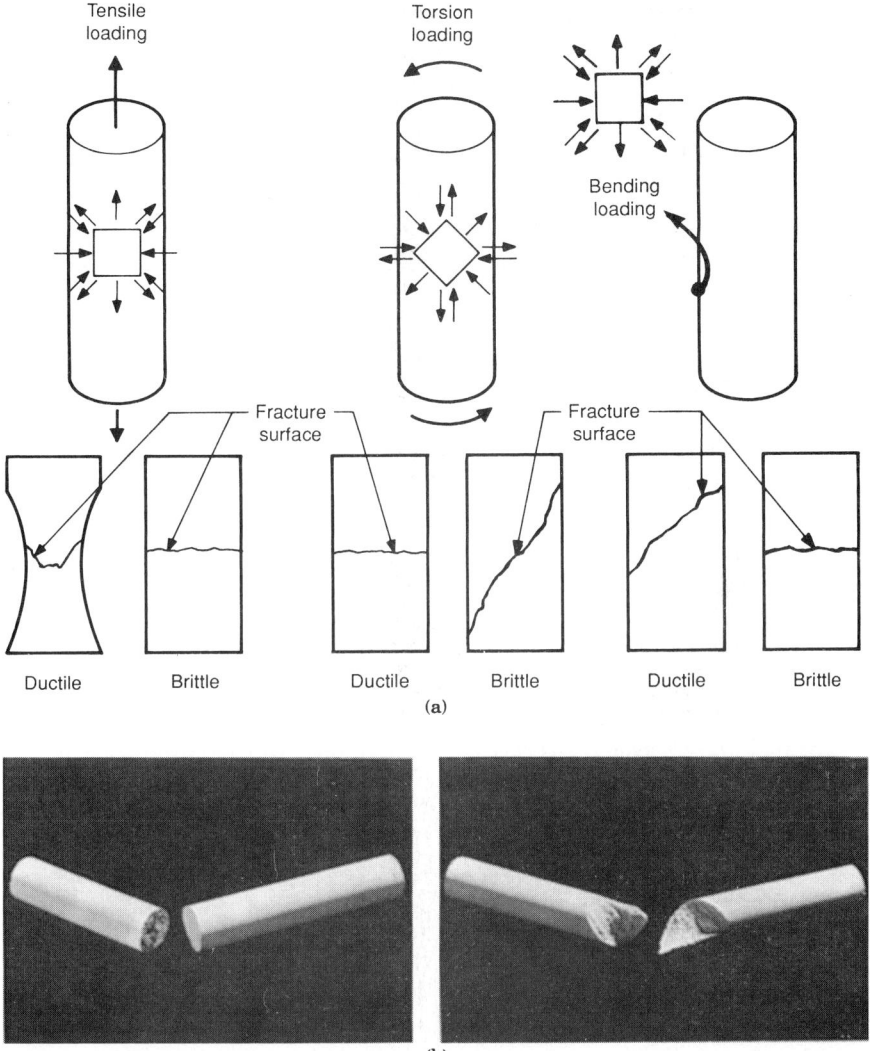

Figure 2.12 (a) Illustration of the orientation of the principal normal and shear stresses at the surface of cylinders loaded in simple tension, torsion, and bending, and the expected orientation of the fracture surface if the material is brittle and ductile. (b) Photographs of a piece of chalk broken by pulling axially (left) and by twisting (right). (*From Feynman et al.*[4])

inclined at 45° to the cylinder axis (see also Fig. 2.11c). Cracking may also be found on a longitudinal plane which has the same shear stress acting on it. For simple bending, the maximum normal stress is parallel to the cylinder axis and varies from a tensile value on one surface to zero at the center and to a compressive value on the other surface. If the material is brittle, fracture then occurs on a plane normal to the cylinder axis. If the material is ductile, a topography similar to that of a tensile fracture of a ductile material occurs (Fig. 2.12a).

2.4 Stress Concentration

So far we have considered geometries with constant cross-sectional areas. Machine components, however, are usually not so simple; even at the surface there are irregularities from machining. Consider a cylinder of a cross-sectional area of 20 in^2 (129 cm^2) loaded axially with 2000 lb (8900 N). If the load is applied uniformly, then each unit of area has the same load on it. For simplicity we can think of each square inch supporting 100 lb (445 N) to give a normal stress of 100 lb/in^2 (689,000 Pa). Also the lines centered on each square inch can be considered to be a line of force (or stress line), as depicted in Fig. 2.13a. Now consider a cylinder

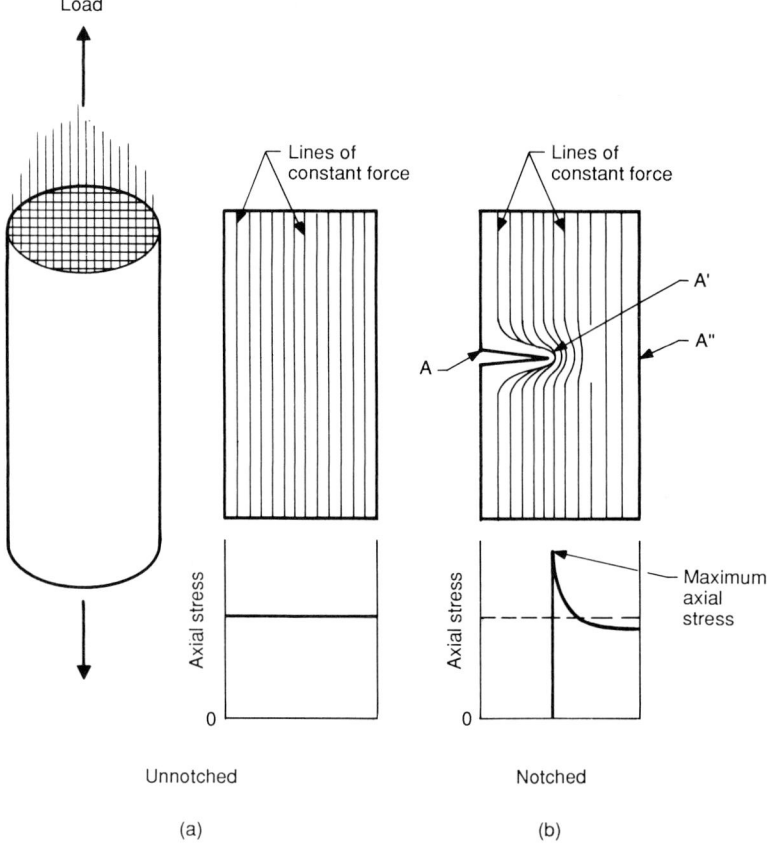

Figure 2.13 Illustration of the concept of lines of force and of the stress concentration at the root of a notch.

which has a circumferential notch in it so that the load-supporting area at the notch location is less than 20 in² (129 cm²) (Fig. 2.13b). There is no load applied to the surface A–A', so the internal stress must go to zero on this surface. The remaining cross section A'–A" must now support 2000 lb (8900 N). Far removed from the notched surface, the load is still distributed essentially uniformly, but as the notch is approached, the lines of force become denser, attaining a maximum density right at the root of the notch. Thus the axial stress increases, as shown in Fig. 2.13b.

The stress distribution depends on the size and shape of the component and the change in geometry (such as hole, circumferential notch). Examples in terms of the lines of force are shown in Fig. 2.14. The locations at which the axial stress is high (higher density of lines of stress or force) are clear, and the increase in the stress intensity with increasing sharpness of the surface discontinuity is clear (compare Fig. 2.14a,b,c).

For simple cases the effect of such geometry changes can be given in terms of a stress concentration factor K_t. This factor is defined as the maximum normal stress (principal stress) divided by the nominal stress in the same direction, and its magnitude depends on the geometry and type of loading. The curves in Fig. 2.15 illustrate this. Consider the case for simple tension in Fig. 2.15a. The maximum axial stress occurs right at the root of the notch. It depends on the geometric factors shown in the figure and increases as the radius of the notch decreases relative to the diameter of the cylinder. For example, if the surface of a 1-in-diameter cylinder has a circumferential notch around it which is 0.01 in in radius, then $h/d = 1$ and $h/r = 3$. This gives $K_t = 1.8$. Thus the axial stress right at the root of the notch is about twice that calculated based on the overall area and the axial load. If

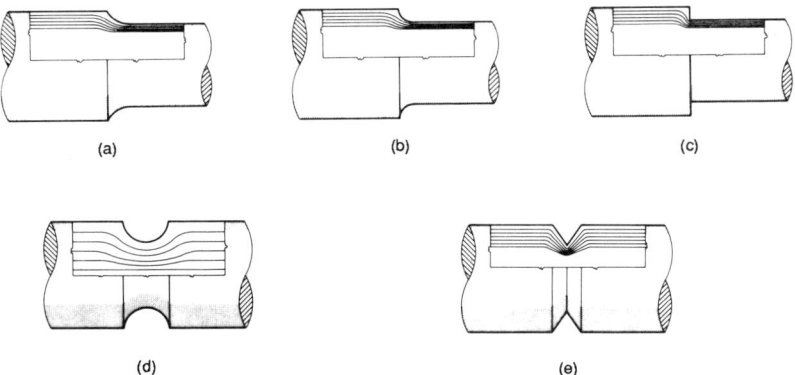

Figure 2.14 Illustration of the effect of geometry on the stress distribution in cylinders which are loaded axially. (*Adapted from* Metals Handbook.[5])

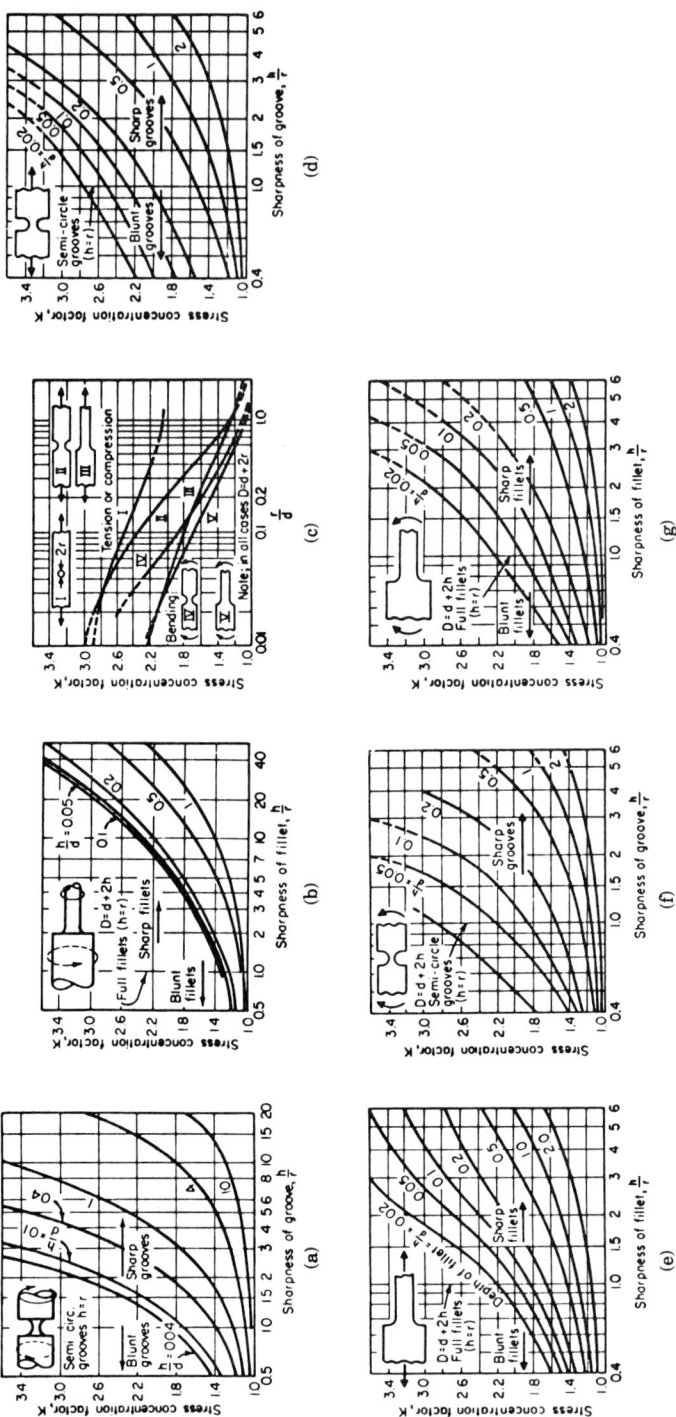

Figure 2.15 Calculated values of the stress concentration factor for various geometries. (*a*) Filleted shaft in tension. (*b*) Filleted shaft in torsion. (*c*) Flat plate with semicircular fillets and grooves or with holes in tension and compression. (*d*) Flat plate with grooves in tension. (*e*) Flat plate with fillets in tension. (*f*) Flat plate with grooves in bending. (*g*) Flat plate with fillets in bending. (*From Avallone and Baumester.*[6])

the notch is about 0.001 in, the maximum stress is considerably higher. This points up the fact that surface irregularities act as *stress concentrators* or *stress risers,* and thus are common locations for fracture to commence. Internal geometry changes, such as holes, also act as stress concentrators, and thus the critical locations depend on the geometry of the component.

The curves in Fig. 2.15 indicate that fracture would be ensured for surface notches (or irregularities) sufficiently acute since K_t increases greatly as the notch size decreases. However, these curves are based on elasticity theory, and in even relatively brittle alloys the region at the root of the notch undergoes some plastic deformation to relieve the stress. Thus in real materials the stresses are modulated by the plastic deformation zone which develops around the notch root.

Another important consideration is the variation in K_t with the location in the component. This is illustrated in Fig. 2.16 for the axial loading of a cylinder and of a plate with a surface notch. It is seen that K_t is a maximum at the root of the notch, as demonstrated before. Thus this is the location of expected initiation of fracture. However, consider the case of a cylinder made of an alloy for which the strength at the surface is sufficient to resist fracture. Then if the strength decreases with radius (such as a carburized steel), fracture may initiate below the surface at a location where the loading stress exceeds the

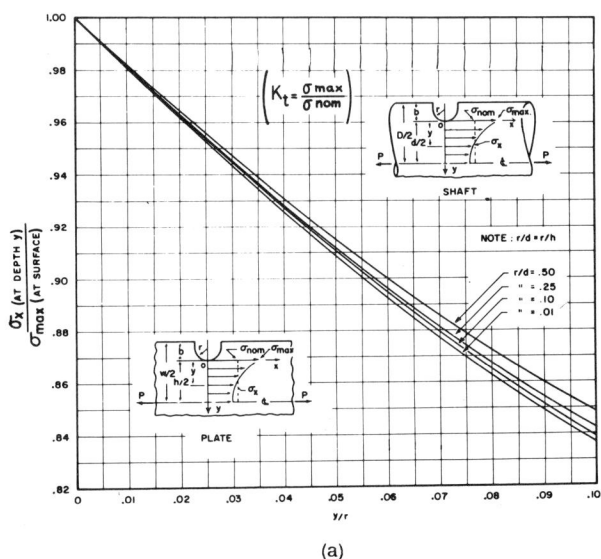

(a)

Figure 2.16 Calculated values of the variation of the maximum normal stress with depth in cylinders and plates loaded in (*a*) tension. (*From Lipson and Juvinall.*[7])

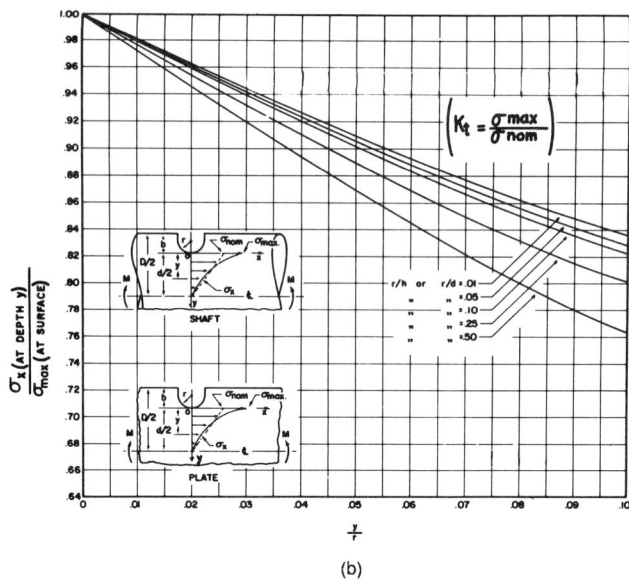

(b)

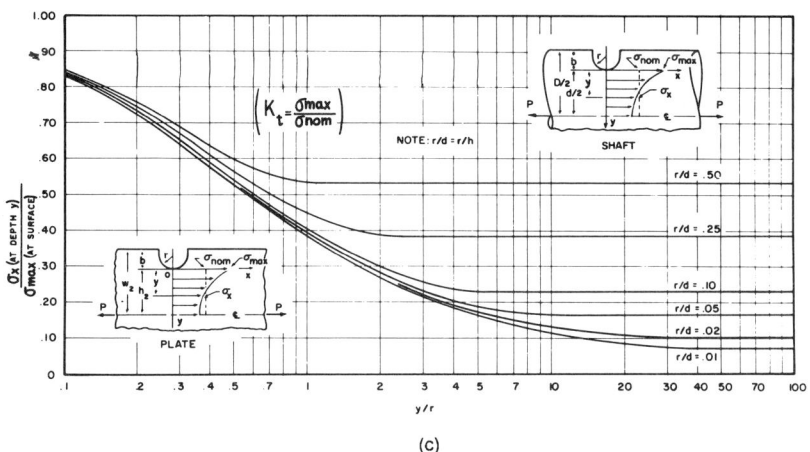

(c)

Figure 2.16 (*Continued*) Calculated values of the variation of the maximum normal stress with depth in cylinders and plates loaded in (*b*) bending and (*c*) torsion. (*From Lipson and Juvinall.*[7])

material's strength. (Note that this possibility is present even if there is no surface discontinuity.)

2.5 Triaxial Stress and Constraint

It is important to recognize that a triaxial stress distribution can develop even if the external load is uniaxial. When a sample is loaded elastically in uniaxial tension (for example, a tensile test), the length of the sample increases. However, there is a concomitant reduction in radius. The ratio of the dimensional change (strain) in the axial direction to that in the radial direction is called *Poisson's ratio*. It is between 0.25 and 0.35 for most alloys.

Now consider a sample loaded in uniaxial tension with a notch present (Fig. 2.17a). We want to examine the stresses in the axial, width, and thickness directions (Y, X, and Z). In the width direction and at the surface of the notch, there is a bending stress due to the

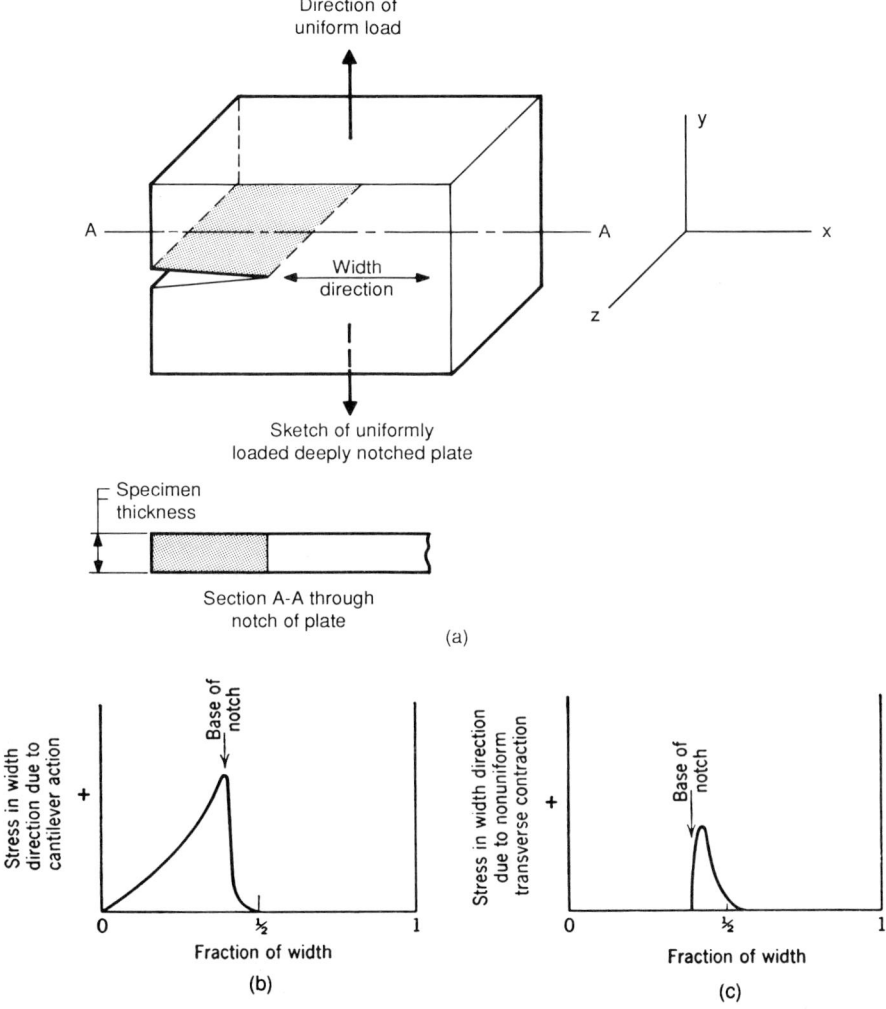

Figure 2.17 Illustration of the development of triaxial stresses due to the presence of a notch. (*Adapted from Parker.*[8])

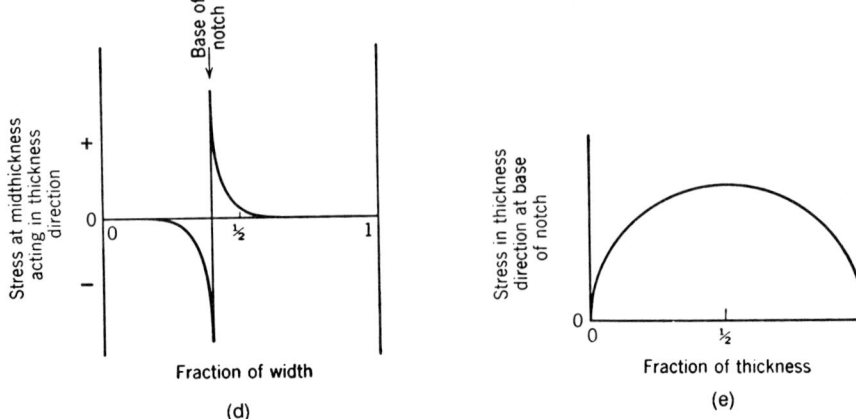

Figure 2.17 (*Continued*) Illustration of the development of triaxial stresses due to the presence of a notch. (*Adapted from Parker.*[8])

cantilever action of the notch, which decreases to zero at the root of the notch. This is shown in Fig. 2.17b. Also, just inside the root of the notch, the axial stress is high and the sample tries to shrink in the X and Z directions. Just outside the notch there is no axial stress applied on the free surface and hence here the material experiences no strain in these two directions. However, these two locations are continuous, and thus the contraction of the region inside the notch is resisted by the lack of contraction outside the notch. This uneven strain in the X and Z directions induces stresses in these directions. Thus even though the external load is axial, the presence of the notch causes a triaxial stress state, namely, stresses along all three axes. The stress distribution in the width direction (X direction) is shown in Fig. 2.17c, and in the thickness direction (at midthickness) in Fig. 2.17d. The stress analysis also reveals that the stress in the thickness direction depends on the location across the thickness. At the root of the notch it varies from zero at the surface to a maximum at midthickness (Fig. 2.17e).

It is thus seen that the notch constrains the material from developing an elastic deformation which would not cause any triaxial stresses, and instead induces a triaxial stress state. This constraint makes it more difficult locally for the material to flow plastically, and hence more susceptible to brittle fracture. An important aspect of such a constraint is that it increases with plate thickness and crack length. Hence brittle fracture may be more likely in thicker plates than in thin plates, and in plates with deeper cracks. It is worth noting that the fracture behavior depends not only on the surface area of the notch but also on that of the remaining material which supports the load. In

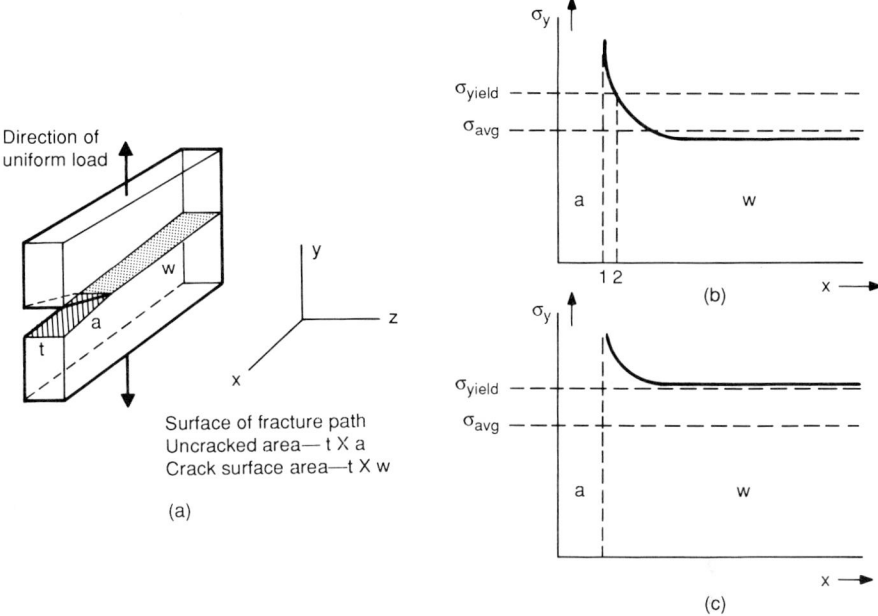

Figure 2.18 Illustration of the effect of reduction in the area supporting the load on the development of gross yielding.

Fig. 2.18a the crack area is $t \times a$ and the uncracked area is $t \times w$. The stress profile in the Y direction is shown in Fig. 2.18b. For X values between 1 and 2, the stress exceeds the yield strength, and in this region of the crack root, plastic deformation occurs. However, for cracks that are large relative to the area of the uncracked material (as will occur as the crack grows), the uncracked area supporting the load ($t \times w$) is small, and the axial stress is high. Over the entire length of the remaining uncracked region, the axial stress may become greater than the yield strength, and hence gross plastic deformation will occur. This situation is shown in Fig. 2.18c.

2.6 Plane Stress

If a small region (or element) in a body is stressed, the stress state can be represented as shown in Fig. 2.19a. There are three principal normal stresses and also shear stresses on each face of the element (shown on only two faces in the figure). Now if the stress σ_Z is small relative to σ_X and σ_Y, then the state of stress is referred to as a *plane stress*, since the normal stresses of finite or significant magnitude lie only in the X–Y plane (Fig. 2.19b).

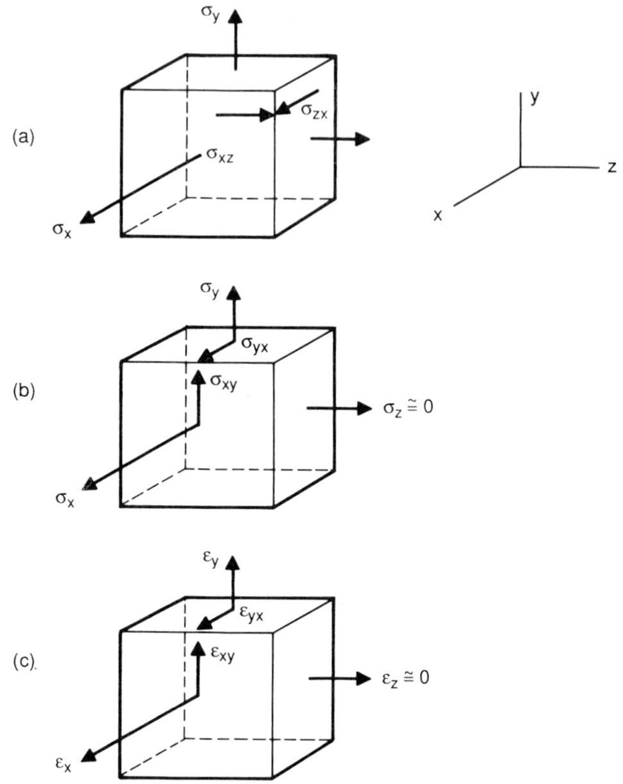

Figure 2.19 (a) Stress state of element in a loaded body. (b) Plane-stress condition. (c) Plane-strain condition.

As pointed out in the preceding section, the constraint at the root of a notch in a uniaxially loaded sample produces a triaxial stress state. Consider a notched sample of rectangular cross section loaded in tension (Fig. 2.20), where the notch surface area is small compared to that of the remaining material supporting the load. At the surface and near the notch root, the stress in the Z direction is zero, since there is

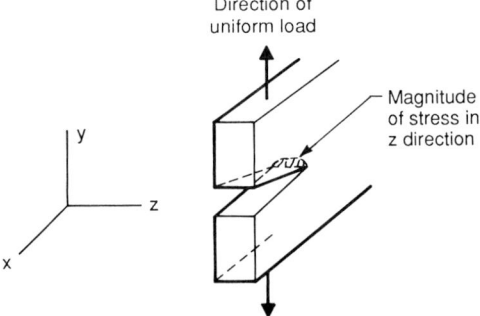

Figure 2.20 Notch in a thin plate, showing the low stress distribution in the thickness (Z) direction at the root of the notch corresponding to the plane-stress condition. The height of the region shows the magnitude of the stress in the Z direction. It is zero at both surfaces.

no load applied to this surface. This stress rises to a maximum value at the center. If the thickness in the Z direction is small, then the magnitude of this stress is low, relative to those in the X and Y directions. Thus a plane-stress condition exists. The plane-stress condition is common to loaded thin sheets or tubes.

For the plane-stress geometry there is no gross section yield because of the low normal stress and the high value of K_t, and this makes the value of the axial stress in the notch root relatively low. However, there is a considerably large maximum shear stress, which lies on planes at 45° to the axial load direction, as shown in Fig. 2.21a. Thus plastic deformation is limited to shear along these planes (Fig. 2.21b), leading to a *shear-lip* fracture surface. This fracture surface is also called a *slant fracture* (Fig. 2.22a).

The significance of a plane-stress condition is that a crack present under such conditions will propagate as a slant fracture, and hence

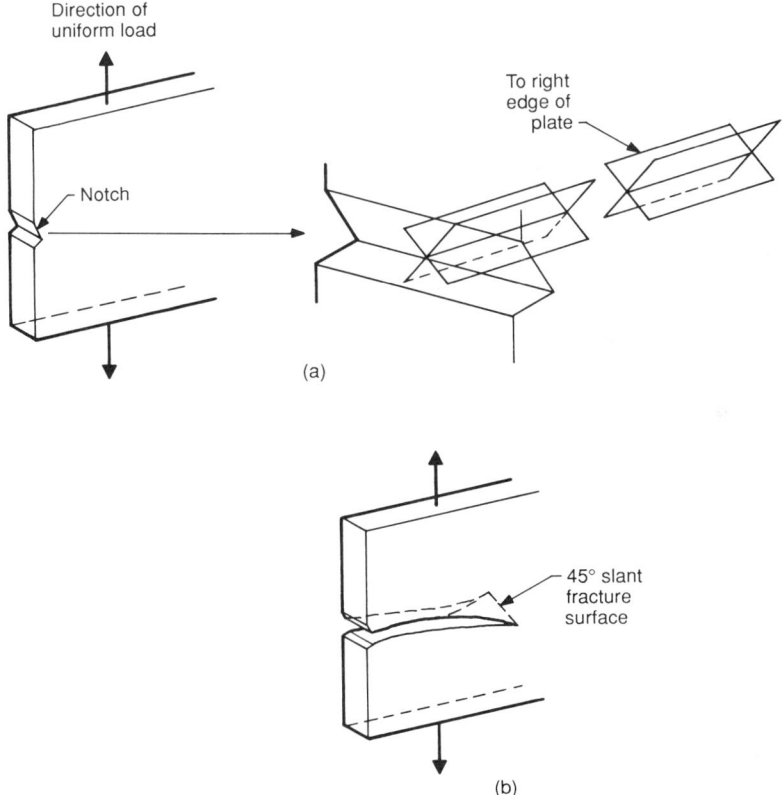

Figure 2.21 (a) Maximum shear stress planes at the root of the notch in plane-stress loading. (b) Overall orientation of the propagating fracture surface at 45° to the load (slant fracture).

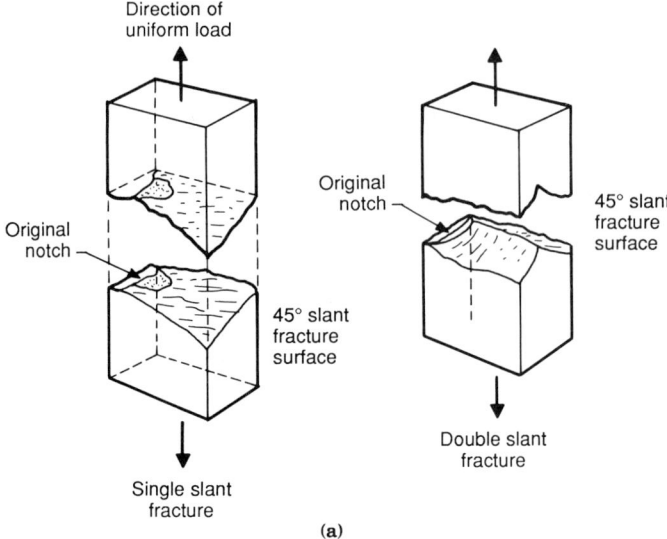

Figure 2.22 (a) Slant fracture surface. (b) Fracture surface of a thin plate. (*From Henry and Horstmann.*[9])

this macroscopic fracture appearance gives information about the fracture process. An example of such a fracture-surface orientation is given in Fig. 2.22b.

2.7 Plane Strain

If the stress state is such that the strain ε in one direction is considerably less than that in the other two directions, the condition is re-

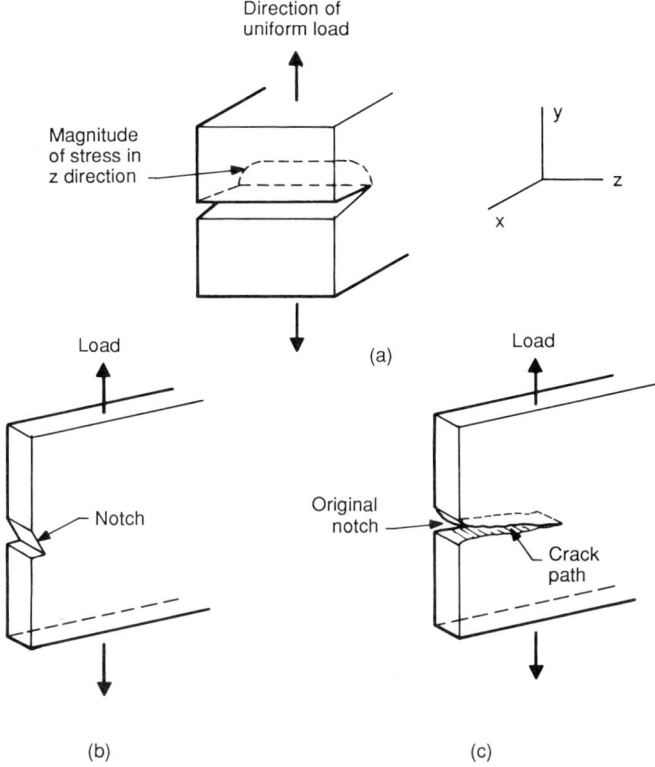

Figure 2.23 (a) Stress in the Z direction at the root of the notch in a thick plate, corresponding to plane-strain condition. (b) Notched plate. (c) Fracture surface propagating from notch has overall orientation perpendicular to the load.

ferred to as *plane strain* (see Fig. 2.19c), since the change in dimensions is limited to two dimensions, defining a plane. When a notched plate loaded in uniaxial tension has significant thickness, the strain in the Z direction is constrained, leading to a plane-strain condition. If strain hardening is sufficiently high, the size of the plastic zone at the crack tip is smaller than in the plane-stress case, and hence the axial stress at the crack root is very large (compare Fig. 2.20 and Fig. 2.23a). Then the maximum shear stresses are relatively low, and the high normal axial stress controls the overall fracture path, which is thus relatively flat and normal to the axis of the external load. This is depicted in Fig. 2.23c, and an example is shown in Fig. 2.24.

2.8 Fracture of Tensile Samples

In a tensile test of a ductile metal, the initial elastic (and essentially linear) deformation is followed by plastic deformation during which the

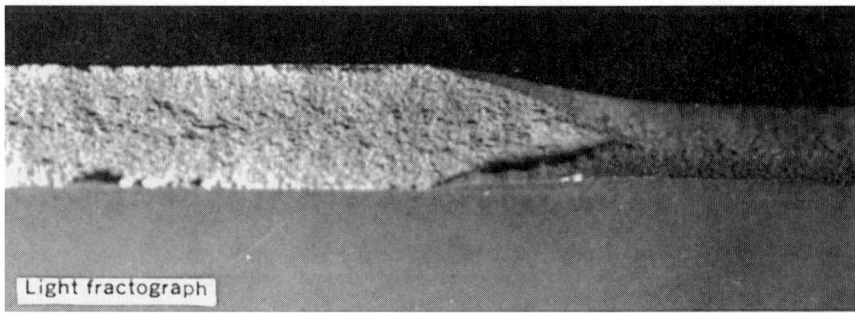

Figure 2.24 Light fractograph of a thick plate showing flat fracture surface developed from a plane-strain condition. (*From* Metals Handbook.[10])

length increases and the cross section decreases (see Sec. 2.2). In the plastic deformation region, usually strain (work) hardening occurs and the material gets stronger, and hence an increasing load is required to continue to strain (elongate) the sample at a constant (approximately) strain rate. Eventually a strain is attained where further plastic deformation becomes very localized. This may initiate at a region of slightly smaller cross section (such as a machining notch) or at a microstructural feature which has work-hardened less. In any case, at this location the cross section becomes locally reduced and hence serves as a surface notch. This can be shown to result in a triaxial state of stress, with the radial and axial tensile stresses being a maximum at the center (Fig. 2.25).

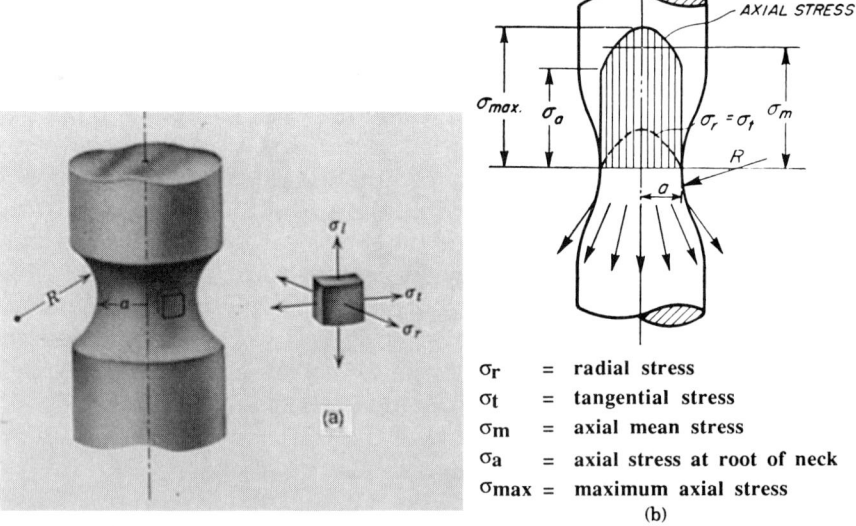

σ_r = radial stress
σ_t = tangential stress
σ_m = axial mean stress
σ_a = axial stress at root of neck
σ_{max} = maximum axial stress

(b)

Figure 2.25 (*a*) Stress distribution at the surface of a notch in a cylinder loaded axially. (*From Hertzberg.*[11]) (*b*) Axial stress across the diameter of a cylindrical tensile sample after necking has occurred. (*Adapted from Nadai.*[12])

This stress distribution makes the center of the tensile bar susceptible to crack initiation, and in many alloys this is where fracture initiates. (However, this depends on the alloy and test conditions; the crack initiation mechanisms are discussed in Chap. 3.) If a crack initiates in the center, it then spreads radially, so that at this stage the tensile bar has an internal notch. As the crack approaches the surface, the remaining material is a thin shell, and hence the plane-stress condition develops. This leads to a change in crack propagation from flat fracture to slant process. Thus the fracture surface consists of a central region, which is macroscopically relatively flat with its surface normal to the tensile load direction (plain-strain region), and a rim inclined at 45° (plane-stress region). This is illustrated in Fig. 2.26 and is called a *cup-and-cone fracture*. If fracture does not initiate in the center, then the material will pull down to a point, as shown in Fig. 2.27. Thus the fracture appearance depends on the effect of internal microstructural features on the initiation of internal cracks.

The relative amount of slant and flat fracture depends on the degree of constraint. In components of rectangular cross section this increases with thickness, so there is more flat fracture. This is illustrated in Fig. 2.28.

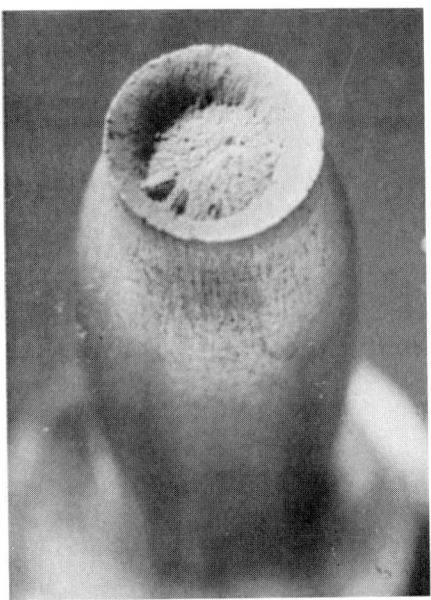

(a)

Figure 2.26 Fracture surfaces of two tensile samples. (*a*) Ductile fracture, with flat fracture in the center and slant fracture on the sides (cup-and-cone fracture). (*From Vander Voort.*[13])

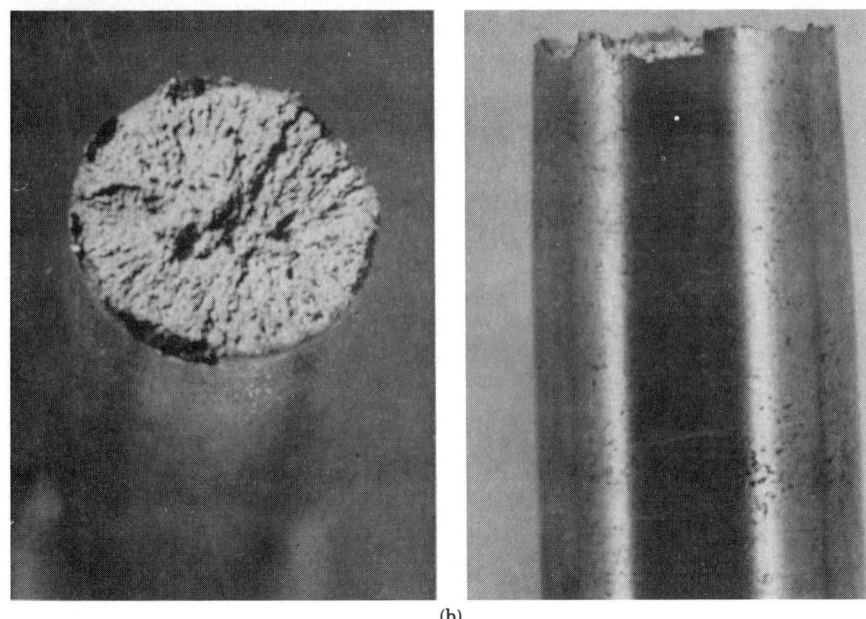

(b)

Figure 2.26 (*Continued*) Fracture surfaces of two tensile samples. (*b*) Brittle fracture. (*From Vander Voort.*[13])

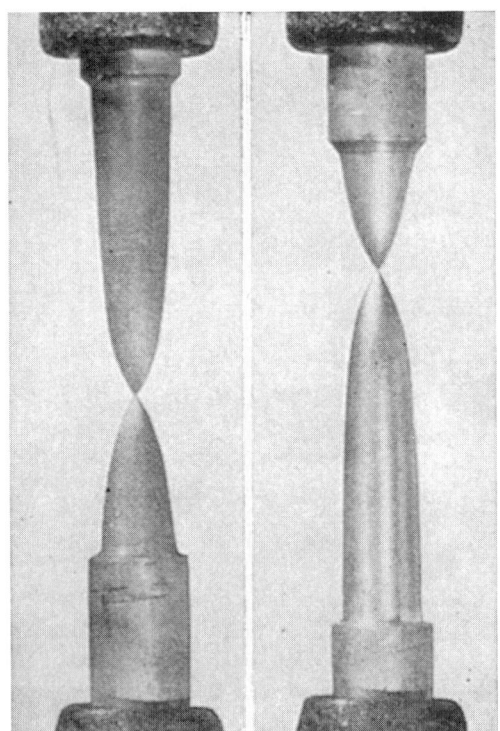

Figure 2.27 Aluminum tensile samples broken at 600°C (1110°F), showing necking to a point at fracture. (*From Henry and Horstmann.*[9])

Mechanical Aspects and Macroscopic Fracture-Surface Orientation   99

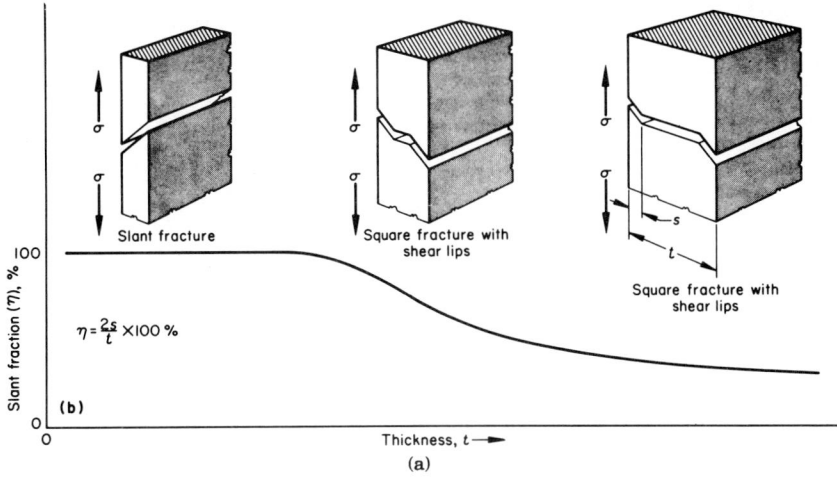

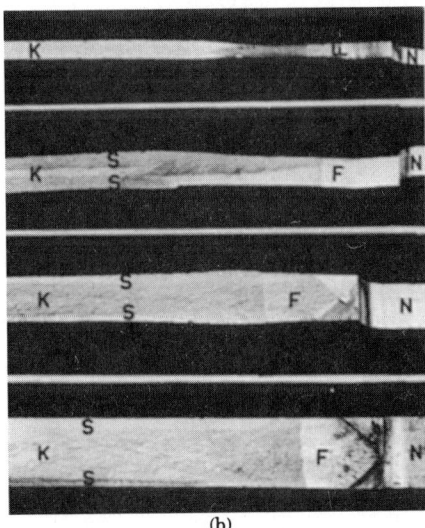

Figure 2.28 (a) Effect of the thickness of plates on the relative amount of slant and flat fracture. (*From* Metals Handbook.[5]) (b) Influence of plate thickness on fracture appearance. Slant fracture in thin sheet (top); square fracture in plate (bottom); transitional behavior (center). N—central starter notch; F—fatigue fracture area; K—final fracture area; S—shear lips. (*From* Broek.[14])

In wide thin sheets, the necking process is different from that in cylinders. A large width-to-thickness ratio (such as 10) produces a constraint which prevents necking on a plane containing the width direction and perpendicular to the load direction. Instead, it is found (and predicted from a plain-strain analysis) that necking is restricted to a band inclined approximately at 55° to the load axis. This, then, is the orientation of the fracture plane, as illustrated schematically in Fig. 2.29.

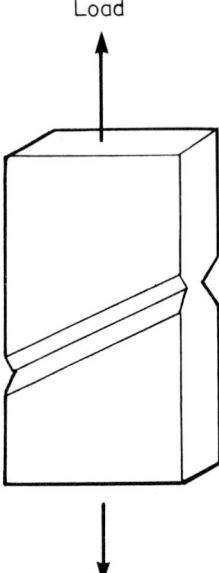

Figure 2.29 Region of localized deformation which develops in thin sheet.

2.9 Effect of Strain Rate and Temperature

The fracture process depends on the strain rate (which depends on the loading rate) and the temperature. The specific response depends on the material, but here only the effect of these two factors will be examined. Most materials become stronger but more brittle with increasing strain rate and lower temperature. The typical effect of temperature on the true-stress–true-strain curve is shown in Fig. 2.30.

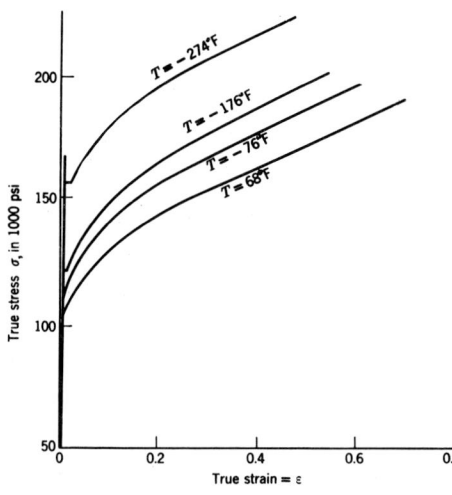

Figure 2.30 Effect of test temperature on the true-stress–true-strain curves in a tensile test of a pearlitic steel. Note that the yield strength increases with decreasing temperature and the ductility (elongation at fracture) decreases. (*Adapted from Zener and Holloman.*[15])

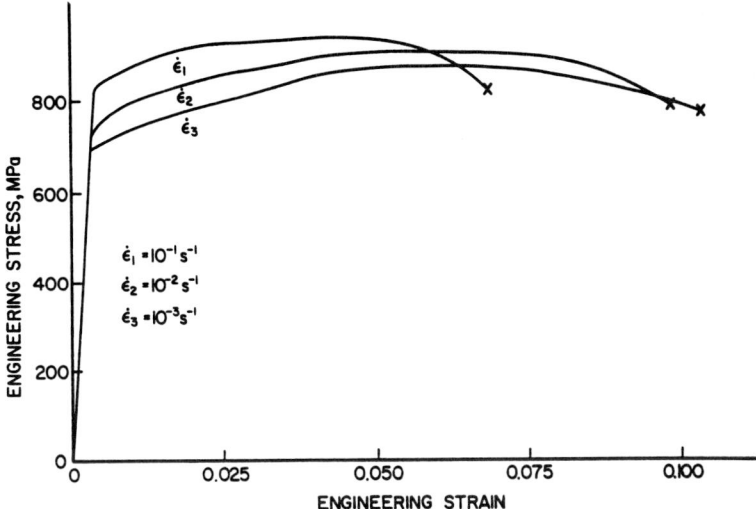

Figure 2.31 Effect of strain rate $\dot{\varepsilon}$ on the engineering-stress–engineering-strain curves in a tensile test of a 1040 steel. Note that as the strain rate increases, the yield and tensile strengths increase and the ductility (elongation at fracture) decreases. (*From Meyers and Chawla.*[16])

Note that the yield strength is increased with decreasing temperature, and that the elongation at fracture is reduced. The fracture surfaces show more flat fracture and less slant fracture as the temperature is reduced. Figure 2.31 shows the typical effect of the strain rate on the engineering-stress–engineering-strain curve. The more rapidly the sample is elongated during the test, the higher the yield strength and the lower the elongation at fracture. The higher-strain-rate tests produce fracture surfaces with less slant fracture and more flat fracture. The effects of temperature and strain rate on the fracture surface are shown in Fig. 2.32. The samples were notched and impact-tested, so that the strain rate was high. Each sample was broken at a different temperature. The lower the temperature, the less the amount of shear lip (slant fracture area). These data demonstrate that the general appearance of the fracture-surface orientation gives clues to the temperature of fracture and the fracture propagation velocity.

2.10 Crack Propagation

The rate of crack growth is usually separated into two categories. One is those that propagate "slowly" [for example, 20 ft/s (6.1 m/s)], and only propagate on application of an external load. This is called *stable crack growth*. The other category is "fast" growth [for example, 3000 ft/s (915 m/s)], and these cracks can proceed without external loading

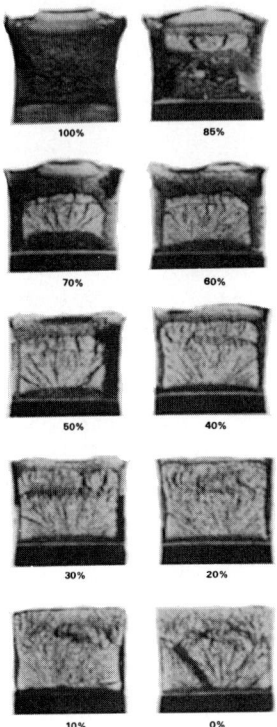

Figure 2.32 Fracture surfaces of a series of impact samples. The material was 4340 steel and the samples were broken in a Charpy V-notch test. The percentage of the fracture surface covered by the shear lip (slant fracture) is shown below each picture. (*From Metals Handbook.*[5])

if there is sufficient internal elastic stressing (such as residual stresses from heat treating). This is referred to as *unstable crack growth*. (The propagation of the fracture in a tensile sample after necking is also unstable crack growth.) The type of fracture that occurs depends on many factors, such as loading conditions, temperature, material, and geometry of the component. It is to be noted that the plane-strain condition is more conducive to unstable crack propagation.

Figure 2.28 shows that the relative amount of flat and slant fracture depends on the thickness of the plate. However, another factor that affects the type of fracture is the stress concentration, which changes as the crack propagates and less uncracked material remains to carry the load. For example, in terms of round notches, r/d changes as the crack becomes deeper since d decreases (see Fig. 2.15). Thus a crack may begin propagating under plane-strain conditions and then change to plane stress (as for the tensile sample fracture described in Sec. 2.8).

An important complicating factor affecting the fracture-surface macroscopic features is the dynamics of crack propagation. This aspect is not well understood, and the common approach to interpreting the

Mechanical Aspects and Macroscopic Fracture-Surface Orientation 103

fracture-surface features is to rely on empirical observations and experience. Two features are cited here and described in more detail in Chap. 4. In many fractures of components of circular or square (or near-square) cross section, radial marks are observed which point back toward the origin of the fracture. Examples are presented in Fig. 2.33. Figure 2.33a shows the fracture-surface of a cylindrical tensile sample. The radial marks point back toward the center of the sample, where fracture initiated (see Sec. 4.2). Upon the development of a sufficiently acute notch, the plain-strain condition caused relatively rapid crack propagation, with the dynamics of crack movement leading to the crack moving from one inclined plane to another as it advanced, developing the radial marks. Figure 2.33b shows the fracture-surface of a tensile sample of the same steel but of a different hardness and tested at a lower temperature. The radial marks are finer and cover more of the fracture. The relationship between this macroscopic fracture-surface topography and the fracture process is discussed in detail in Chap. 4.

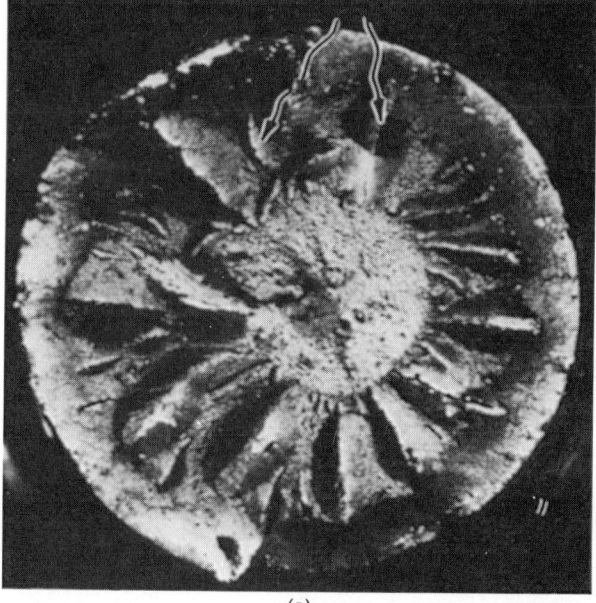

(a)

Figure 2.33 Radial marks on fracture surfaces of tensile samples of 4340 steel having circular cross section. (a) Structure of tempered martensite with a hardness of Rockwell C 28. The inner fibrous zone is ridged circumferentially; the intermediate zone has coarse radial marks; the outer ring is the shear-lip zone. (*Adapted from* Metals Handbook.[10])

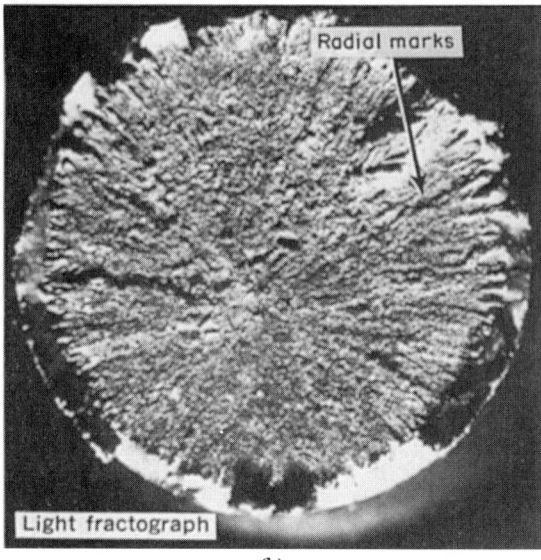

(b)

Figure 2.33 (*Continued*) Radial marks on fracture surfaces of tensile samples of 4340 steel having circular cross section. (*b*) Sample was fractured at −196°C (−321°F). The structure was tempered martensite; the hardness was Rockwell C 35. This fracture has no fibrous zone. The radial marks occupy the entire area inside the narrow shear-lip zone. (*Adapted from* Metals Handbook.[10])

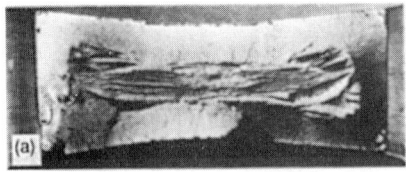

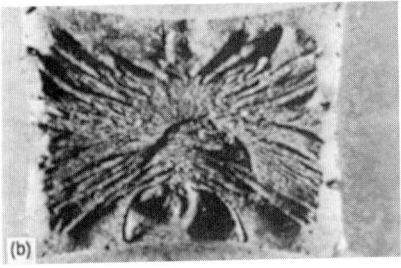

Figure 2.34 Fracture surfaces of a rectangular and a square tensile steel sample showing radial marks. (*From* Metals Handbook,[10] *as adapted from* Nunes et al.[17])

The fracture surfaces of tensile samples of square and rectangular cross section are shown in Fig. 2.34. In the square cross section the radial marks are evident, but in the rectangular cross section these are parallel to the long axis of the sample. As the thickness of such a sample decreases, the radial marks become symmetrical about the midline of the plate, forming *chevron* marks (also called *herringbone* marks), as shown in Fig. 2.35. One representation of the dynamics of crack propagation is given in Fig. 2.36. It is hypothesized that cracks form just ahead of the advancing crack front, and the intersection of these cracks with the front forms the chevron marks.

2.11 Meaning of Ductile and Brittle Fracture

The terms *ductile* and *brittle* are commonly taken as self-descriptive, but care should be taken in using them. They are used to indicate the general plasticity at fracture, relying on such quantities as the elongation at fracture (ductility) and the presence of necking to indicate whether the material is ductile or brittle. A difficulty here is to decide how much ductility is required to place the material in a ductile class; there is no common value to be used. More pertinent to our interests is that the terms ductile and brittle are used to distinguish ductile and brittle fracture, both the macroscopic and the microscopic aspects. As is discussed in Chap. 3, the void coalescence mechanism (on a microscopic scale) of crack formation involves plastic deformation by slip. Thus this mechanism is sometimes referred to as ductile. However, it is possible to have cracks form and fracture occur by this mechanism, yet the material shows no obvious or little macroscopic plastic deformation. On this basis, the fracture would be categorized as brittle.

In this book, the terms ductile and brittle are used to describe the macroscopic aspects of fracture and are not based on the microscopic mechanisms. The following characteristics of the two processes allow

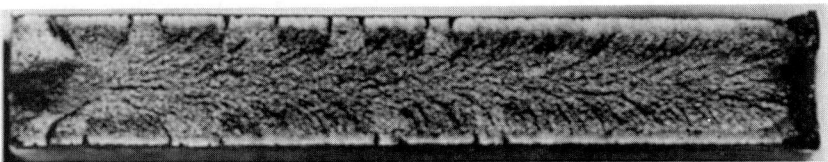

Figure 2.35 Fracture surface of a steel plate, showing well-developed chevron marks which point back toward the origin of fracture (at left), which is adjoined by a small fibrous zone. This is a high-velocity fracture. (*From* Metals Handbook.[10])

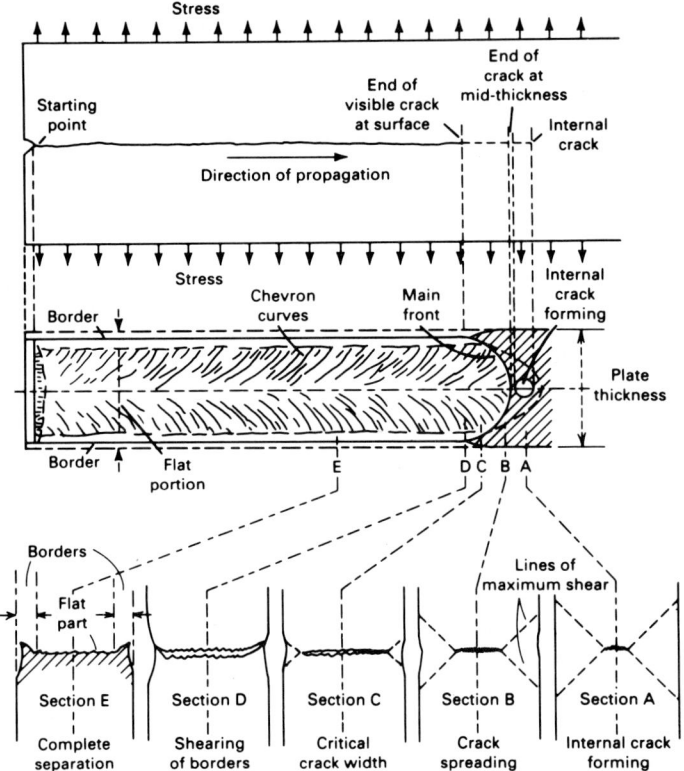

Figure 2.36 Formation of chevron marks on a fracture surface. The lower portion shows an analogy between chevron formation in a plate and cup-and-cone fracture. (*Adapted from Boyd.*[18])

distinguishing them (adapted from Vander Voort[13]). They include macroscopic fracture-surface characteristics.

1. *Ductile fracture*

- A relatively large amount of plastic deformation precedes the fracture.
- Shear lips are usually observed at the fracture termination areas.
- The fracture surface may appear to be fibrous or may have a matte or silky texture, depending on the material.

- The cross section at the fracture is usually reduced by necking.
- Crack growth is slow.

2. *Brittle fracture*

- Little or no visible plastic deformation precedes the fracture.
- The fracture is generally flat and perpendicular to the surface of the component.
- The fracture may appear granular or crystalline and is often highly reflective to light. Facets may also be observed, particularly in coarse-grain steels.
- Herringbone (chevron) patterns may be present.
- Cracks grow rapidly, often accompanied by a loud noise.

A convenient distinction can be made based on the amount of macroscopic change in shape or size. For example, in tensile testing, the amount of elongation at fracture is usually taken as a measure of ductility. The problem is what value makes a usable distinction between ductile and brittle behavior. Note from the preceding that a characteristic for brittle fracture is little or no visible plastic deformation. This implies that none or little shear-lip region is present, but an exact distinction between brittle and ductile may not be possible.

In spite of the difficulty of establishing an exact distinction between brittle and ductile fracture, this usually does not present a practical problem. A recognition of the amount of flat and slant fracture surface and the degree of macroscopic plastic deformation required for a suitable working distinction usually can be discerned.

2.12 Fracture Mechanics and Failure

A major concern in the design of structure and machine components is the prevention of sudden catastrophic fracture. There are many documented cases in which the systems had functioned properly for many years, then suddenly broke. As experience increased, certain factors, such as low temperature and environment, appeared to exacerbate the problem. It was often found that the fracture originated from already existing cracks or surface (internal and external) discontinuities.

An example of such behavior is usually found in fatigue fractures (see Sec. 2.13). If a component is loaded elastically below the yield strength, and is not operating under conditions where creep (see Sec. 2.14) would occur, then it is expected that it will perform safely. However, if the load is cyclic (but not exceeding the yield strength), it is often found that small cracks will form, especially at surface discontinuities (such as machining notches) due to stress concentration at these locations (see Sec. 2.4). These cracks will expand with each cycle, and the remaining material must support the load. This decreasing support area eventually becomes insufficient to sustain the next increase in load, and fracture is expected. The important aspect of the fracture is that it usually occurs rapidly (catastrophically) and with little plastic deformation, that is, the final fracture is brittle. Thus there is something about the crack (notch) which eventually made the material break in this fashion. Two important characteristics of the notch are its size and its sharpness.

Thus loaded components may fracture suddenly and brittlely if cracks or notches have a certain size. Note that such behavior is not predictable from the results of a tensile test. Although a material which is relatively brittle (for example, low elongation at fracture in a tensile test) would be suspect, a relatively ductile material may fracture catastrophically. What is desired is a material property which can be measured and from which such behavior can be predicted. Fracture mechanics allows treatment of this problem. Here we briefly outline the ideas involved; more realistic treatments can be found in the references listed at the end of the chapter.

Consider a parallelepiped of the dimensions shown in Fig. 2.37a. If it is loaded elastically along its axis, the force-length curve is that

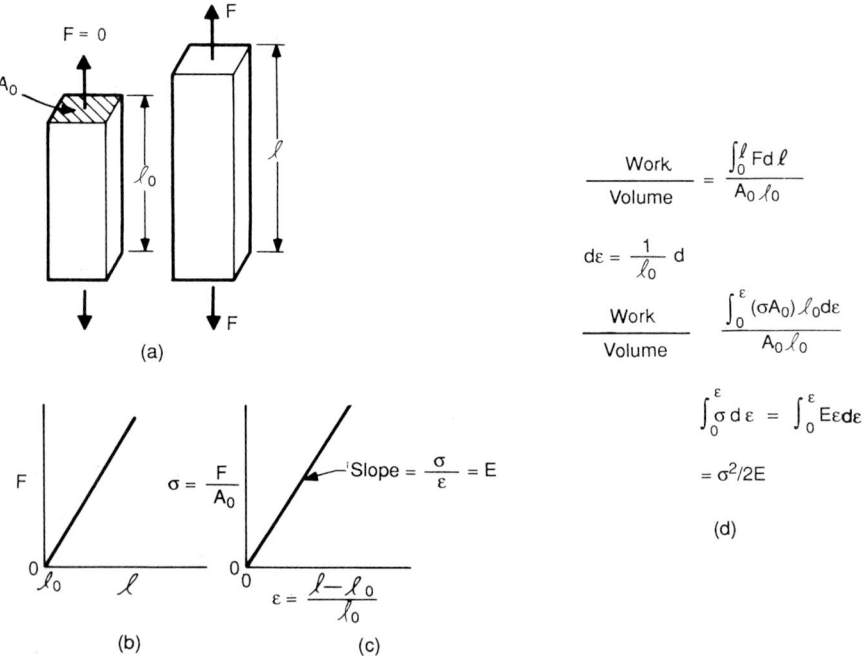

Figure 2.37 Derivation of the elastic energy in a cylinder when loaded axially to stress σ.

shown in Fig. 2.37b, which is assumed (realistically) to be linear. This force-length curve can be converted to the engineering-stress–engineering-strain curve shown in Fig. 2.37c. The slope of the line in the elastic region is the elastic modulus E. The work to stretch the bar to a length l at force F is given by $\int F\, dl$. This work is the stored elastic energy U and is shown in Fig. 2.37d to be $\sigma^2/2E$ per volume of material.

Now consider a small region taken out of the center of the bar, as shown in Fig. 2.38. If the stress on the bar is still σ, then the elastic energy of this bar has been reduced since there is less material present. This reduction is shown in Fig. 2.38 to be $\sigma^2 a^2 l_0/2E$.

If the solid bar has a crack or notch in the center, as shown in Fig. 2.39, and the bar is still loaded to stress σ, the material in the region of the notch supports no load, so the presence of the notch lowers the energy of the bar. Assume that the volume which supports no load is given by the elliptical shape shown. Then the reduction in energy is $\pi\sigma^2 a^2 B/E$. Thus the elastic energy of the bar decreases with the area of the notch, assuming that the notch is small compared to the bar cross section. In Fig. 2.39, the crack extends through the bar thickness, so the energy reduction is proportional to a^2.

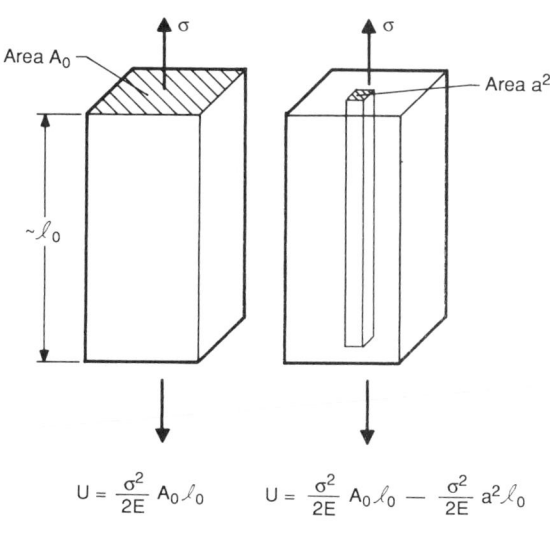

Figure 2.38 Derivation of the reduction in energy upon removal of a small bar from the sample.

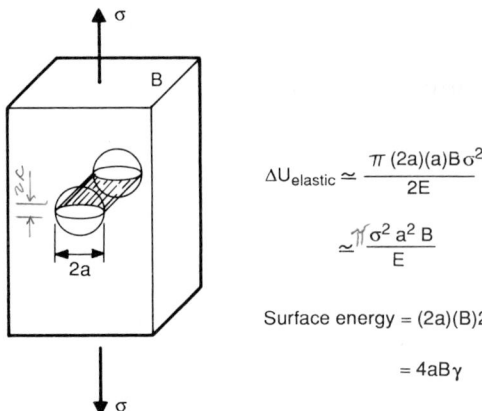

$$\Delta U_{elastic} \simeq \frac{\pi (2a)(a)B\sigma^2}{2E}$$

$$\simeq \frac{\pi \sigma^2 a^2 B}{E}$$

Surface energy $= (2a)(B)2\gamma$

$= 4aB\gamma$

Figure 2.39 Derivation of the reduction in energy upon forming a small crack in a parallelepiped.

However, creating a crack forms a free surface which has an energy per surface area of γ. For the case in Fig. 2.39, the notch surface area is $2 \times 2aB$, and so the energy of the bar has been increased to $4aB\gamma$. This increase in energy is proportional to a. The total energy then is the sum of the reduction in elastic energy and the increase in surface energy. This is depicted graphically in Fig. 2.40.

It is seen that once the crack attains the size a^*, a further increase in its size is accompanied by a decrease in energy of the bar, which

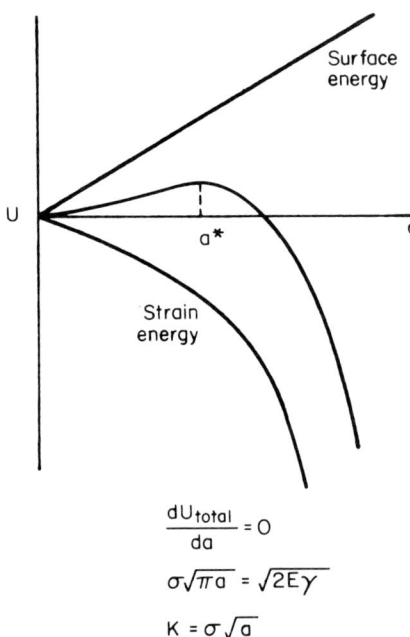

$$\frac{dU_{total}}{da} = 0$$

$$\sigma\sqrt{\pi a} = \sqrt{2E\gamma}$$

$$K = \sigma\sqrt{a}$$

Figure 2.40 Variation of elastic strain energy (for a given stress) and surface energy with crack size. The maximum in the sum of these two curves defines the critical crack size a^*.

implies that a crack of this size will grow. This is the explanation why such catastrophic rapid crack propagation occurs.

The relationship of critical crack size to stress is obtained by differentiation of the relation for the total energy and equating it to zero (corresponding to the maximum in the total-energy curve; Fig. 2.40). This gives

$$\sigma\sqrt{\pi a} = \sqrt{2E\gamma} \qquad (2.1)$$

The right side contains only material constants, and hence we can write

$$\sigma\sqrt{\pi a} = K \qquad (2.2)$$

A parallel treatment takes the approach of calculating (from elasticity theory) the stress field ahead of the crack tip and expressing the result in terms of a critical stress intensity factor K_I. It is similar to K in Eq. (2.2), but the exact form depends on the size and shape of the crack and plate. K_I is taken as a property of the material (that is, chemical composition and microstructure) and is called the *fracture toughness*.

The treatment of fracture toughness must be expanded to account for local plastic deformation. Also, the fracture toughness K depends on whether the loading conditions are plain strain or plain stress. Thus there are different types of K values. Also, the K value may be sensitive to the environment, and values are defined for crack propagation under stress corrosion cracking conditions, hydrogen embrittlement, and so on.

The toughness also depends on the material, its microstructure, and the temperature. These are illustrated in Fig. 2.41. In general, the

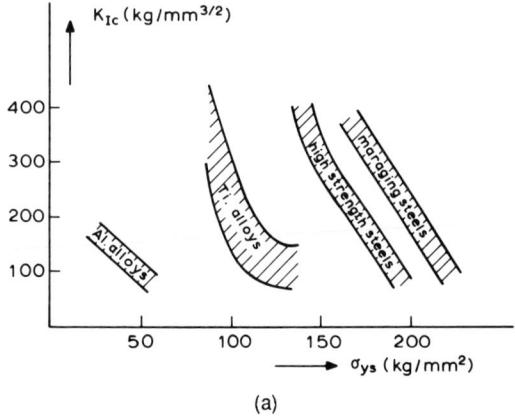

Figure 2.41 Fracture toughness (*a*) as a function of yield strength. (*From Broek.*[14])

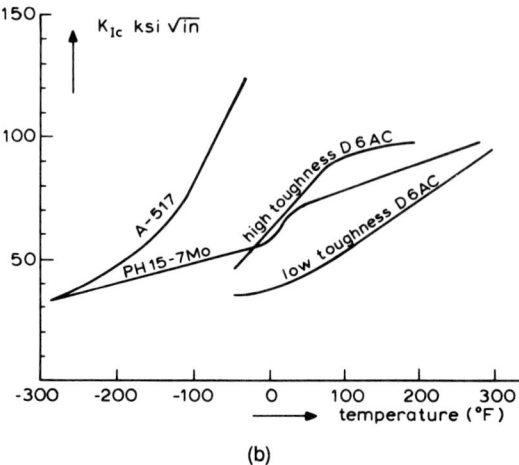

Figure 2.41 (*Continued*) Fracture toughness (*b*) as a function of temperature. (*From Broek.*[14])

stronger the material or the lower the temperature, the lower the toughness.

In failure analysis of a broken part, a fracture mechanics analysis of the failure will often reveal considerable information about the failure. For example, on a fracture surface the distance a from the origin of the crack to the location of the beginning of rapid crack propagation can often be measured. If K_I is known, then the gross stress under which the component was at fracture can be estimated via relations similar to Eq. (2.2). If both K_I and the stress σ are known, and the calculated critical crack size is equal to the measured a, then the crack must have grown slowly to the final size, probably by fatigue (see Sec. 2.13).

2.13 Fatigue Loading

Consider a cylinder subjected to a time-dependent load, as shown in Fig. 2.42. The amplitude of the load at the maximum of the cycle is only sufficient to induce an elastic stress. That is, the cylinder is stressed at a value below the yield strength based on a standard tensile test, and thus it appears that the cylinder should not fail. However, due to stress concentration at discontinuities (such as surface notches or internal cracks), the local stress will exceed that required for yielding, and deformation and local fracture occurs (see Sec. 3.9 for details of mechanisms). With each cycle the crack advances, which is

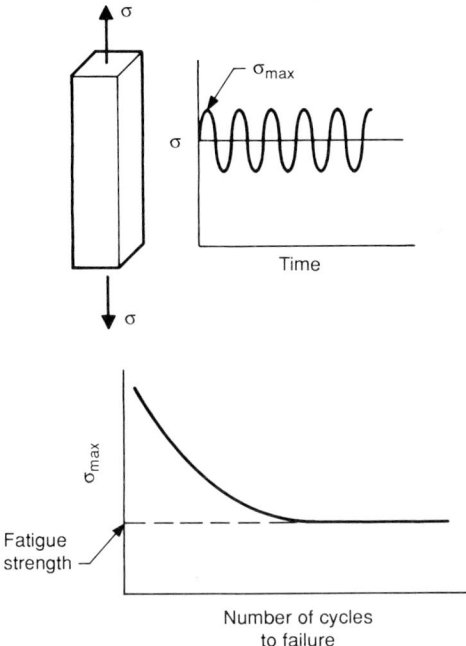

Figure 2.42 Depiction of a fatigue test.

subcritical crack growth. Eventually the crack exceeds the critical value for the toughness of the material, and it then propagates unstably, leading to fracture.

The time or number of cycles to reach fracture increases with a reduction in the magnitude of the load. For some materials there is a stress below which the crack does not propagate. This is called the *fatigue limit* or *fatigue strength,* as illustrated in Fig. 2.43. Usually the results are not sensitive to the frequency of cycling.

The rate of propagation of the crack depends on whether the load, the stress, or the strain is kept constant during the test. These must be defined to properly interpret the test results. The difficulty with using data such as those in Fig. 2.43 is that there is no simple way to predict when a growing crack will attain the critical value which leads to rapid crack propagation. This is achieved by measuring the effect of the stress intensity factor on the change in the crack propagation distance. The concept is illustrated in Fig. 2.44, and Fig. 2.45 shows some data. The curve in Fig. 2.45 can be separated into a region of no crack propagation, where ΔK is very low, a region of linear crack propagation, and a very rapid crack-propagation region at high ΔK values.

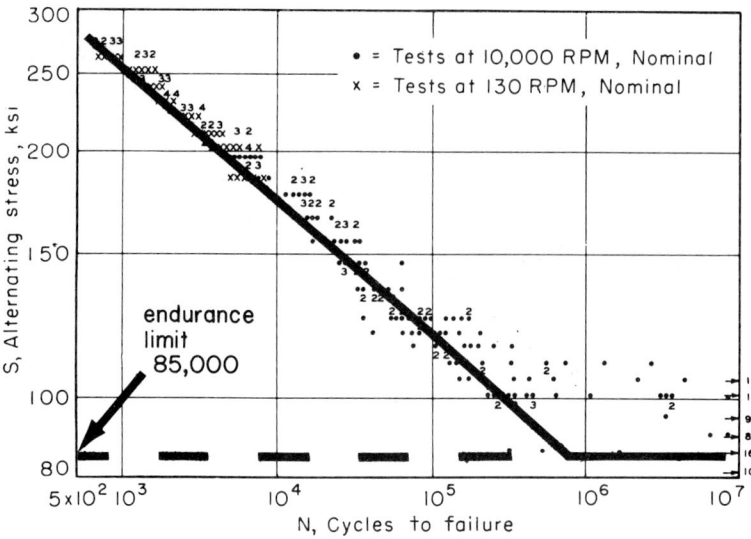

Figure 2.43 Typical fatigue curve. The 4340 steel was heat-treated to a tensile strength of 260,000 lb/in^2 (250,000 lb/in^2 yield strength, hardness 53 R_c). The numbers next to the data points are the number of tests giving the same data point. The numbers next to the arrows indicate the number of specimens at the indicated stress which did not break at 10^7 cycles. The specimens were cylinders tested in a rotating-beam configuration. Note that the results are valid for two widely different rotational speeds. (*Adapted from Cummings et al.*[19])

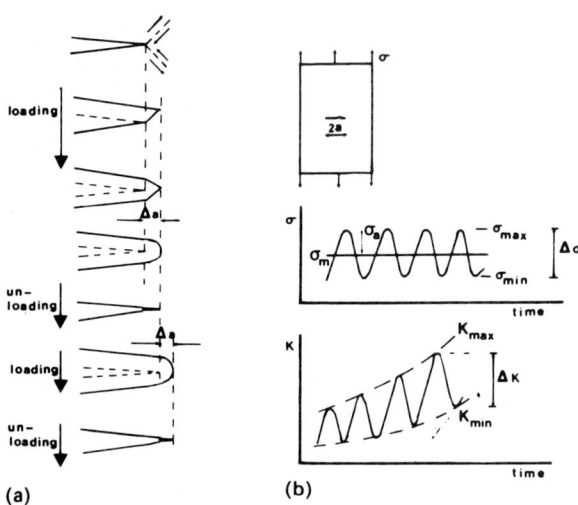

Figure 2.44 Schematic illustration of parameters used in obtaining a fatigue crack growth curve. (*Adapted from Broek.*[20])

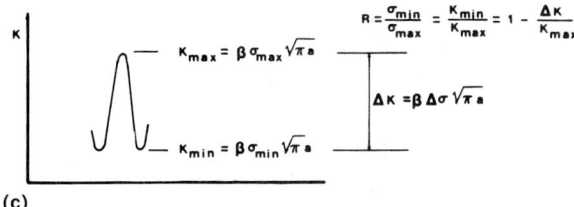

(c)

Figure 2.44 (Continued) Schematic illustration of parameters used in obtaining a fatigue crack growth curve. (Adapted from Broek.[20])

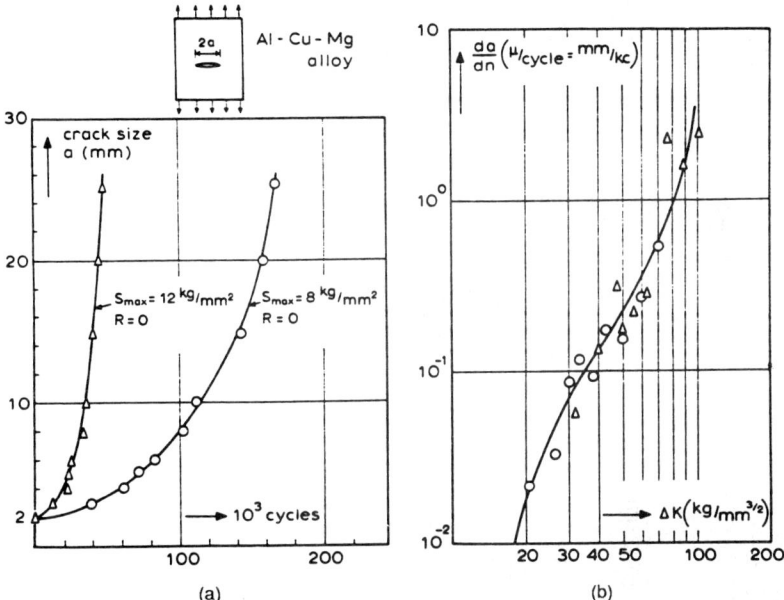

Figure 2.45 Fatigue crack propagation curves. (a) Crack-growth curves. (b) Crack-propagation rate. (From Broek.[14])

2.14 Creep Deformation

If a cylinder is loaded axially at a stress less than that of the yield strength, then yielding and fracture are not expected. However, there are situations in which a slow deformation will occur. This means that the response of materials to loading is time-dependent, but may not be observed in short-time tensile tests (but see Fig. 2.31). Figure 2.46 shows the elongation with time for a fixed axial stress. This elongation process is called *creep*. The elongation rate (*creep rate*) increases with increasing temperature and stress, and the time to fracture

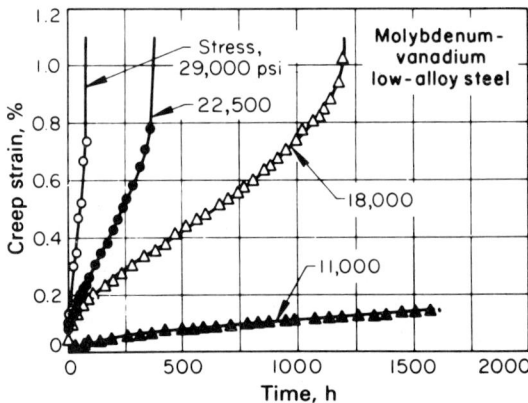

Figure 2.46 Typical creep curve. The sample was a Mo-V low-alloy steel under tension at 600°C (1110°F) at the four stress levels noted. (*From* Metals Handbook.[21])

(*rapture life*) decreases. Note that the fracture elongation is relatively low (10 percent or less).

The fracture surface of a creep fracture is generally normal to the maximum normal stress. An example of creep rupture in a pipe is shown in Fig. 2.47. The fracture strain is usually low (see Fig. 2.46), so dimensional changes in the vicinity of the fracture are slight, although local cracking can lead to apparent plastic deformation.

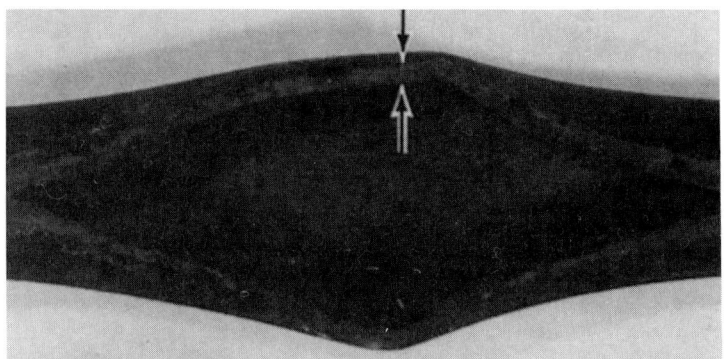

Figure 2.47 Creep fracture in a 321 stainless-steel (ASME SA-213, grade TP321H) superheater tube that failed by thick-lip stress rupture. The fracture surface is normal to the circumferential (tangential) stress. Note that the change in tube-wall thickness at the arrows is slight. This fracture appearance is sometimes called *fishmouth* rupture. (*Adapted from* Metals Handbook.[21])

References

1. W. F. Smith, *Principles of Materials Science and Engineering*, 2d ed., McGraw-Hill, New York, 1986.
2. C. O. Smith, *The Science of Engineering Materials*, 2d ed., Prentice-Hall, Englewood Cliffs, N.J., 1977.
3. D. J. Wulpi, *Understanding How Components Fail*, American Society for Metals, Metals Park, Ohio, 1985.
4. R. P. Feynman, R. B. Leighton, and M. Sands, *The Feynman Lectures on Physics*, California Institute of Technology, 1964, reprinted with permission of Addison-Wesley, Reading, Mass.
5. *Metals Handbook*, 8th ed., vol. 10: *Failure Analysis and Prevention*, American Society for Metals, Metals Park, Ohio, 1975.
6. F. A. Avallone and T. Baumester III, *Mark's Standard Handbook for Mechanical Engineering*, 9th ed., McGraw-Hill, New York, 1987.
7. C. Lipson and R. C. Juvinall, *Handbook of Stress and Strength*, Macmillan, New York, 1963.
8. E. R. Parker, *Brittle Behavior of Engineering Structures*, Wiley, New York, 1957.
9. G. Henry and D. Horstmann, *De Ferri Metallographia*, vol. V: *Fractography and Microfractography*, Verlag Stahleisen, Düsseldorf, Germany, 1979.
10. *Metals Handbook*, 8th ed., vol. 9: *Fractography and Atlas of Fractographs*, American Society for Metals, Metals Park, Ohio, 1974.
11. R. W. Hertzberg, *Deformation and Fracture Mechanics of Engineering Materials*, 2d ed., Wiley, New York, 1983.
12. A. Nadai, *Theory of Flow and Fracture of Solids*, 2d ed., McGraw-Hill, New York, 1950.
13. G. F. Vander Voort, "Visual Examination and Light Microscopy," in *Metals Handbook*, 9th ed., vol. 12: *Fractography*, ASM International, Metals Park, Ohio, 1987.
14. D. Broek, *Elementary Engineering Fracture Mechanics*, Sijthoff and Noordhoff, The Netherlands, 1978.
15. C. Zener and J. H. Holloman, "Effect of Strain Rate upon Plastic Flow of Steel," *J. Appl. Phys.*, vol. 15, p. 22, 1944.
16. M. A. Meyers and K. K. Chawla, *Mechanical Metallurgy—Principles and Applications*, Prentice-Hall, Englewood Cliffs, N.J., 1984.
17. J. Nunes, F. L. Carr, and F. R. Larsen, "Macrofractographic Features," in R. F. Bunshah (ed.), *Techniques of Metals Research*, vol. 2, pt. 1, Wiley, New York, 1968, p. 379.
18. G. M. Boyd, "The Propagation of Fracture in Mild Steel Plates," *Engineering*, vol. 175, pp. 65 and 100, 1953.
19. H. M. Cummings, F. B. Stulen, and W. C. Schulte, "Relation of Inclusions to the Fatigue Properties of SAE 4340 Steel," *Trans. ASM*, vol. 49, p. 482, 1957.
20. D. Broek, *The Practical Use of Fracture Mechanics*, Kluwer, Boston, Mass., 1989.
21. *Metals Handbook*, 9th ed., vol. 11: *Failure Analysis and Prevention*, American Society for Metals, Metals Park, Ohio, 1986.

Bibliography

Campbell, J. E., W. W. Gerberich, and J. H. Underwood (eds.): *Application of Fracture Mechanics*, American Society for Metals, Metals Park, Ohio, 1982.
Ewalds, H. L., and R. J. H. Wanhill: *Fracture Mechanics*, Edward Arnold, Baltimore, Md., 1984.
Hertzberg, R. W.: *Deformation and Fracture Mechanics of Engineering Materials*, Wiley, New York, 1983.
Tetelman, A. S., and A. J. McEvily: *Fracture of Structural Materials*, Wiley, New York, 1967.

Chapter 3

Fracture Mechanisms and Microfractographic Features

"Fractography" is the name we have applied to this technique. Reticence in introducing a new term is counteracted by the advantage in substituting a word for a phrase as long as "the micrographic study of cleavage facets on fractured metal specimens," which is the definition of fractography. In addition, the word fractography has some etymological correctness in stemming from roots similar to those affording metallography, for "fracto-" is the combining form of the Latin "fractus," meaning fracture.
> C. A. ZAPFFE AND M. CLOGG, JR.
> *Transactions of the American Society for Metals, 1945*

3.1 Introduction

In this chapter we review the common processes which occur on a relatively fine scale when a metallic material is subjected to increasing loads, and which lead to material separation (fracture). Since at this scale the human eye cannot resolve individual features, the use of microscopes is required. These processes can be referred to as *fracture micromechanisms,* but we will use the term *fracture mechanisms.* This distinguishes the fine-scale processes from the coarser-scale processes, or fracture macromechanisms, which we call modes in Chap. 4.

Since the fracture mechanisms determine the fine topology of the fracture surface, its appearance gives clues, and frequently direct confirmation, of the fracture mechanism. Examination of this fine surface topography is called *microfractography.* Examination of the coarser fracture-surface topography, generally resolvable with the eye or low-magnification (such as 25×) optical instruments, is called macrofractography, which is reviewed in Chap. 4.

We begin this chapter by reviewing the common, basic mechanisms of the response of a metallic material to loading, namely, slip, twinning, and cleavage. We do not examine the underlying atomistic mechanisms, such as dislocation generation and movement, but instead restrict our comments to those features which most directly affect the fracture-surface topography. We then relate these basic mechanisms to the fracture mechanisms and to the resulting fracture-surface topography.

3.2 Slip and Cleavage

A common mechanism of plastic deformation is *slip*, which is the parallel movement of layers of material in a crystal past adjacent layers. This can be illustrated by the response of a single crystal to a shear stress, as depicted in Fig. 3.1. Here it is assumed that plastic deformation is initiated when the shear stress on the plane shown exceeds a critical value, *the critical resolved shear stress*. Continued application of this stress will cause separation and hence fracture.

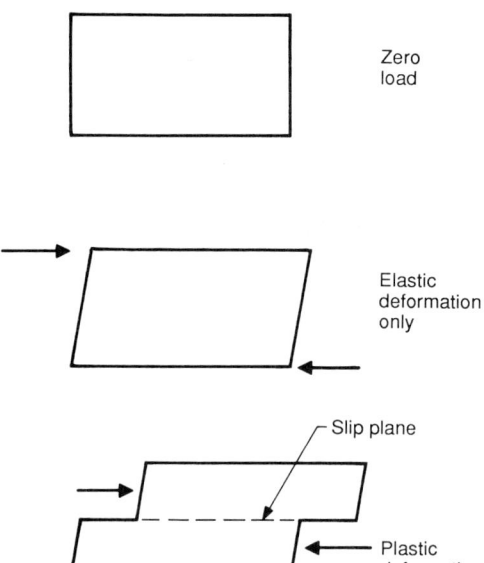

Figure 3.1 Illustration of plastic deformation of a single crystal by slip.

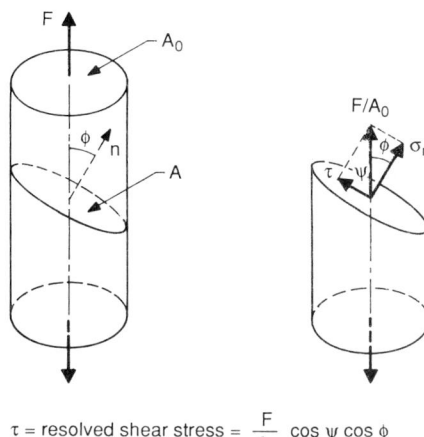

τ = resolved shear stress = $\dfrac{F}{A_0} \cos \psi \cos \phi$

σ = resolved normal stress = $\dfrac{F}{A_0} \cos \psi \sin \phi$

Figure 3.2 Illustration showing resolved shear stress τ [= (F/A_0) cos ψ cos ϕ] and resolved normal stress σ_n [= (F/A_0) cos ψ sin ϕ] for a cylindrical sample loaded axially.

A more realistic case is illustrated in Fig. 3.2. Here a cylindrical single crystal is loaded axially with an axial stress of F/A_0, where A_0 is the cross-sectional area of the cylinder. The plane on which slip occurs is oblique to the cylinder axis, with its normal at an angle ϕ to the cylinder direction (the loading axis). Slip occurs along only one direction in this plane, the slip direction, which makes an angle ψ with the loading axis. Thus the component of the axial load F resolved onto this plane (the shear force F_s) is F cos ψ. The area of the shear plane is given by $A_0/\cos \phi$. The resolved shear stress τ is (F/A_0) cos ψ cos ϕ.

Upon subjecting the cylindrical single crystal to an increasing load, slip begins when the resolved shear stress on the slip plane reaches the critical resolved shear stress value. It is observed to occur at many locations, with elliptically shaped discs moving past each other, as depicted in Fig. 3.3b. If the crystal is relatively large (such as 1 cm in diameter) and the surface initially well polished, these offsets are visible to the unaided eye with sufficient deformation (Fig. 3.3c). Examination with a microscope shows that each coarse offset (slip bands) is made up of much finer steps (slip lines) (Fig. 3.4).

This effect can be seen on a polycrystalline sample. If the sample is metallographically polished, then loaded as shown in Fig. 3.5a, in each grain a slip system is activated and surface offset occurs when the critical resolved shear stress is exceeded. An example is given in Fig. 3.5b. Note that the slip direction changes at the grain boundaries,

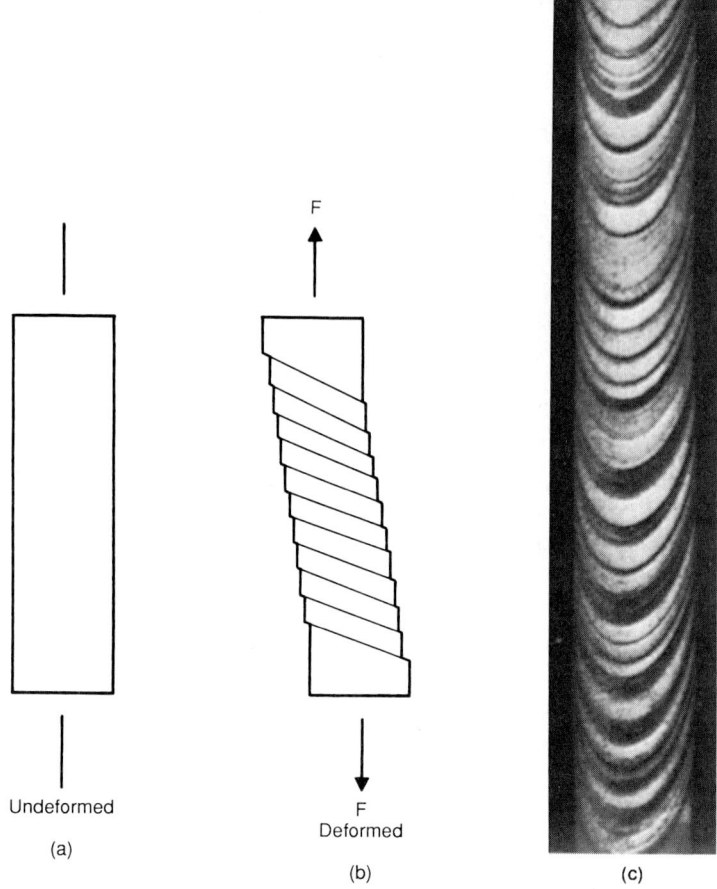

Figure 3.3 Illustration of plastic deformation of a single crystal. (*a*) Undeformed. (*b*) Deformed showing slip. (*c*) Deformed cylindrical single crystal of zinc, showing elliptical offsets due to slip. (*Adapted from Boas and Schmid.*[1])

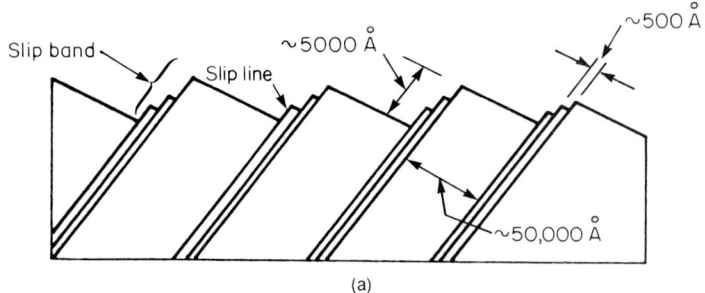

Figure 3.4 Illustration of fine offsets on a free surface produced by slip. (*a*) Appearance of the surface cross section at high magnification. (*Courtesy of Eric Lifshin, Research and Development Center, General Electric Company, from Guy.*[2])

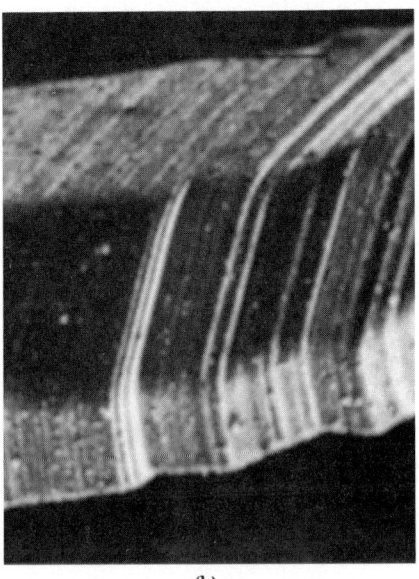

(b)

Figure 3.4 (*Continued*) Illustration of fine offsets on a free surface produced by slip. (*b*) SEM of slip markings on the surface of a deformed single crystal of cobalt. (*Courtesy of Eric Lifshin, Research and Development Center, General Electric Company, from Guy.*[2])

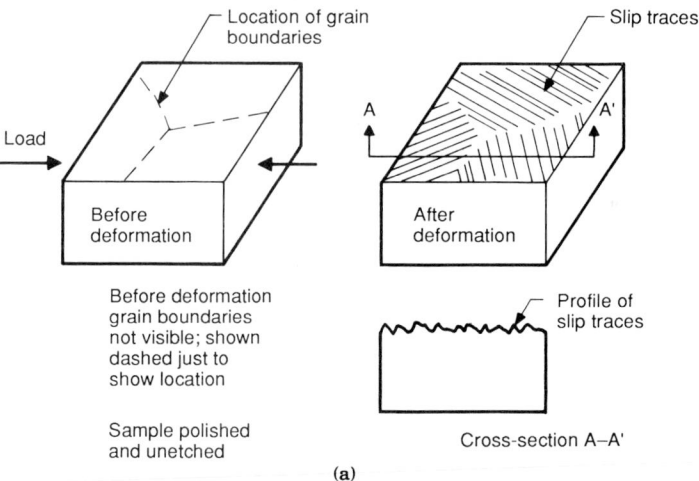

(a)

Figure 3.5 (*a*) Loading of a metallographically polished (but unetched) sample to produce slip markings on the polished surface.

and hence the locations of these boundaries are revealed even though the sample is unetched.

The planes and directions on which slip occurs depend on the crystal structure of the material; usually these are the atomistically close-packed planes and directions. Thus in face-centered cubic crystals

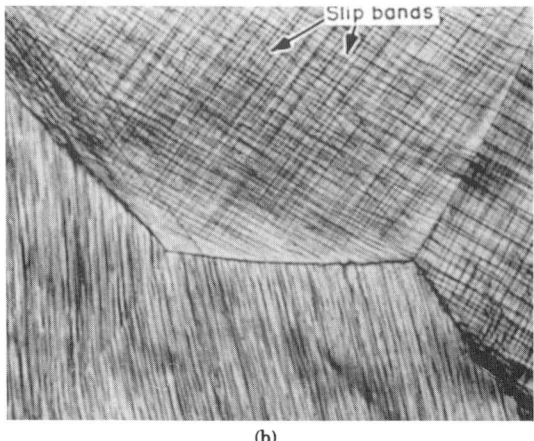

Figure 3.5 (*Continued*) (*b*) Micrograph of such a sample of aluminum showing slip traces. (*Adapted from Verhoeven.*[3])

these are the {111}-type planes and the ⟨110⟩-type directions. A slip plane and a slip direction in that plane constitute a slip system. In face-centered cubic metals there are four {111}-type planes, and in each {111}-type plane three ⟨110⟩-type directions, so there are 12 slip systems. In the other close-packed structure, which is hexagonally close-packed, there is only one close-packed plane, but three identical slip directions, or only three slip systems.

When a crystal is loaded, the shear stress increases on all slip systems, but slip initiates only on the system which first attains the critical resolved shear stress for the material. Which system this is depends on the orientation of the crystal. It is possible to have an orientation such that two slip systems have identical resolved shear stresses, and that these two attain the critical resolved shear stress first. This leads to simultaneous slip on these two slip systems, which is called *cross slip*.

In the hexagonal close-packed structure it is possible to have an orientation in which the resolved shear stress is zero. This is illustrated in Fig. 3.6. The basal plane, which is the slip plane, is normal to the loading force. Thus as the loading force is increased, it might be expected that slip will occur on another type of slip plane (not a close-packed plane). However, the critical resolved shear stress for this is very high. Another possibility is that the resolved *normal* stress on a plane will exceed that which intrinsically bonds the planes together, and these planes will separate, or cleave. Note from Fig. 3.2 that a normal stress σ_n can be defined for a specified

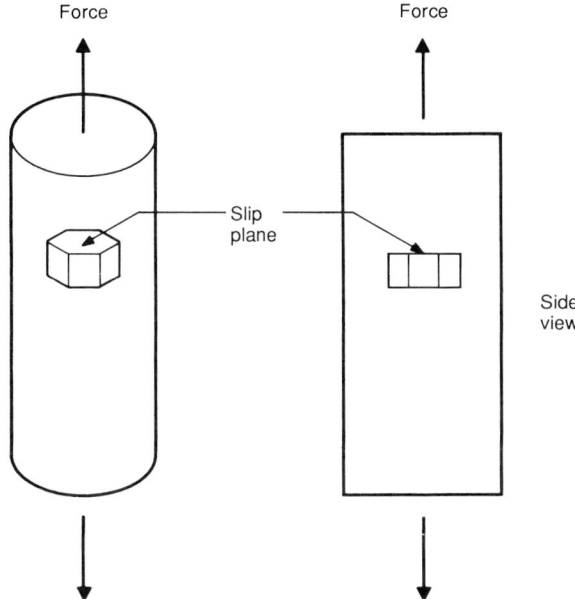

Figure 3.6 Illustration of the orientation of a single crystal of a hexagonal close-packed metal, with the basal (slip) plane normal to the loading axis, so that there is no resolved shear stress on it.

plane (specified by the angle ϕ), which is given by $(F/A_0) \sin \phi \cos \psi$. Thus cleavage will occur when the critical resolved normal stress for cleavage is exceeded.

If a crystal of the hexagonal close-packed metal zinc is loaded with the configuration shown in Fig. 3.6, slip cannot occur. Instead cleavage occurs on the basal plane, and the fracture surface will be relatively smooth, with a mirrorlike appearance. Such cleavage usually cannot occur with a face-centered cubic crystal. If the crystal is oriented so that a {111} slip plane is normal to the loading axis, there are still three other {111} planes at oblique angles to this axis, and hence these planes, each containing three <110> slip directions, have finite resolved shear stresses, and slip will occur on some of these.

Thus there is a competition between slip and cleavage. As the external load on a crystal increases, the shear stresses on the slip systems increase and the normal stresses on the cleavage planes increase. Which occurs first, slip or cleavage, depends on which stress attains the critical value first. This is depicted in Fig. 3.7. If the loading is such that τ and σ_n increase along line A, then slip occurs first; if

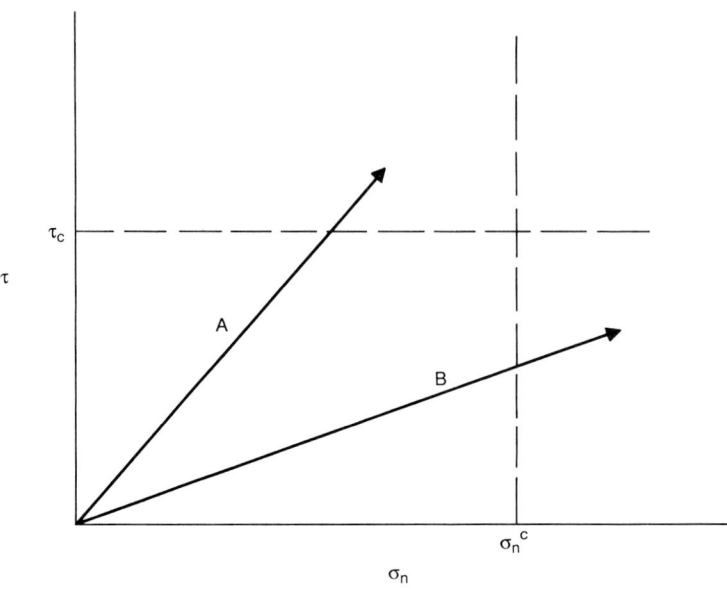

Figure 3.7 Schematic illustration that whether slip or cleavage occurs depends on the relations between resolved shear and resolved normal stress as the load increases, and values of critical resolved shear stress for slip τ_c and critical resolved normal stress for cleavage σ_n^c.

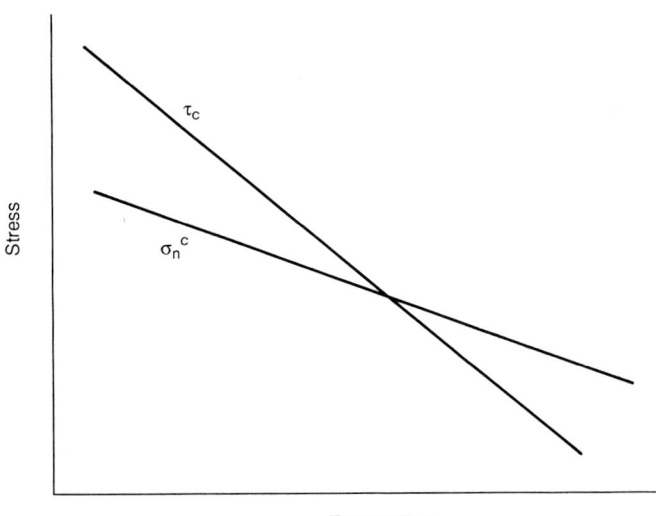

Figure 3.8 Temperature dependence of critical resolved shear stress and critical resolved normal stress for a body-centered cubic metal.

along line B, then cleavage occurs. Clearly slip is more favored if τ_c is much lower than σ_n^c, which is the case for most close-packed metals and most body-centered cubic metals at sufficiently high temperature. The critical values of τ and σ_n are temperature-dependent, both increasing with decreasing temperature. However, in some cases, such as body-centered cubic metals, lower temperatures are more favorable for cleavage (Fig. 3.8). These critical values of τ and σ_n are also affected by the strain rate, so that a metal may exhibit cleavage if the loading which causes fracture is sufficiently rapid.

3.3 Twinning

In some metals and alloys, crystals can respond to increasing loading by twinning. The atomistic rearrangement for the process is depicted schematically in Fig. 3.9. Note that the shift of the lattice along the twinning plane produces the same crystal structure, but as a mirror image across this plane. Hence it is said that the crystal on one side of this plane is a twin of the crystal on the other side.

Twinning is a rather complex process. It rarely occurs in face-centered cubic crystals, but does occur in both hexagonal close-packed and body-centered cubic crystals. Its occurrence may be favored by low temperature and high strain rates. Figure 3.10 shows a microstructure containing typical mechanical twins. The line separating

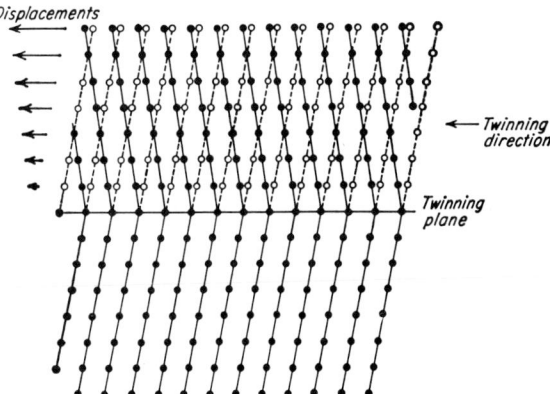

Figure 3.9 Schematic illustration of mechanical twinning. The twinned part lies above the twinning plane. Open circles—initial positions; closed circles—final positions, which are mirror images of the points below the twinning plane. (*From Birchenall.*[4])

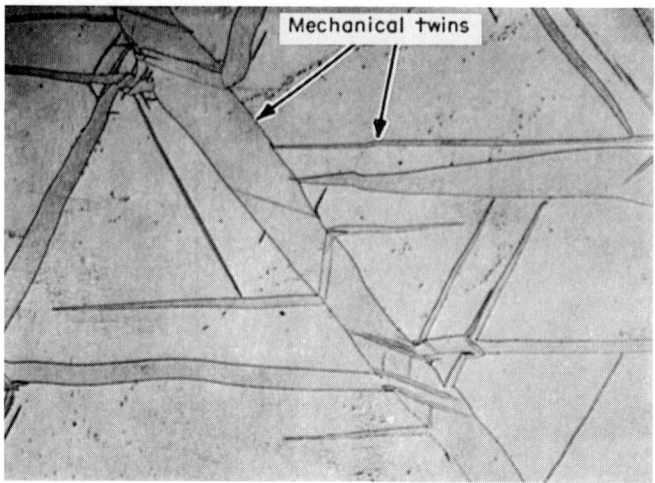

Figure 3.10 Microstructure of plastically deformed magnesium showing mechanical twins. (*Adapted from Roberts.*[5])

the original crystal from the twinned material is the trace of the twinning plane. In fracture of body-centered cubic metals, twinning may play a role in crack nucleation. [Mechanical twins should not be confused with annealing twins. The latter are also crystallographic twins, but they form by other mechanisms. They are especially prominent in many face-centered cubic metals and appear as straight lines in the microstructure (Fig. 3.11).]

3.4 Cleavage Fracture Topography

From the preceding description of cleavage it would appear that the fracture surface of a single crystal would be atomistically smooth and hence featureless. Indeed, at low magnification such surfaces are mirrorlike, and polycrystalline samples which have fractured completely by cleavage may appear to the eye to consist of bright facets due to light reflection. However, even in single crystals of pure metals there are lattice defects which cause the propagating cleavage crack to deviate and hence generate topography. Also, the cleavage cracks may nucleate at more than one location, and thus the cracks propagate at different levels even when on the same type of parallel planes. Where these levels meet, fracture must occur by

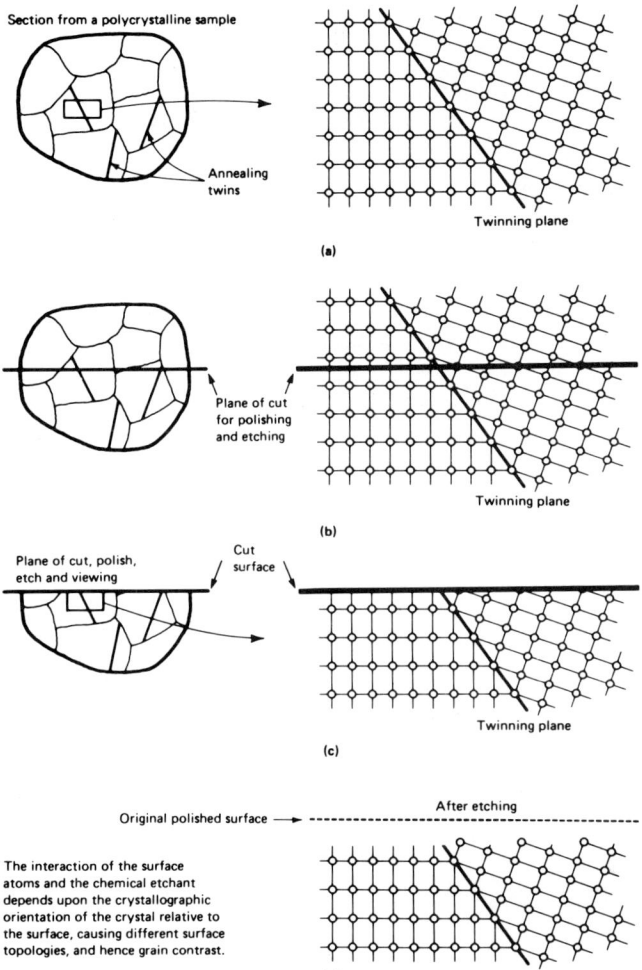

Figure 3.11 (a)–(d) Illustration of why annealing twins are revealed upon etching a polished surface.

either slip or cleavage on another set of planes, so that the fracture surface has offsets on it.

The formation of the fractographic feature where the fronts of cleavage cracks moving on adjacent planes meet is illustrated in Fig. 3.12a. The connecting material fractures either by cleavage on another cleavage plane or by plastic deformation. The characteristics of the steps which form are revealed by cross-sectional views of the fracture

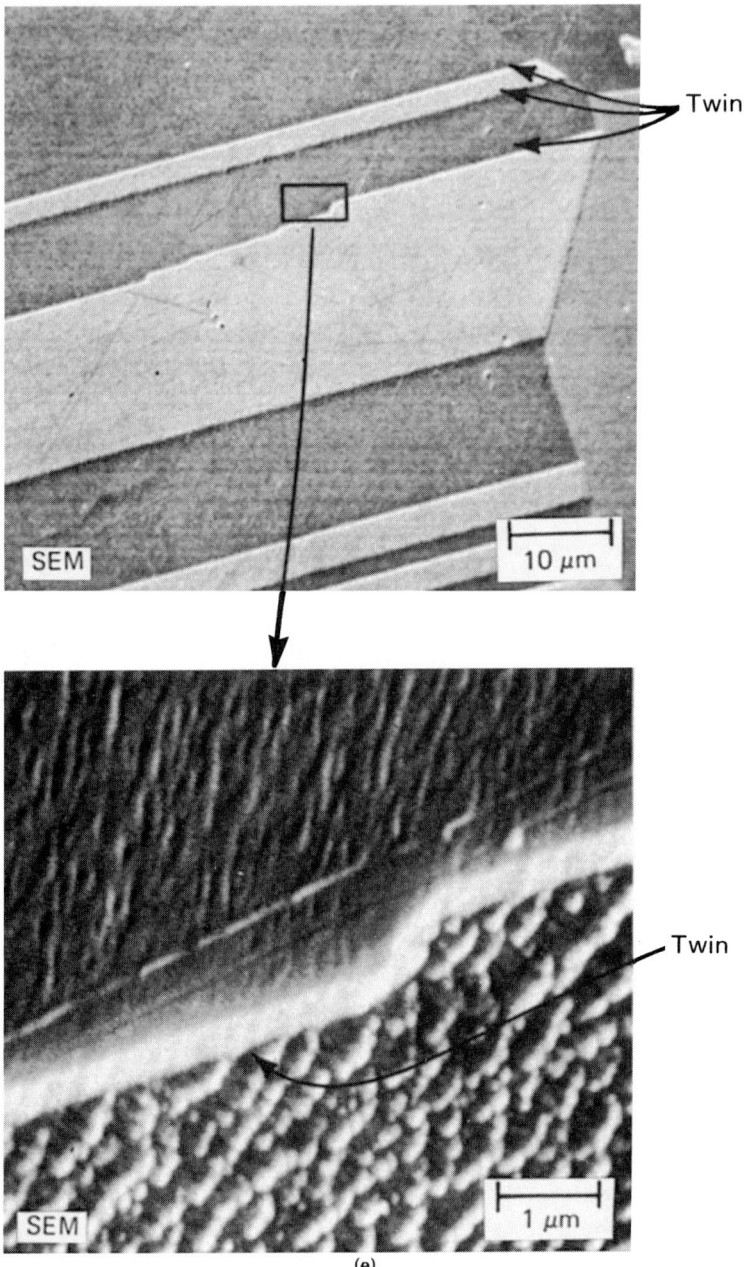

Figure 3.11 (*Continued*) (*e*) SEM of Cu–5% Zn alloy in the annealed condition. (*From Brooks.*[6])

Fracture Mechanisms and Microfractographic Features   131

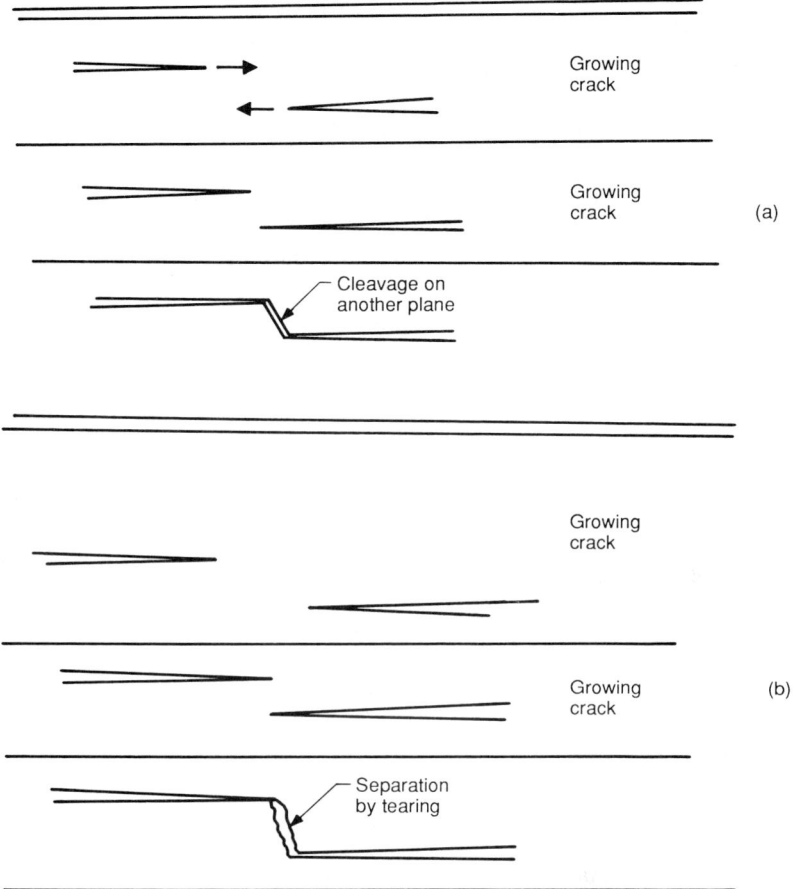

Figure 3.12 Schematic illustration of separation between cleavage planes. (*a*) By cleavage. (*b*) By tearing.

surface. An example is presented in Fig. 3.13, where the steps connecting the parallel cleavage planes are apparent. Note that the straight-line profiles of these steps indicate that they fractured by cleavage. In some cases the connecting material may "tear" by plastic deformation (slip). In the example of Fig. 3.14, the connecting material does not show the straight lines associated with cleavage, but instead is bent.

In body-centered cubic and hexagonal close-packed structures, the presence of mechanical twins, or more commonly the formation of them during the deformation, allows the crack front to deviate by separation of the plane between the twin and the matrix. This is illustrated in Fig. 3.15. In many body-centered cubic metals (par-

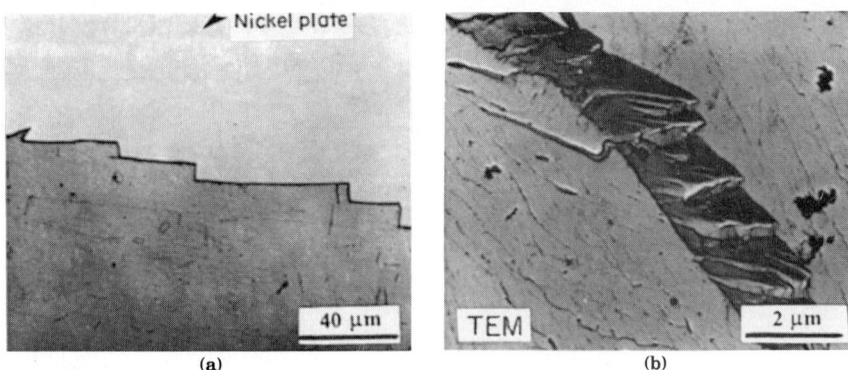

Figure 3.13 (a) OM of the cross section of a cleavage profile (nickel-plated rupture surface) in an Fe–Si single crystal. This shows secondary ruptures between different levels of fracture on the cleavage plane. (b) Fractograph showing cleavage steps and secondary rupture along a (100) plane. Material was mild steel, impact-fractured at −196°C. (*From Henry and Horstmann.*[7])

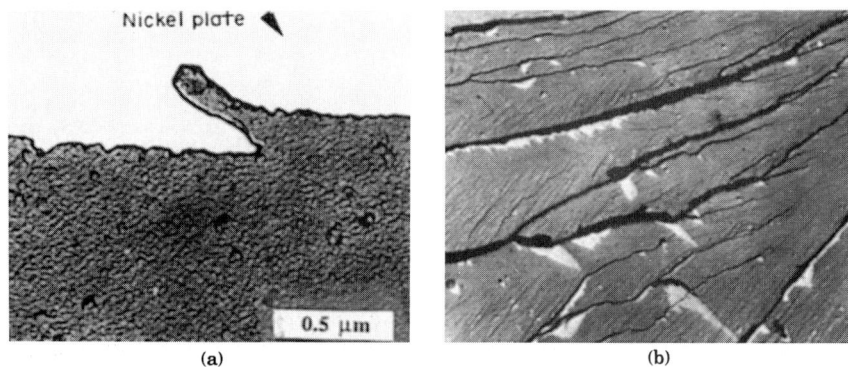

Figure 3.14 (a) Cross section through a cleavage fracture surface, showing tearing of the connecting material between cleavage surfaces. Material was Fe–3% Si. (b) TEM fractograph showing tear ridges which make up a river pattern. Their profile is similar to that in (a). Material was extramild steel. (*From Henry and Horstmann.*[7])

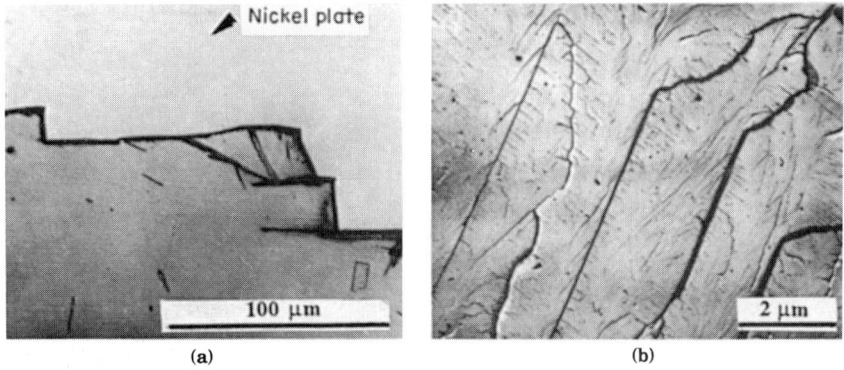

Figure 3.15 (a) Cross section through a cleavage fracture surface, showing interaction with mechanical twins. Material was an Fe–Si single crystal. (b) TEM fractograph showing preferential propagation along a [100] direction, and features due to fracture of twinned regions. Material was an Fe–25% Cr alloy. (*From Henry and Horstmann.*[7])

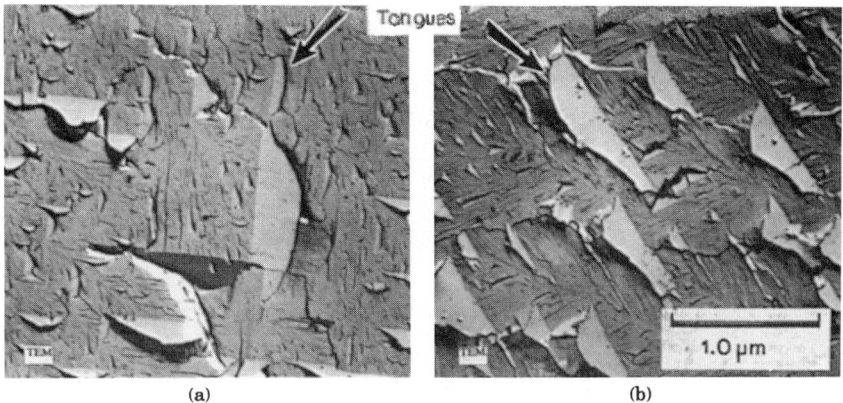

Figure 3.16 Fractographs showing the general appearance of "tongues." Samples were impact-fractured at −196°C. Note that the intersections of the cleavage plane and the tongues are oriented along one or the other of two perpendicular directions. The tongue plane is a (112) twin plane. (a) Extramild steel. (b) Fe–1.8% Si steel. (*From Henry and Horstmann.*[7])

ticularly ferrite in steels), the protrusions or matching depressions shown in Fig. 3.16 are sometimes seen on cleavage fracture surfaces. These are called *tongues*. Cross-sectional views (Fig. 3.17), which allow measurements of the geometry of these regions, have shown that the tongue plane is a {112} type, and it is generally agreed, therefore, that these are created by fracture along the twin-

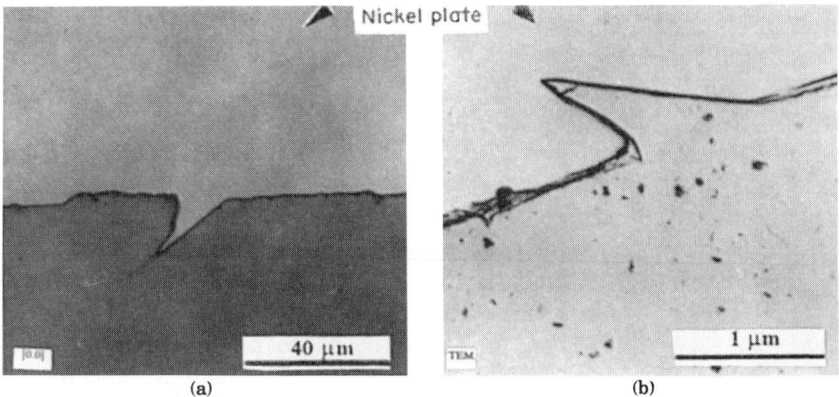

Figure 3.17 Cross section through fracture surface, showing separation along the interface between mechanical twins and the matrix. (a) OM of an Fe–Si single crystal, impact-fractured at −196°C. The fracture surface was nickel-plated (top). In the metal (bottom), a small twin which initiated the tongue is noticed. (b) TEM replica micrograph of fracture in Armco iron, showing profile of a tongue. (*From Henry and Horstmann.*[7])

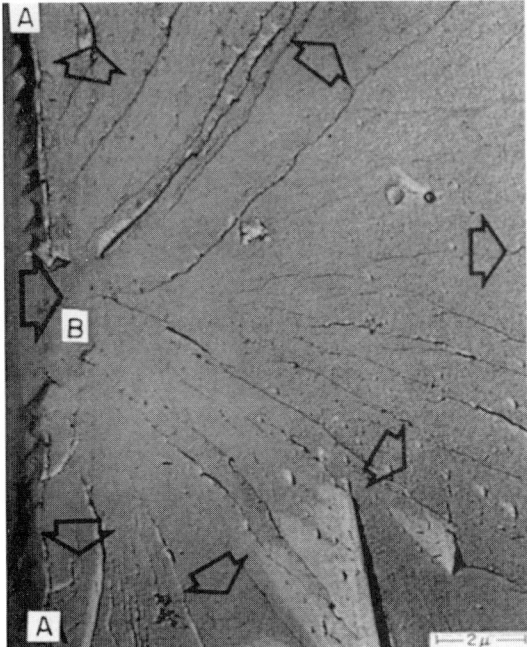

Figure 3.18 TEM micrograph of a two-stage replica, showing feature called a fan. Material was iron. Arrows indicate the crack propagation direction. The fracture surface was slightly etched. (*From Beachem.*[8])

matrix interface. It appears that these twins form ahead of the advancing cleavage crack.

The fractures described appear to illustrate the main mechanism of cleavage crack formation. However, the topography is usually more diverse than this would indicate. This is due to the complex microstructure of most materials through which the crack must propagate and to the tendency for the crack to take a path which minimizes energy expenditure. Although the topography is complicated, fortunately there are usually distinctive features associated with cleavage. One feature is a fan, shown in Fig. 3.18. In this case the advancing crack (from left to right) has encountered a grain boundary (marked A–A), whereupon locally a fresh crack was initiated at B. This cleavage crack then propagated in a fanlike fashion onto planes of slightly different elevation. The local crack propagation direction is noted on the fractograph by arrows.

In another example, shown in Fig. 3.19, the crack advanced from left to right. Along the line A–A it met a low-angle boundary.

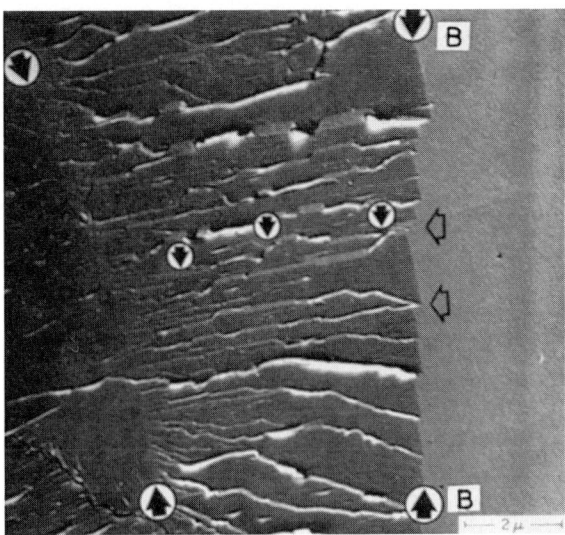

Figure 3.19 TEM micrograph of a two-stage replica, showing cleavage steps, river patterns, grain boundary (left), and mechanical twin interface (right) in cleaved iron. (*From Beachem.*[8])

This causes the initiation of many cracks on the new cleavage plane, but at slightly different elevations. As the crack then propagated, these levels combined to form a crack advancing on fewer parallel planes. The reason is that there is less area associated with the fracture of the connecting material, and hence less expenditure of energy as the crack propagates. Such a mechanism generates a pattern similar to a river and its branches, and hence is called a river pattern. Note that in this sense the local crack propagation direction is "downstream." (Another example is shown in Fig. 3.15.) As pointed out, cleavage in iron and ferrite can also occur on the mechanical twin interface ({112}-type plane), and this is what happened at *B–B*. To the right is an extremely smooth fracture facet along such an interface.

The interaction of the fracture surface with twins sometimes creates the feature shown in Fig. 3.20, called a herringbone pattern. Cleavage on the basic {100} plane has changed on both sides of the midrib to cleavage on the {112} twin interface planes.

The processes described are shown schematically in Fig. 3.21. Examples of fracture surfaces containing these features are illustrated in Fig. 3.22. In some cases it is possible to see slip bands or lines on the

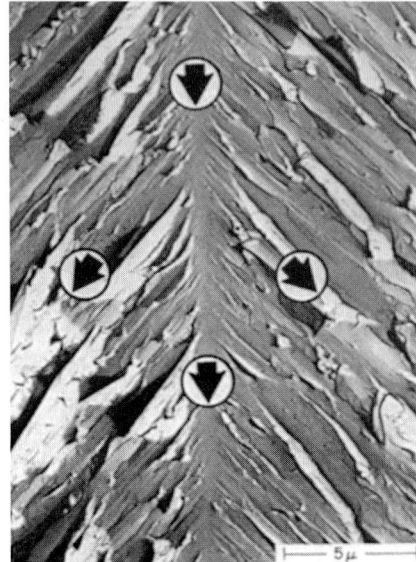

Figure 3.20 TEM micrograph of a two-stage replica, showing feature called a herringbone. Material was an Fe–Cr–Al alloy. (*From Beachem.*[8])

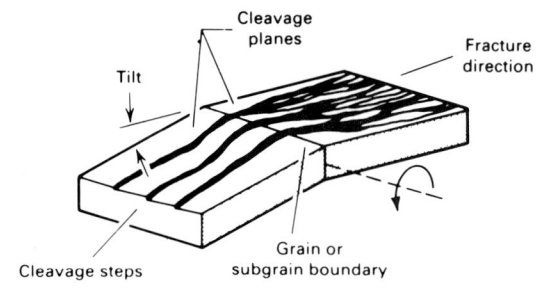

Figure 3.21 Cleavage surface formation, showing the effect of subgrains and grain boundaries. (*From Kerlins and Phillips.*[9])

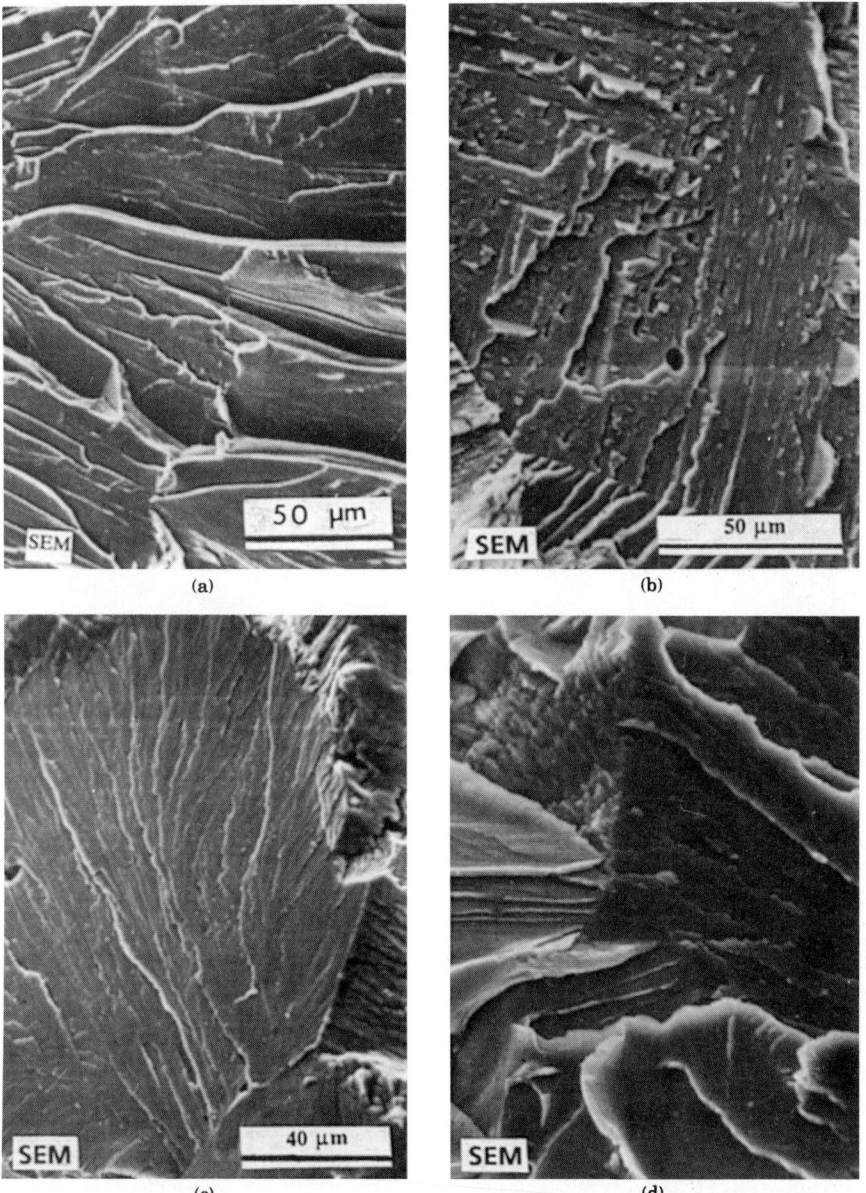

Figure 3.22 Examples of cleavage fracture surfaces. (a) Armco iron, impact-ruptured at −196°C. (*From Henry and Horstmann.*[7]) (b) Weld metal of 30% chromium steel. Owing to the varying crack propagation direction on the cleavage surface of this single grain, a featherlike structure has resulted, consisting of fine steps delineating the individual fracture paths which indicate the direction of propagation. The direction of crack advancement is from bottom to top. (c) Thanks to the presence of numerous twins, various crack propagation directions in a single plane can be deduced. The crack has divided in two along a straight front with a small section moving obliquely away from the main cleavage surface, subsequently returning to it by means of an irregular curved path. Where the crack has crossed into neighboring grains, many new subsidiary cracks form, as seen at the bottom of the picture. (d) 0.5% Mn steel pin. In the center of the picture the crack propagating from the right-hand side crosses a tilted grain boundary. As a result, many new crack surfaces are reinitiated in the grain on the left-hand side of the picture. In this grain a twin can be seen, running obliquely upward from the surface, this having the typically semicircular tonguelike shape. (*b–d*) (*From Engel and Klingele.*[10])

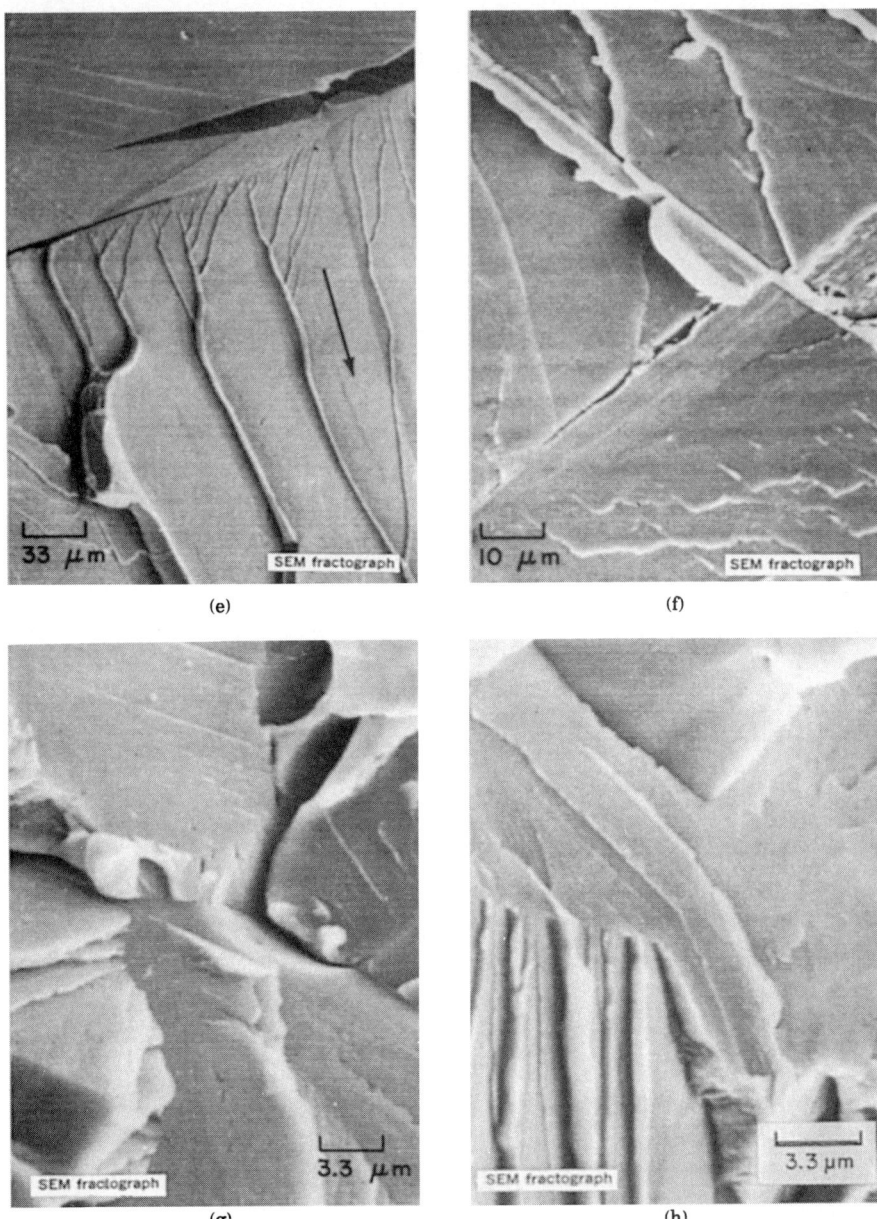

Figure 3.22 (*Continued*) Examples of cleavage fracture surfaces. (*e*) Fe, 0.01% C, 0.24% Mn, and 0.02% Si, heat treated at 950°C (1742°F) ½ h, air-cooled. The structure is ferrite. The fracture was generated by impact at −196°C (−321°F). Cleavage steps beginning at the twin at top form a sharply defined river pattern. Crack propagation was in direction of arrow. (*f*) Higher-magnification view of the specimen(s), showing the blocklike cleavage pattern that is characteristic of low-temperature impact of low-carbon iron. The well-defined tongue at the center of the fractograph is typical. Secondary cracks are visible as well. (*g*) 1021 steel heat treated at 900°C (1652°F) for ½h, air-cooled. Ferrite-pearlite structure. Hardness, 120 dph. The specimen was notched, and was broken by impact at −196°C (−321°F). The cleavage occurred on various {100} planes having different orientations. (*h*) Another SEM view of the fracture in (*g*), showing similar flat cleavage facets along crystallographic planes, but also very pronounced steps that form a gross example of a river pattern beginning at either a twist or a tilt boundary. [(*e*)–(*h*) *From* Metals Handbook.[11]]

Fracture Mechanisms and Microfractographic Features   139

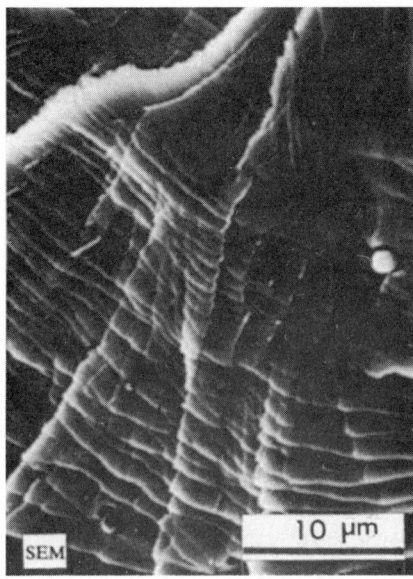

Figure 3.23 Fractograph of cleavage fracture surface showing slip traces. Slip lines reveal plastic deformation which occurs following the formation of ductile zones. (*From Henry and Horstmann.*[7])

cleavage facets. An example is given in Fig. 3.23. This apparently occurs after cleavage, when the connecting material is beginning to fracture by void coalescence. (See dimples, which are described in Sec. 3.5.)

3.5 Void Coalescence

If the metal begins to slip instead of cleavage occurring, then a different fracture mechanism occurs. A single crystal might deform due to cross slip, as shown in Fig. 3.24, with fracture occurring due to eventual "pinching off" of the final connecting material. A similar process can occur in polycrystalline material.

However, a more common mechanism associated with slip is due to the separation of material internally, forming voids which then join to develop the fracture surface. This process is called void coalescence and is depicted schematically in Fig. 3.25. Note that once the void forms, the connecting material continues to deform by slip, allowing the voids to expand until they begin to connect. This fracture mechanism develops a fracture topography consisting of cusps on each fracture surface. Such a surface is called dimpled, and an example is given in Fig. 3.26.

In most engineering alloys the voids form at second-phase particles.

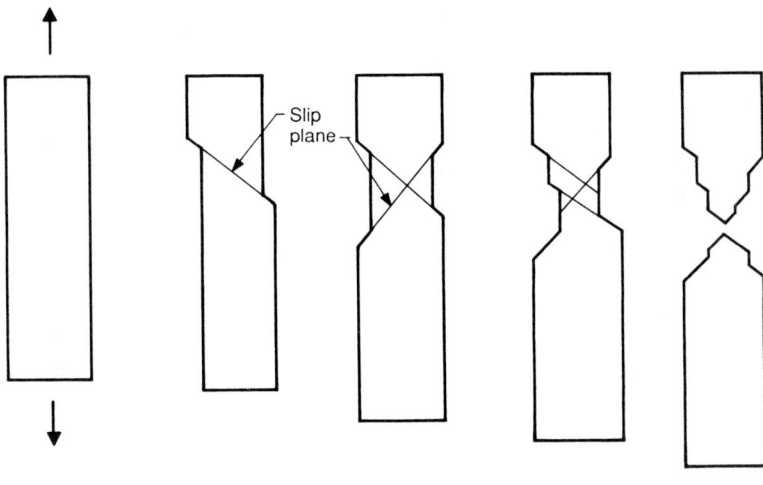

Figure 3.24 Schematic illustration of fracture of a single crystal by slip.

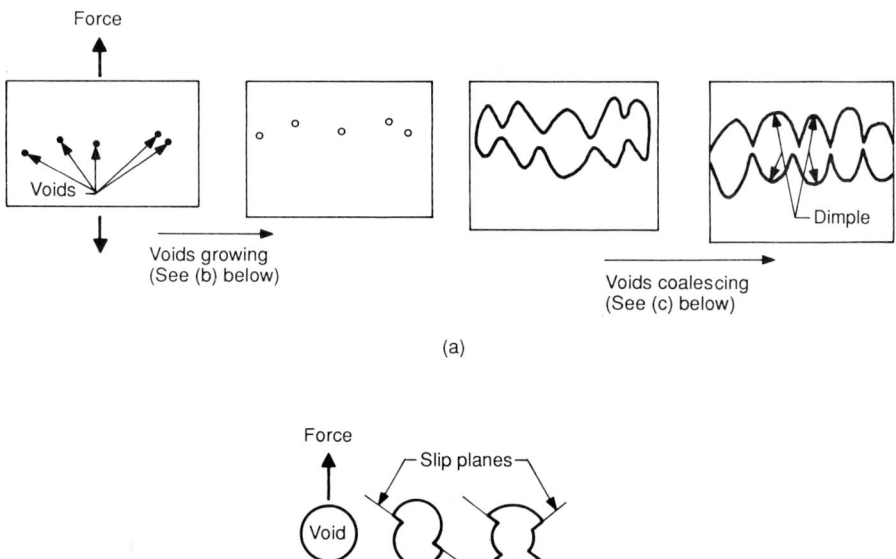

Figure 3.25 Schematic illustration of fracture by void coalescence.

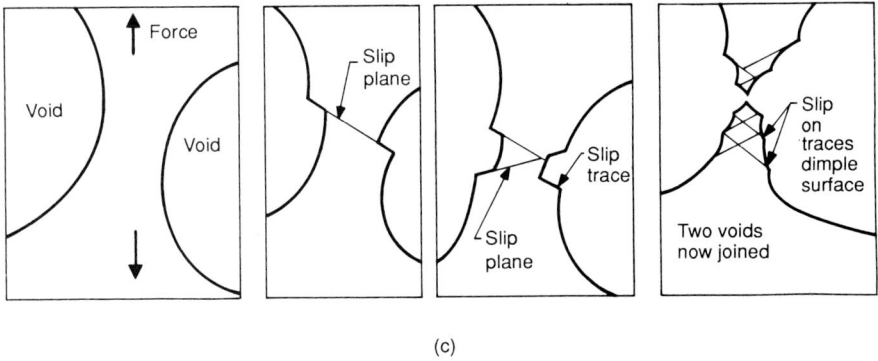

(c)

Figure 3.25 (*Continued*) Schematic illustration of fracture by void coalescence.

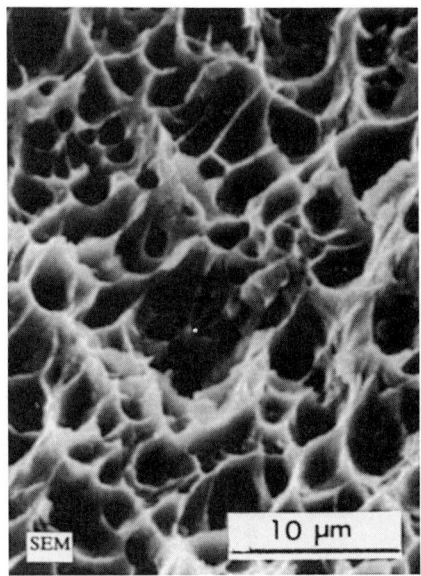

Figure 3.26 Fractograph showing dimples. Material was a steel containing 0.10% C, 0.02% Si, 0.30% Mn, 0.032% P, 0.024% S, and 0.003% N. The sample was notched and broken in bending. (*From Henry and Horstmann.*[7])

This is shown schematically in Fig. 3.27, where the voids are initiated either by decohesion of the particle-matrix interface or by fracture of the particles. Examination of the microstructure of samples in the early stages of plastic deformation confirms this model, as shown in Fig. 3.28. An example of a fracture surface showing such particles is presented in Fig. 3.29. Some particles are in the dimples on the mating fracture surface, or they fell out during separation, so that in the fractograph some dimples are empty.

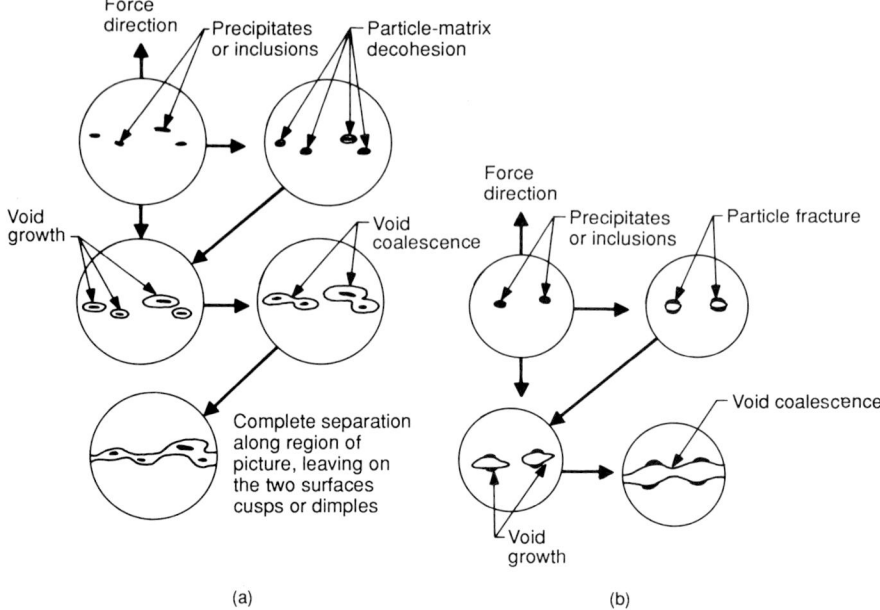

Figure 3.27 Schematic illustration of void nucleation at particles. (*a*) Initiation by particle-matrix interface decohesion. (*b*) Initiation by particle fracture.

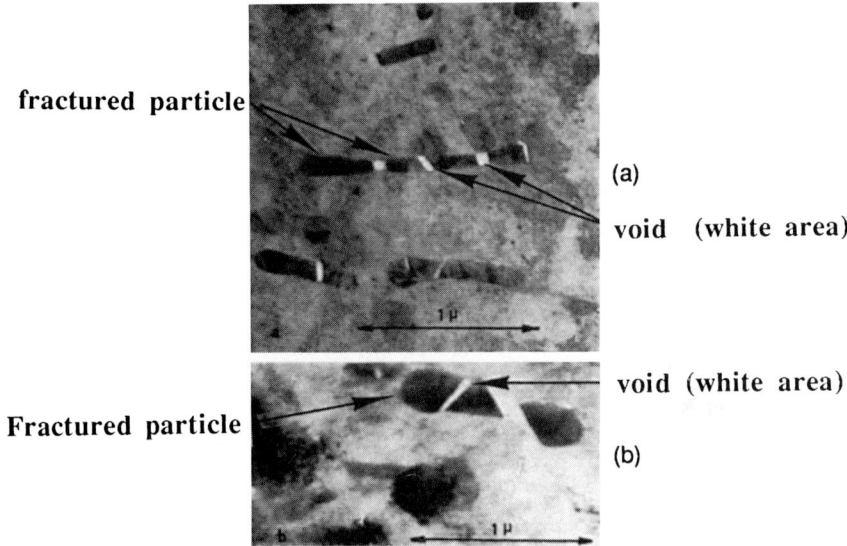

Figure 3.28 Microstructures showing void nucleation by particle fracture or particle-matrix decohesion. Material was aluminum alloy 7079. (*Adapted from Broek.*[12])

particle 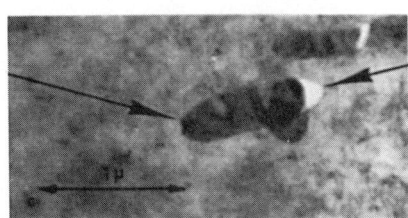 void formed by
decohesion

(c)

Figure 3.28 (*Continued*) Microstructures showing void nucleation by particle fracture or particle-matrix decohesion. Material was aluminum alloy 7079. (*Adapted from Broek.*[12])

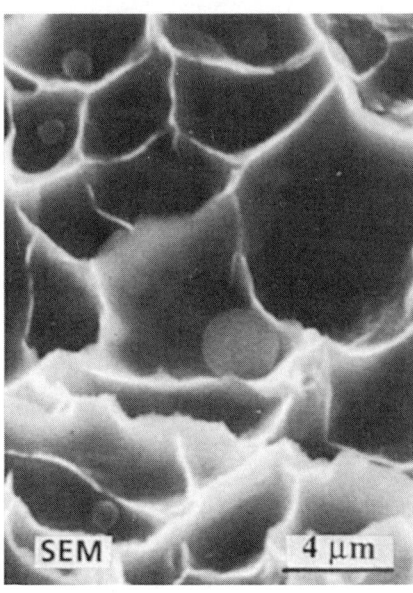

Figure 3.29 Fractograph showing dimples with particles in them. These silicate inclusions determined the form and size of the dimples in this weld-seam fracture of an unalloyed carbon steel. (*From Engel and Klingele.*[10])

The shape of the dimples depends on the loading conditions. This is illustrated schematically in Fig. 3.30. Void coalescence by a stress normal to the overall plane of fracture creates dimples which are equiaxed. (However, this geometry will appear distorted if the surface is examined at an angle.) A shear or tear loading (Fig. 3.30) will create elongated dimples. The two cases can be distinguished by examination of both fracture surfaces, knowing the original alignment of the two pieces. In the case of shear, the elongated dimples on the two surfaces will be in opposite directions. Examples of shear dimples are shown in Fig. 3.31 and of tear dimples in Fig. 3.32.

In torsion failure by void coalescence, the dimples appear similar to those caused by shear, but examination of the surface at suffi-

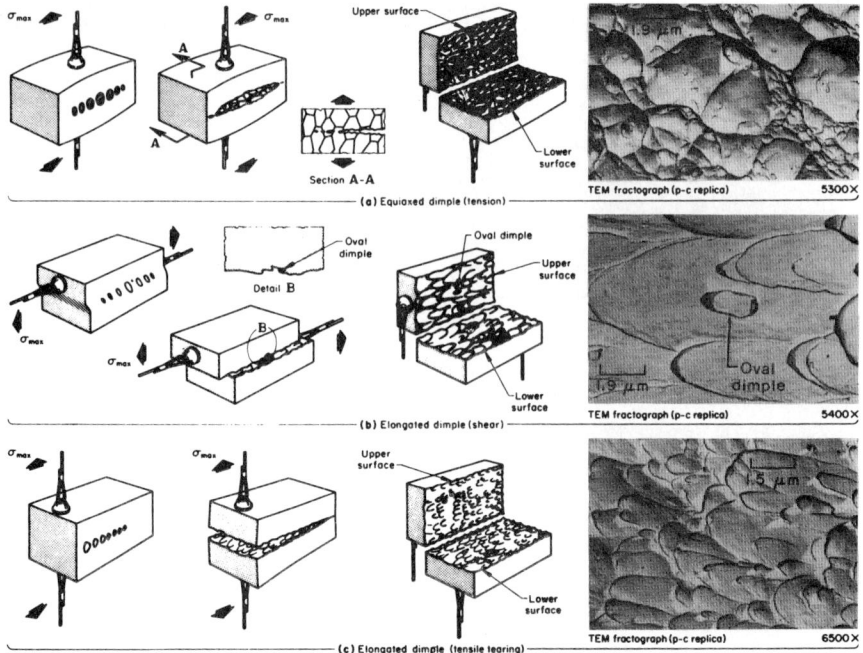

Figure 3.30 Schematic illustration of the influence of loading conditions on the geometry of the dimples found on the fracture surface. The fractographs show the shapes of the dimples formed. (a) In tension, equiaxed dimples are formed on both fracture surfaces. (b) In shear, elongated dimples point in opposite directions on matching fracture surfaces. (c) In tensile tearing, elongated dimples point toward fracture origin on matching fracture surfaces. (*From* Metals Handbook.[11])

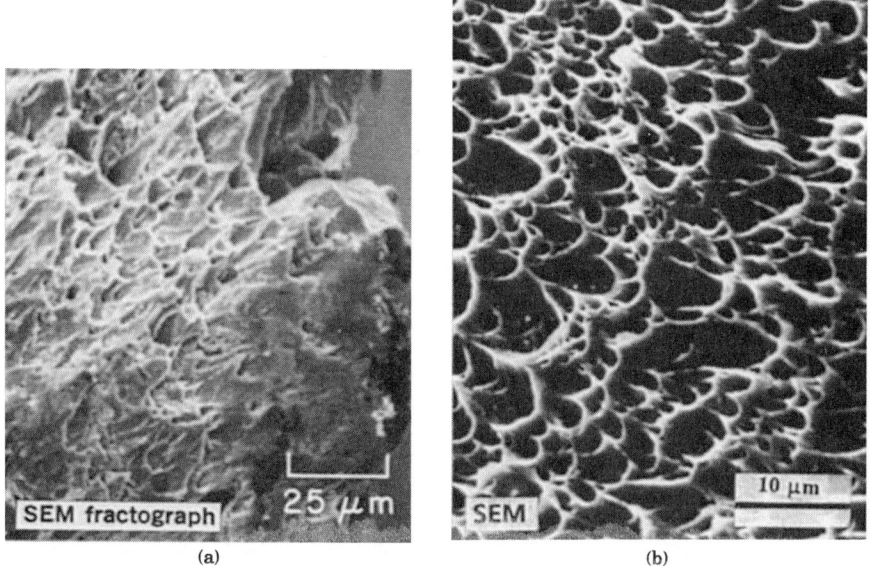

Figure 3.31 Fractographs showing shear dimples. [(a) *From* Metals Handbook.[11] (b) *From* Engel and Klingele.[10]]

Fracture Mechanisms and Microfractographic Features 145

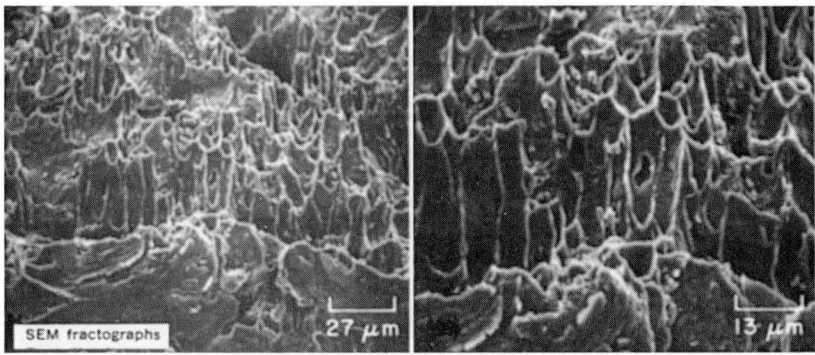

Figure 3.32 Fractographs showing tear dimples. These were taken from the fracture region at the root of the notch of a Charpy impact sample of a 1040 steel. Tear dimples resemble slim parabolas and point toward the site of crack initiation at the notch root, which is shown in the lower third of these fractographs. Unlike shear dimples, the tear dimples on the mating surface of this fracture would point in the same direction—toward the notch. Occasionally, when the overall shape of the fracture surface is unknown, a comparison of the mating fracture surfaces to observe the relative directions of dimple orientation may be the only method of determining whether the dimples present were formed by shear or by tearing. (*From* Metals Handbook.[11])

ciently low magnification will show a rotation or circular pattern, which reveals the direction of torsion. An example is given in Fig. 3.33.

The size and shape of the dimples are directly related to the size, shape, and dispersion of the second-phase particles. This is seen in

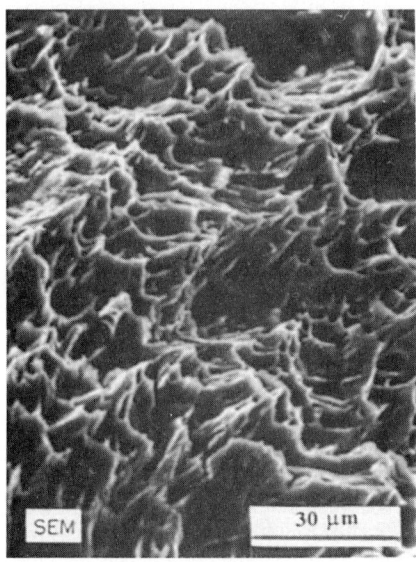

Figure 3.33 Fractograph showing torsional dimples. The wire sample was fractured by means of a monotonic torsional loading. The resulting arrangement of the dimples in a clockwise fashion indicates the direction of torsion. (*From Engel and Klingele.*[10])

Fig. 3.29, where the larger voids have the larger particles in them. Figure 3.34a shows the fracture surface of a steel with elongated MnS inclusions. Note the correlation between the geometry of the elongated dimples and the inclusions. In between the regions containing the inclusions, equiaxed dimples are present. If the

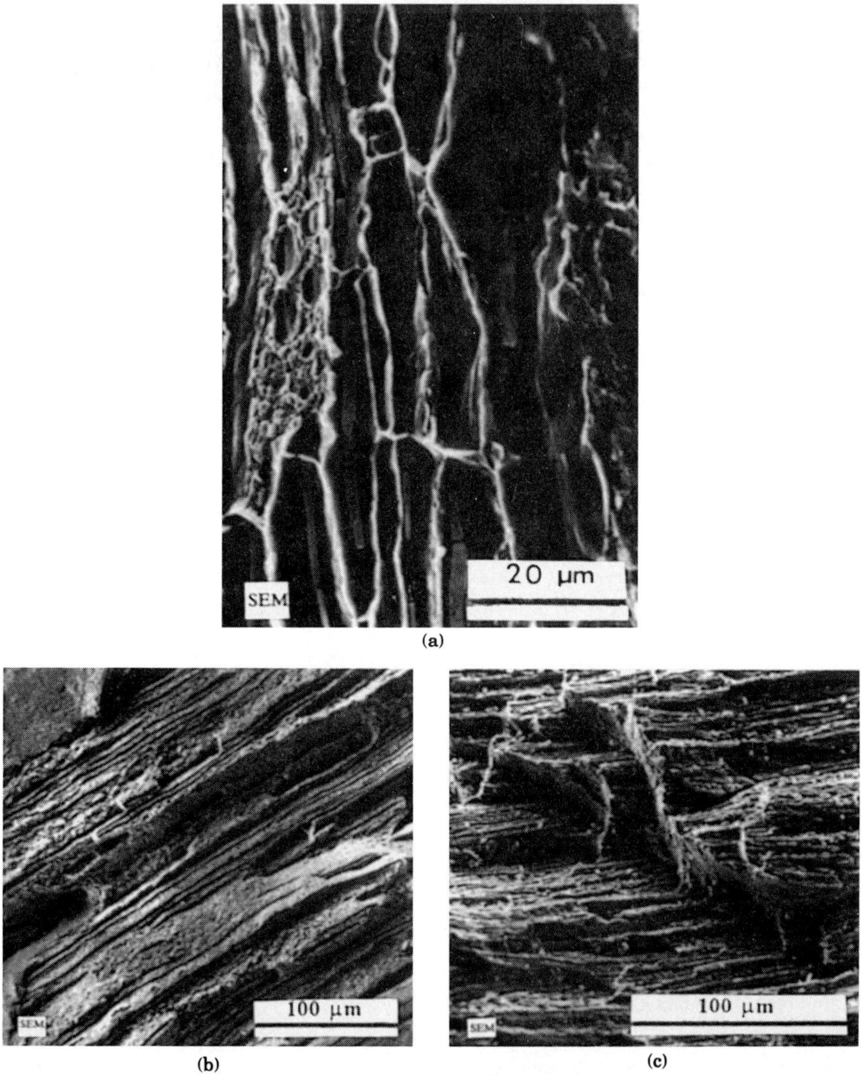

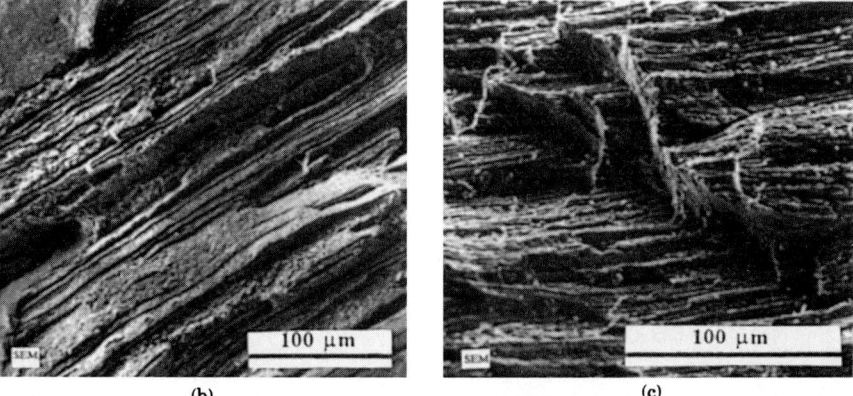

Figure 3.34 (a) Fracture showing elongated dimples due to elongated inclusions. Material is steel containing 0.12%C, 0.02% Si, 0.44% Mn, 0.059% P, and 0.068% S. Cracks are initiated along sulfide inclusion lines. (b) Fracture showing woody structure. 0.35% C, 0.62% Mn, 0.028% S structural steel, water quenched, and tempered at 550°C. Tensile rupture at 20°C (transverse direction). (c) Fracture showing woody structure. 0.44% C, 1.5% Mn, 0.30% S free cutting steel, water quenched and tempered at 550°C. Tensile rupture at 20°C (transverse direction). (*From Henry and Horstmann.*[7])

Fracture Mechanisms and Microfractographic Features   147

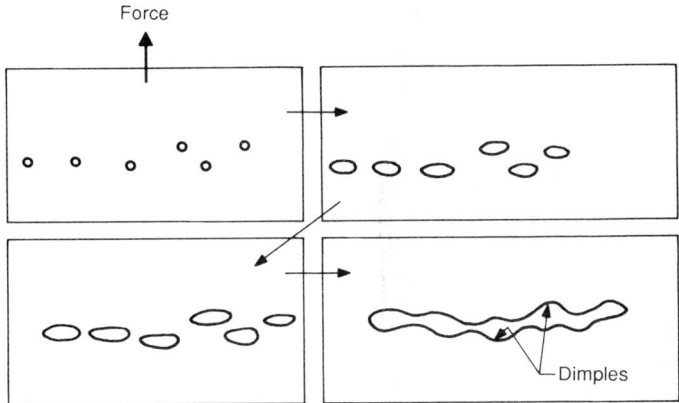

Figure 3.35 Schematic illustration of the formation of shallow (flat) dimples. (Compare to Fig. 3.25.)

amount of such inclusions is high or if they are relatively large, the fracture at low magnification may appear similar to that of fractured wood and is called *woody fracture*. Examples are shown in Fig. 3.34b and c.

If the strength of the material is high, but fracture still occurs by void coalescence, then the dimples will be shallow due to the limited ductility of the material connecting the voids. This is illustrated schematically in Fig. 3.35. Figure 3.36 is a fractograph showing shallow dimples.

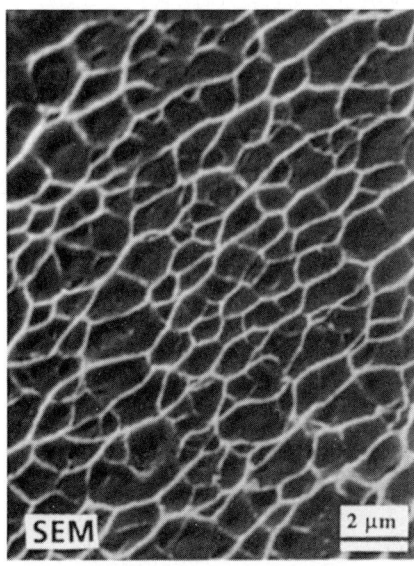

Figure 3.36 Fractograph showing shallow dimples. The sample was a maraging steel (0.02% C, 0.1% Si, 0.1% Mn, 18% Ni, 9% Co, 5% Mo, 0.9% Ti) whose formability had been exhausted due to severe cold working. The specimen failed under static loading with a minimum amount of plastic deformation. (*From Engel and Klingele.*[10])

If the grain boundaries contain features that make this region initiate voids, then appearances illustrated in Fig. 3.37a may be found. This might occur if the boundary contains particles, or if there is a region of particles adjacent and parallel to the boundary. (Note that this type of fracture is not considered to be intergranular in the sense that fracture occurs by separation of the boundary between two planes. This type of fracture is covered in Sec. 3.8.) The effect of the structure on and near the grain boundary on the topography of the intergranular fracture surface is shown in Fig. 3.37b.

The plastic deformation of the material between the voids occurs by slip, as depicted in Fig. 3.25. Since this slip extends to the free surface of the void, then slip traces form; an example is shown in Fig. 3.38. However, they may be difficult to resolve. Also, their appearance depends on the local crystallographic orientation of the region slipping relative to the free surface. In some cases, slip traces appear as wavy lines, probably associated with cross slip. This feature is called serpentine glide; an example is shown in Fig. 3.39. If the slip plane is quite oblique to the free surface of the cusp, the slip traces are spread out so that the surface appears very smooth, but with faint traces of the slip offset (Fig. 3.40). These features are called ripples, and the mechanism is called stretching. [The mechanism has also been called glide plane decohesion (because the slip planes eventually are highly offset, see Fig. 3.24), and it has been called ductile cleavage.]

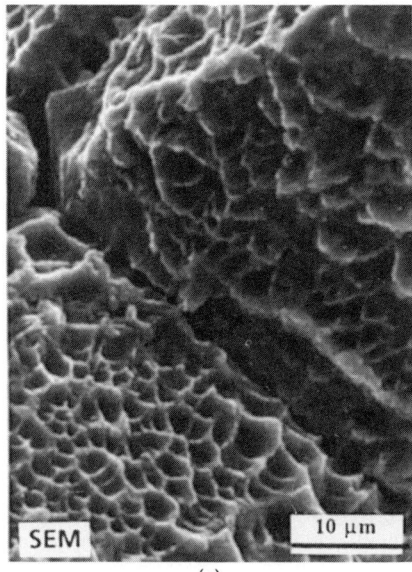

Figure 3.37 (a) Intergranular fracture surface showing dimples on the surface. (*From Engel and Klingele.*[10])

(a)

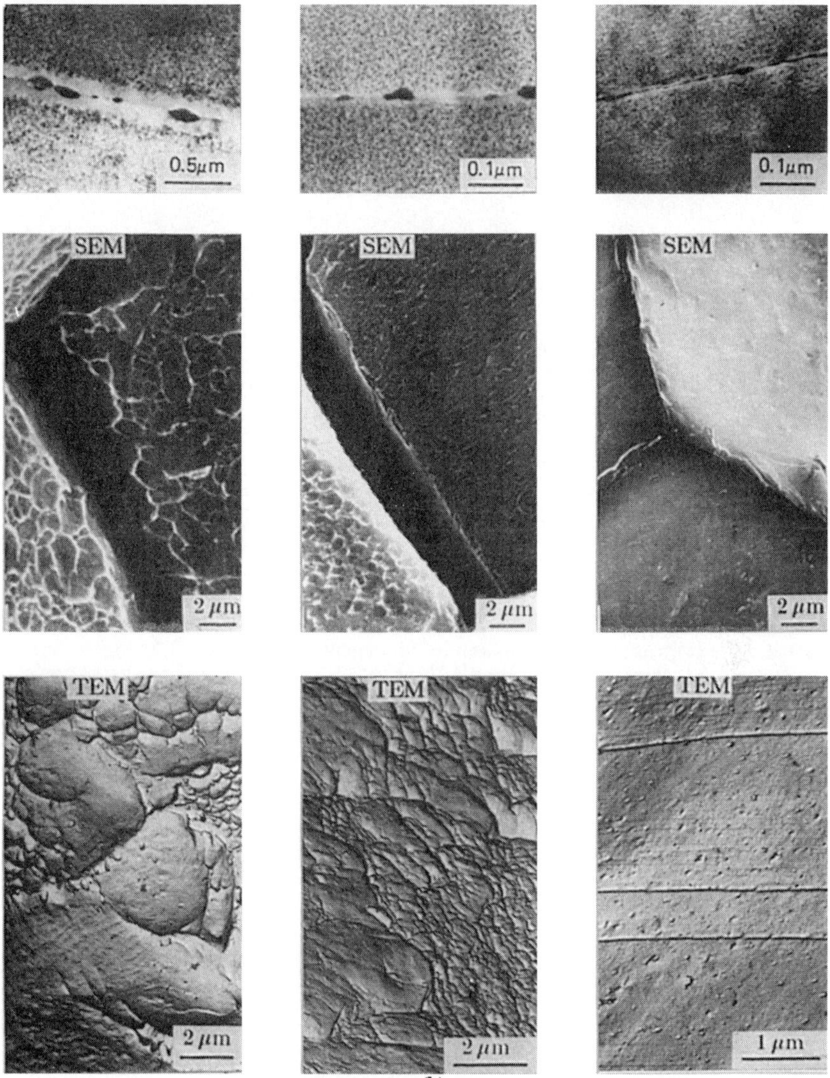

Figure 3.37 (*Continued*) (*b*) Effect of grain-boundary structure on fine topology of the intergranular fracture surface. (*From Lynch.*[13])

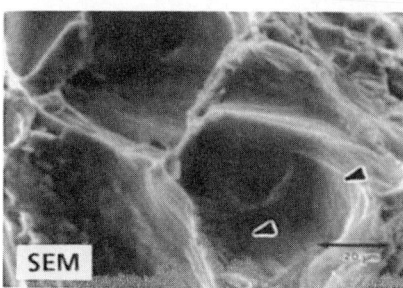

Figure 3.38 Fractograph showing slip traces (arrows) on the side of the dimples. (*From Gabriel.*[14])

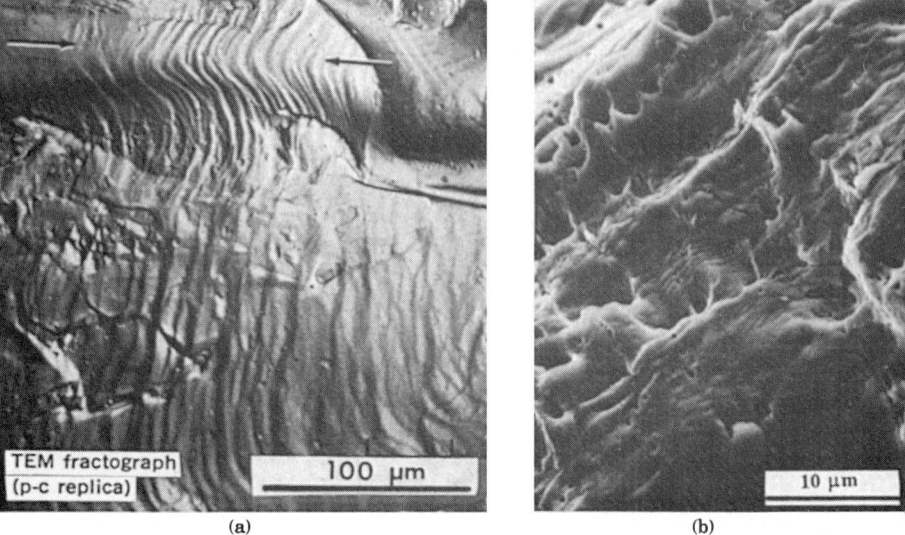

Figure 3.39 Fractographs showing examples of serpentine glide. (a) TEM replica of fracture surface of Armco iron. (*From* Metals Handbook.[11]) (b) Fracture in oxygen-free copper tube, produced by forcing open the crack in a precracked specimen. (*From Engel and Klingele.*[10])

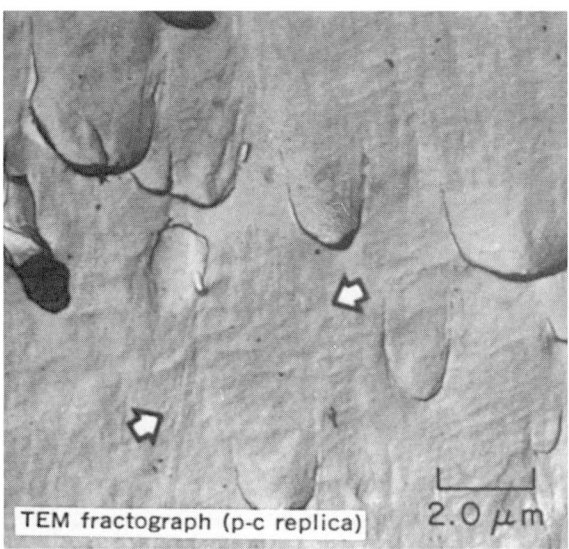

Figure 3.40 Two-stage carbon TEM replica showing ripples (at arrows). The sample was a 302 stainless steel. (*From* Metals Handbook)

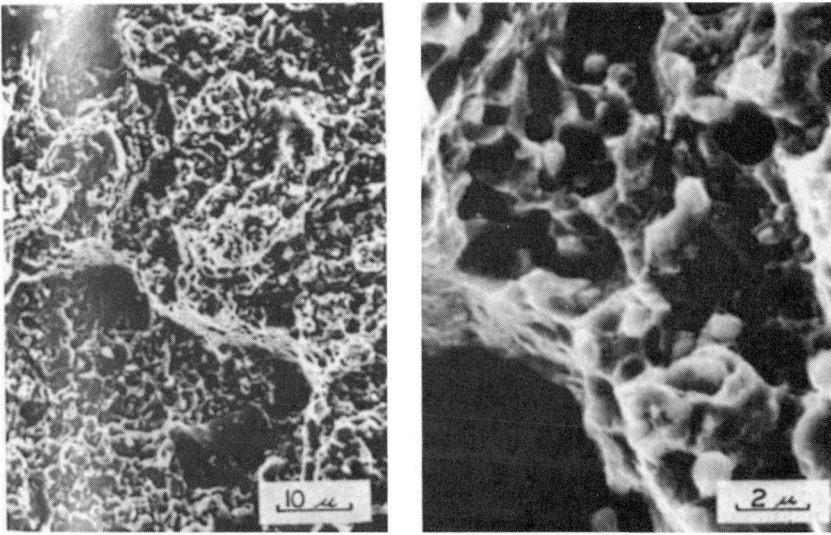

Figure 3.41 SEM fractographs showing shallow dimples in a material which failed in a brittle mode. (*From Syniuta and Corrow.*[15])

Usually macroscopically ductile fractures, which show gross dimensional changes, occur microscopically by void coalescence. However, it is important to realize that macroscopically brittle fracture, with little dimensional change, can also occur by void coalescence. This is favored by a high-strength material and by a fine dispersion of particles. An example of such a fracture is shown in Fig. 3.41. Note that the dimples are relatively flat.

3.6 Mixed Mechanism and Quasicleavage Fracture

In many materials, fracture may first occur by cleavage, then fracture of the connecting material by void coalescence. Examples are shown in Fig. 3.42. This is the case particularly in body-centered cubic materials (such as ferrite), when fracture occurs in the vicinity of the transition temperature (which depends on the strain rate). This is an example of a mixed mechanism of fracture.

In some materials, especially steels, a fracture surface forms which consists of regions showing ill-defined or relatively faint signs of cleavage (sometimes called *rosettes*) separated by regions showing extensive deformation. These latter regions are *tear ridges*. Examples of such surfaces are given in Fig. 3.43. They ap-

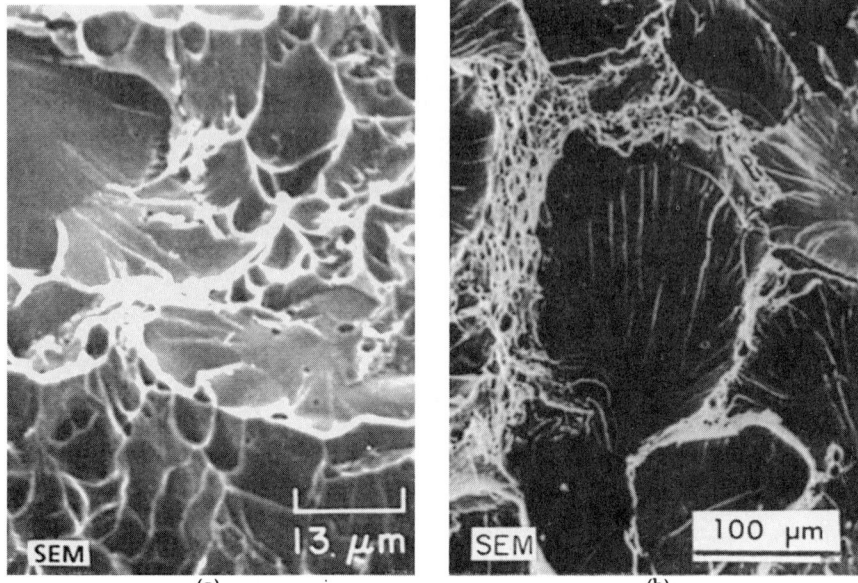

Figure 3.42 Examples of cleavage and void coalescence mixed mechanisms of fracture. (a) 1040 hot-rolled steel. (*From* Metals Handbook.[11]) (b) Tensile sample broken at 0°C. Steel contained 0.08% C, 0.14% Si, 0.34% Mn, 0.021% P, 0.018% S, and 0.034% Al. (*From Henry and Horstmann.*[7])

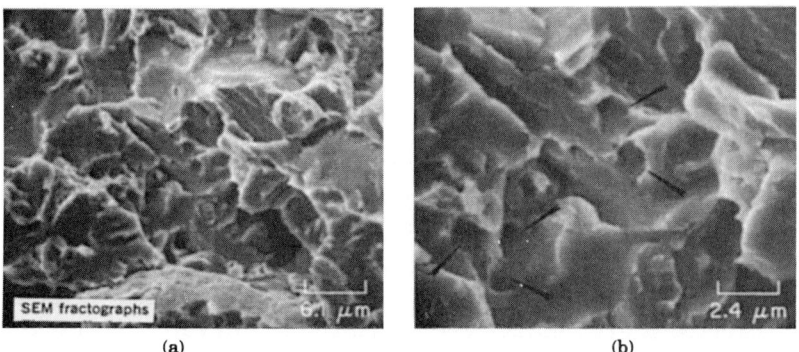

Figure 3.43 Examples of quasicleavage. (a), (b) Fractographs of an impact sample of a 4340 steel. The small cleavage facets in martensite platelets contain river patterns and are separated by tear ridges. Shallow dimples, marked by arrowheads, are also visible. The direction of crack propagation is from bottom to top. The specimen was heat-treated at 843°C (1550°F) for 1 h, oil-quenched, and tempered at 427°C (800°F) for 1 h. Fracture was by Charpy impact at −196°C (−321°F). (*From* Metals Handbook.[11])

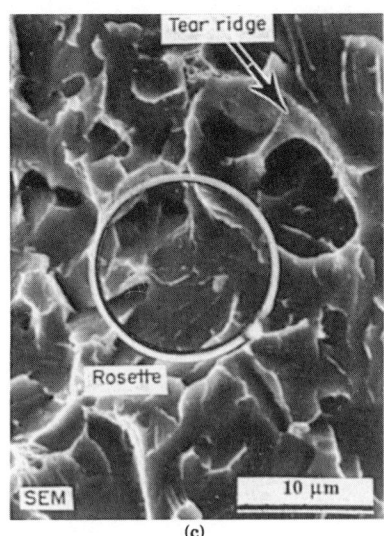

Figure 3.43 (*Continued*) Examples of quasicleavage. (*c*) Material was CrMoV steel (0.20% C, 1.0% Cr, 0.7% Mo, 0.15% V) with a tempered structure formed in the upper bainite range. (*Adapted from Engel and Klingele.*[10])

pear to develop by the formation of microcracks and then tearing (plastic deformation) of the connecting material, as shown schematically in Fig. 3.44. This mechanism is called *quasicleavage*. It is to be noted that this is actually cleavage, accompanied by tearing or dimples, where the fine microstructural details make the identification less apparent than described earlier for cleavage. In this sense, it is really a mixed-mechanism fracture mechanism.

3.7 Tearing Topography Surface

In some cases, a fractographic topography as shown in Fig. 3.45 is found. It consists of very fine "facets" connected by tear ridges and is referred to as *tearing topography surface* (TTS). This appearance seems to be caused by the particular complex microstructures involved in the fracture. The underlying mechanisms are probably those already described.

3.8 Intergranular Separation

When the grain boundaries are the weakest locations in the microstructure, then decohesion occurs on these surfaces before cleavage or

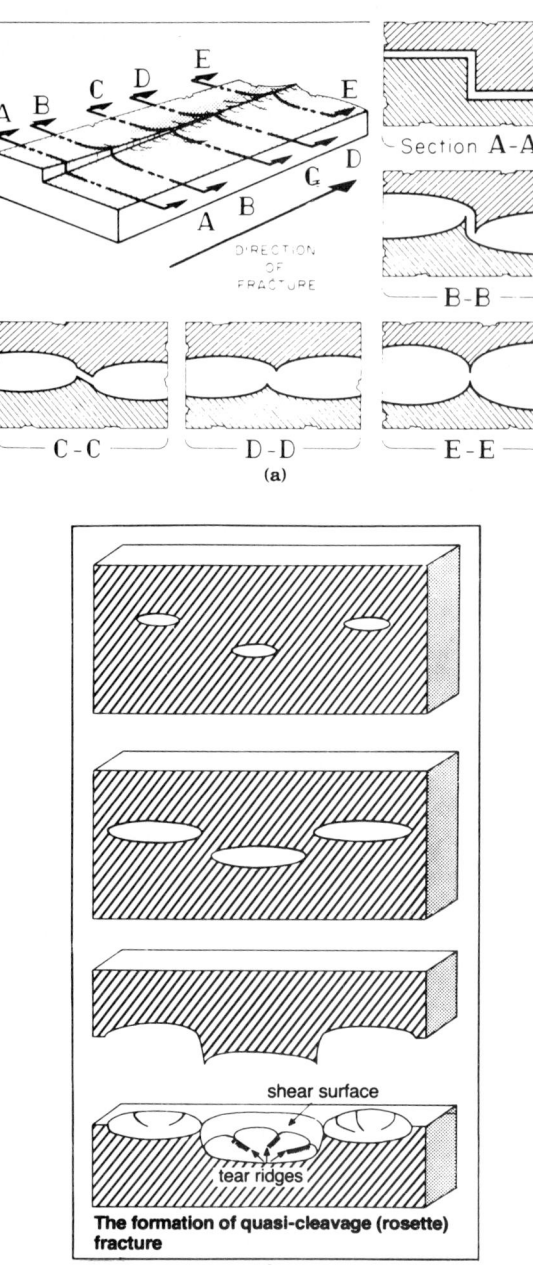

Figure 3.44 Schematic illustration of the formation of fracture by quasicleavage. (*a*) Model showing a cleavage step blending in with a tear ridge. At the top left is the lower surface of a fracture, showing a step at the lower left and a ridge at the upper right. At right and bottom are sections through the fractured member, showing profiles of both the upper and the lower fracture surfaces. (*From* Metals Handbook.[11]) (*b*) Model of the formation of quasicleavage rosettes. (*From Engel and Klingele*.[10])

Fracture Mechanisms and Microfractographic Features

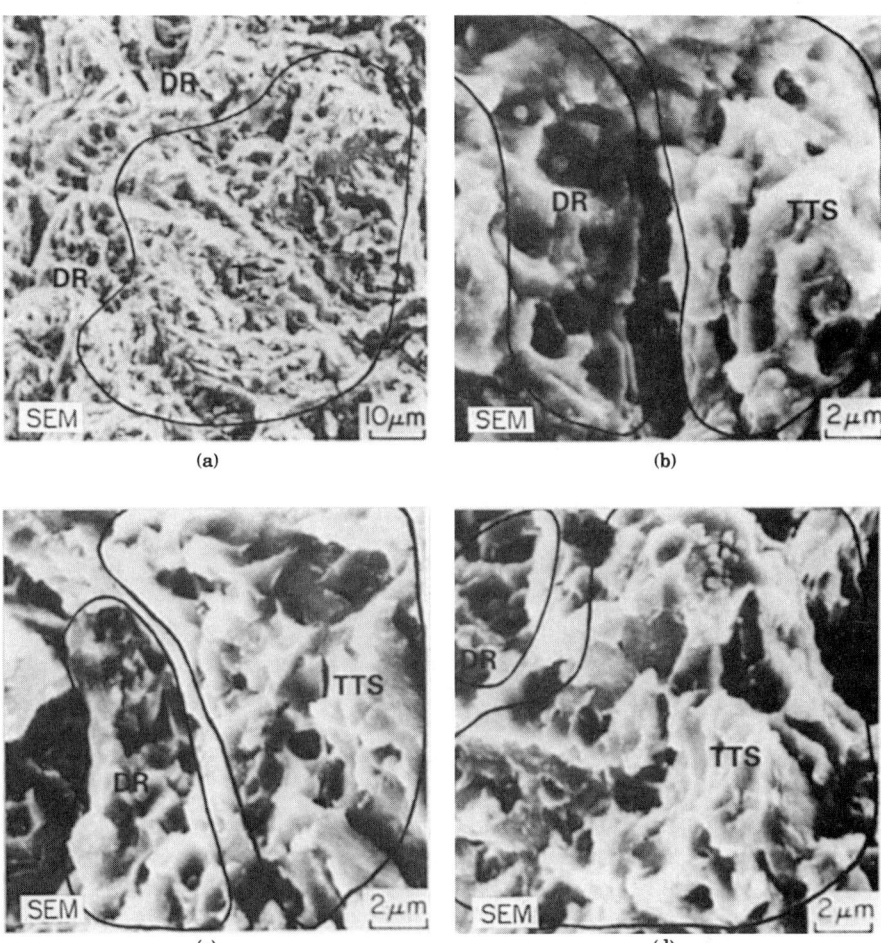

Figure 3.45 Examples of tearing topography surfaces (TTS). In these cases dimple rupture (DR) is also found. (a)–(d) HY-130 steel. (a) Areas of complex tearing (T) and dimple rupture (DR). (b) Detail from upper left corner of (a) showing particle-nucleated dimples (DR) and region of TTS. (c), (d) Additional examples of TTS. (*From Thompson and Chesnutt.*[16])

slip. In the strict sense, this decohesion mechanism of fracture is called intergranular fracture. However, in complex microstructures, the fracture surface may appear intergranular, even though the fracture path may have been, for example, along particles on the boundary (by decohesion or separation along the particle-matrix interface), by void coalescence associated with particles on the boundary, or by void coalescence and tearing in a weak zone parallel and adjacent to the boundary. Even in these cases, the fracture topography is referred to

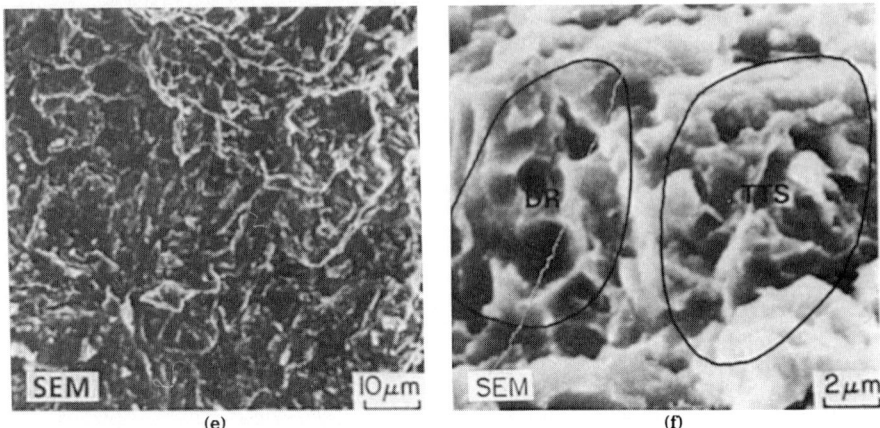

Figure 3.45 (*Continued*) Examples of tearing topograph surfaces (TTS). In these cases dimple rupture (DR) is also found. (*e*) TTS in an essentially 10% pearlitic eutectoid steel (similar to AISI 1080) where fracture propagates across pearlite colonies. (*f*) Fractograph showing dimple rupture (DR) and TTS fracture in a quenched-and-tempered (martensitic) HY-130 steel. (*From Thompson and Chesnutt.*[16])

as intergranular fracture. In such cases the fracture surface will show topographical details, whereas in decohesion the surface will be smooth if the boundary contains no particles. The shape of the intergranular fracture topography is determined by the grain shape, as illustrated in Fig. 3.46. Typical intergranular fracture fractographs are shown in Fig. 3.47. In some cases the structure on or near the grain boundaries causes the fracture surface to be relatively rough on a fine scale, as shown in Fig. 3.37.

It is to be noted that intergranular fracture has a variety of causes. In some alloys, hydrogen embrittlement causes grain-boundary fracture during loading. In some corrosion environments for some alloys,

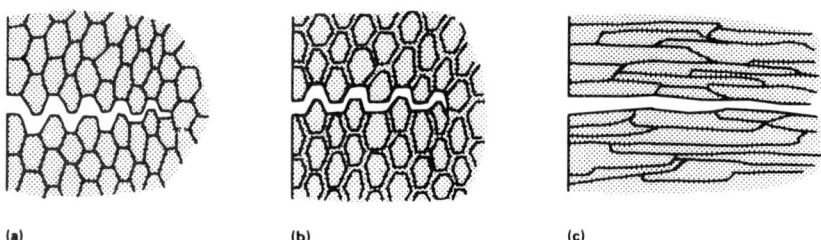

Figure 3.46 Schematic illustration of grain-boundary fracture. (*a*) Decohesion along grain boundaries of equiaxed grains. (*b*) Decohesion through a weak grain-boundary plane. (*c*) Decohesion along grain boundaries of elongated grains. (*From Kerlins and Phillips.*[9])

Fracture Mechanisms and Microfractographic Features 157

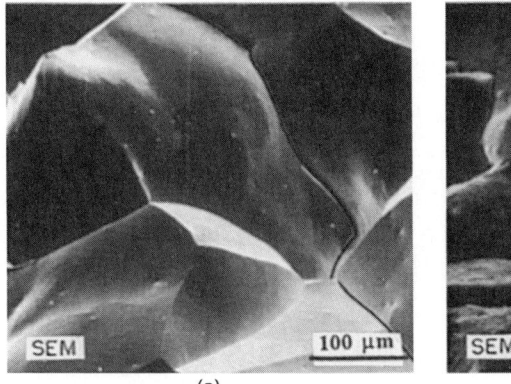
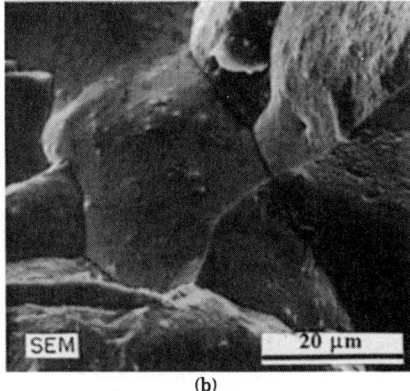

(a)   (b)

Figure 3.47 Examples of intergranular fracture. (a) Fracture by tensile testing at 20°C of coarse-grained annealed iron containing 0.68% P. (b) Fracture of a steel containing 0.58% C, 0.30% Si, 0.82% Mn, 0.024% P, and 0.017% S. (*From Henry and Horstmann.*[7])

stress corrosion cracking is intergranular. Some steels show temper embrittlement fracture along the former austenite grain boundaries, which are the high-angle boundaries of the ferrite matrix. In some steels this is due to segregation of minor impurities (such as P or Bi) to these boundaries, and in some cases it is due to carbide network on these boundaries.

3.9 Fatigue Fracture Topography

Consider a material which has a certain yield strength, obtained from testing a cylinder of this material in a simple tensile test. Thus if a cylindrical sample of this material is loaded to an axial stress two-thirds of the yield strength, then plastic deformation and fracture are not expected to occur. However, if the axial load is cycled between zero and this two-thirds value, after sufficient cycles fracture may occur. This is called a fatigue fracture. Since most structural and machine components operate under alternating loading, fatigue failures are relatively common, and being able to recognize them from the fracture-surface topography is important.

In fatigue fractures the loading level is usually macroscopically below that required to initiate gross or macroscopic plastic deformation. However, due to surface irregularities, microstructural features, and so on, the local microscopic stress exceeds the yield strength, in which case, upon repeated fatigue loading, plastic deformation occurs locally. On highly polished free surfaces the local plastic deformation can be

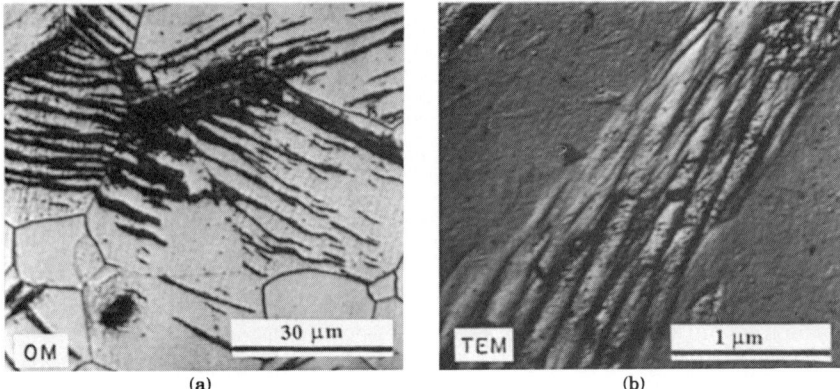

Figure 3.48 Formation of deformation bands during fatigue loading. (*a*) OM of a mild steel specimen surface after 525,000 cycles of alternate bending under ±13 daN/mm^2. (*b*) TEM replica micrograph of the surface of a mild steel after 125,000 cycles of alternate bending under ±10 daN/mm^2. (*From Henry and Horstmann.*[7])

seen as the development of slip lines and bands, which eventually reach a size such that they can act as the initiator of the fatigue crack. An example of such deformation bands is shown in Fig. 3.48. Generally this local deformation is associated with a shear stress at approximately 45° to the tensile stress created by the load. In some alloys the initial stages of plastic deformation create extrusions and intrusions at the surface at these sites (Fig. 3.49). These are shown on the free

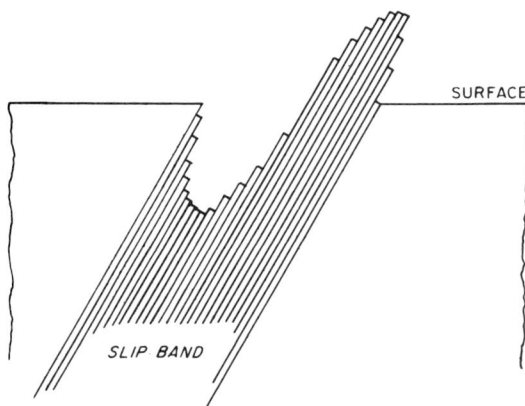

Figure 3.49 Schematic illustration of the formation of extrusions and intrusions in stage I of fatigue. (*From Grosskreutz.*[17])

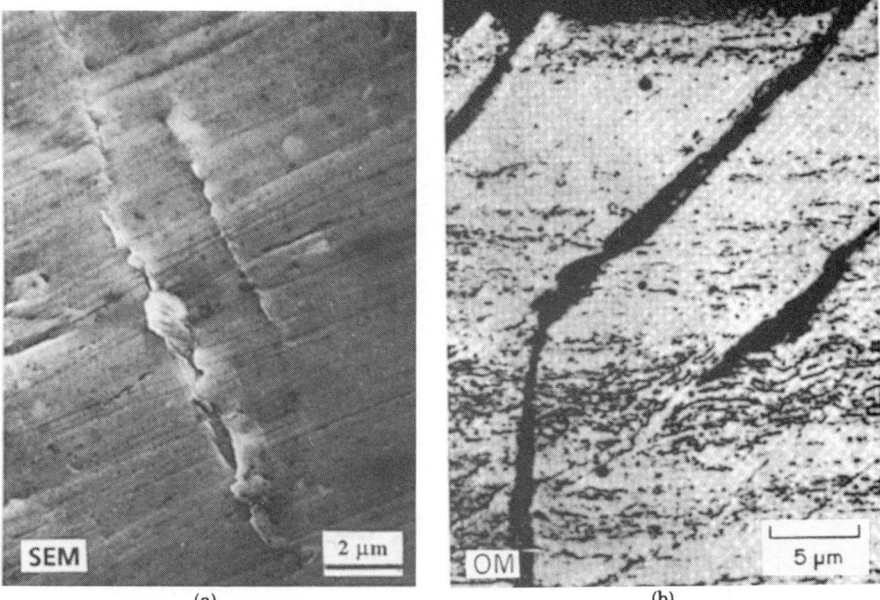

Figure 3.50 (a) Example of extrusions on the surface of a fatigue test sample. Material was an Al alloy containing 4% Cu, 1.2% Mg, 0.3% Mn. Test time was 10,000 cycles. (*From Engel and Klingele.*[10]) (b) Cross section of the surface of a fatigue sample showing stage I cracks at the surface and the beginning of stage II. Material was pure aluminum. (*Courtesy of Forsyth; from Henry and Horstmann.*[7])

surface in Fig. 3.50a and in the cross-sectional view in Fig. 3.50b. Such surface discontinuities serve as the fatigue crack initiator. The fracture surface in this initiation stage appears cleavagelike, as shown by the fractographs in Fig. 3.51.

The crack initiation stage of fatigue is called stage I. It occurs for high-cycle–low-stress fatigue loading, but if the stress becomes too high, it is not observed. In most machine components the surface is sufficiently rough that this stage of fatigue crack formation occurs at surface irregularities or defects. Stage I may not comprise a significant portion of the total crack propagation. For these reasons, recognition of stage I may require careful examination of the fracture surface in the vicinity of the crack initiation as deduced from macroscopic observations.

When the crack has progressed a small distance from the surface by the mechanism just described (stage I), the crack begins to propagate by a different mechanism, called stage II. The general crack path becomes perpendicular to the tensile stress, as illustrated in

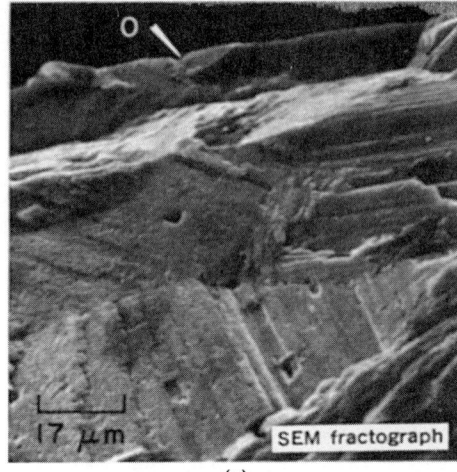
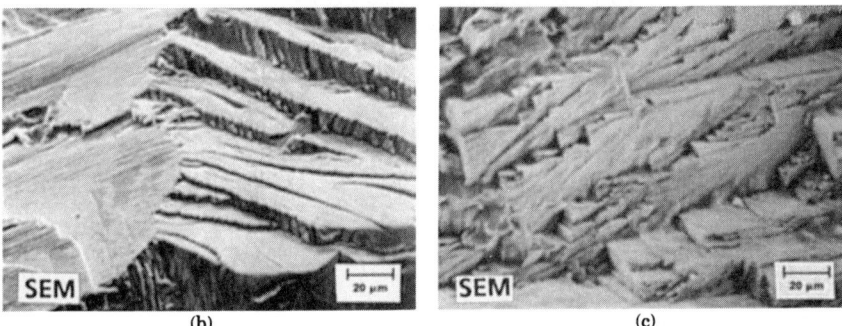

Figure 3.51 Examples of the appearance of the fracture surface of stage I. (a) Stage I of fatigue fracture progress in a smooth (unnotched) specimen of aluminum alloy 2024-T3. The fracture origin is at 0, near the top of the fractograph, in a free surface of the specimen. The change of orientation of the fracture plane from grain to grain is well demonstrated. Within a given grain, the fracture is along a {111}-type plane. The ridge patterns within each grain are parallel to the direction of crack propagation. Each ridge is probably in a ⟨110⟩ direction, which is the intersection of two (111) slip planes. The presence of inclusions does not affect the fracture path. Stage I of fatigue-fracture progress extended over 3 or 4 grains; stage II began at bottom right. (*From Metals Handbook.*[11]) (b) Cleavagelike, crystallographically oriented stage I fatigue fracture in a cast Ni–14Cr–4.5Mo–1Ti–6Al–1.5Fe–2.0(Nb + Ta) alloy. (c) Stair-step fracture surface indicative of stage I fatigue fracture in cast ASTM F75 cobalt-base alloy. [(b), (c) *Courtesy of R. Abrams, Howmedica, Div. Pfizer Hospital Products Group Inc., from Kerlins and Phillips.*[9]]

Fig. 3.52. The crack propagates only during the tensile stress part of the cycle. This is illustrated in Fig. 3.53. Here a notched sample was loaded in tension-compression, and the polished surface near the notch was observed in a scanning electron microscope. Note the opening of the crack during the tensile loading (1 to 2), the closing during compressive load-

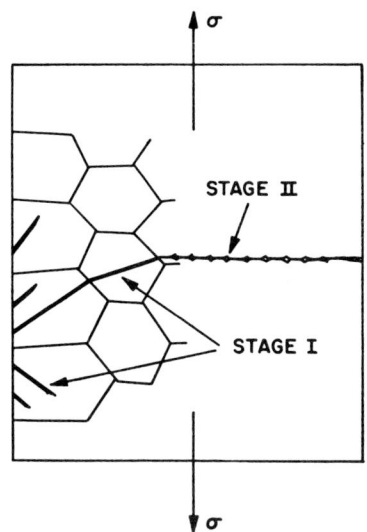

Figure 3.52 Schematic illustration of stage I and stage II fatigue crack formation. (*From Laird.*[18])

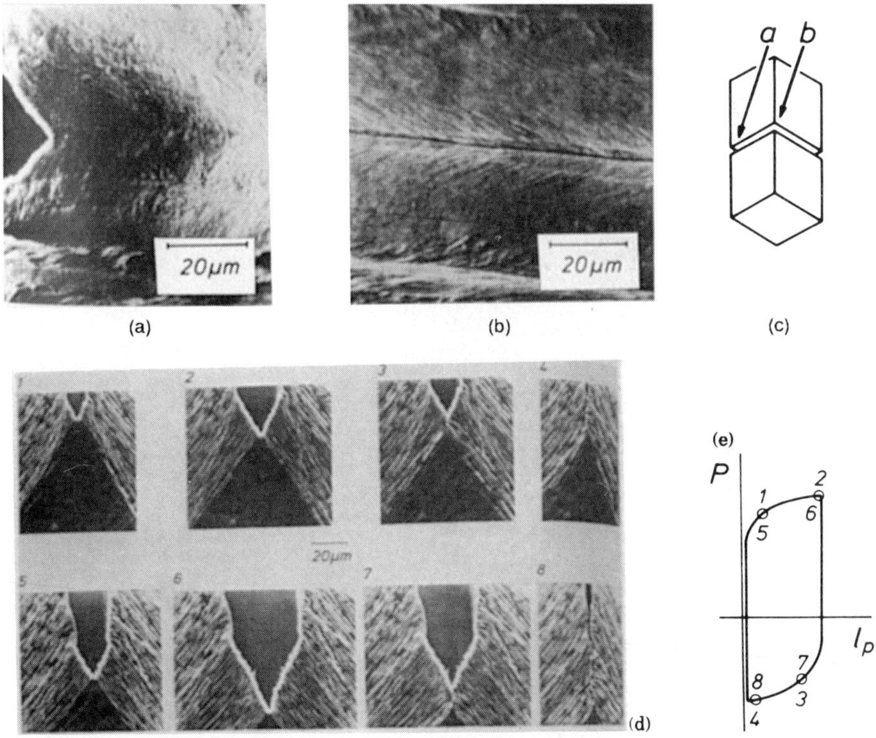

Figure 3.53 Microstructural observations (SEM) of in situ fatigue crack growth. (*a*) Bottom of the ductile crack under tension near the specimen's side face and near the center of the specimen. The dark triangle is the V-shaped crack tip intersecting the specimen side faces (*b*). (*c*)Viewing direction under which (*a*) and (*b*) were taken. (*d*) Sequence of two load cycles as observed on the specimen's side face during straining in an SEM. The crack propagation direction is from top to bottom. (*e*) Approximate location of each picture on the load cycle. (*From Verhoff and Newmann.*[19])

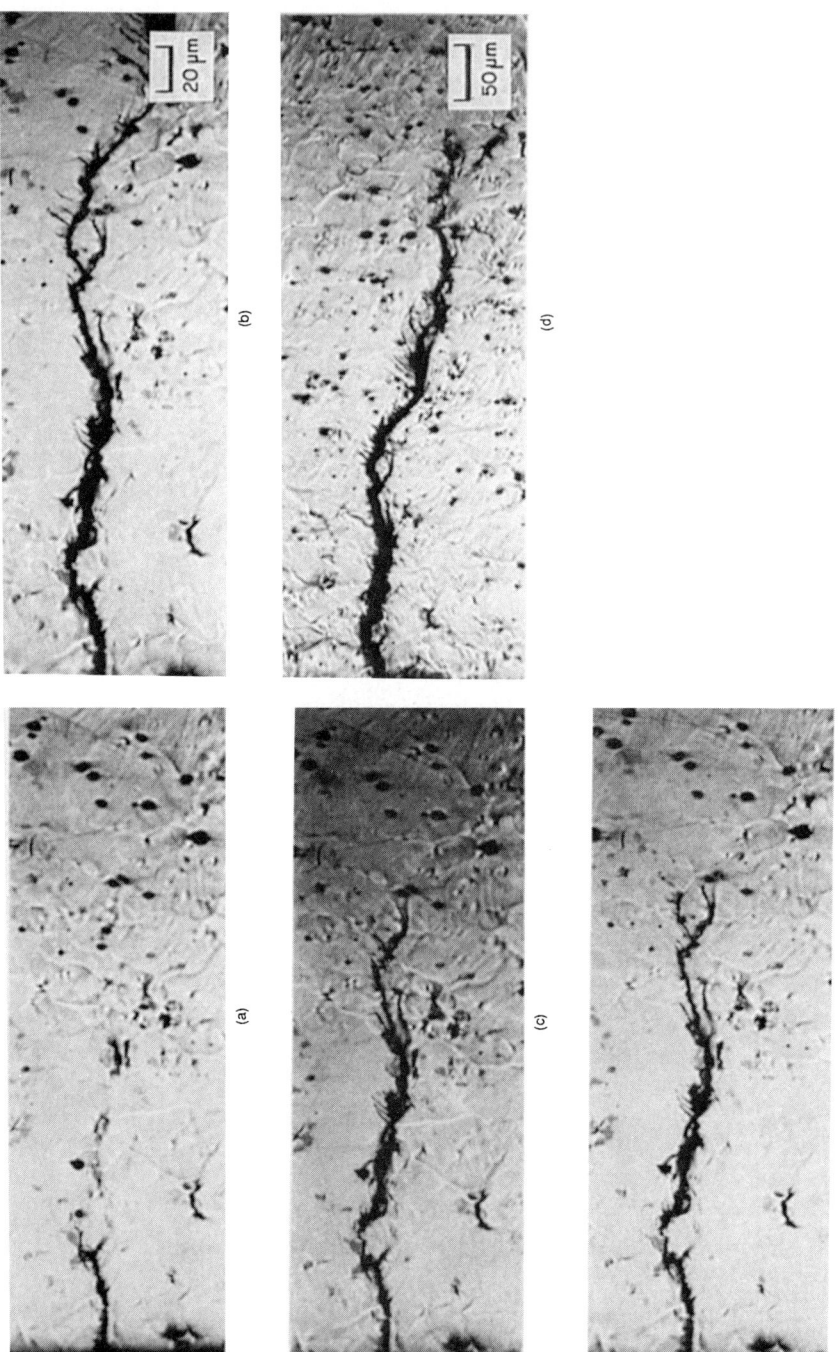

Figure 3.54 OMs of the growth of a fatigue crack on the surface of a fatigue sample. (a) 96,000 cycles, crack length 60 μm. (b) 96,300 cycles, crack length 235 μm. (c) 96,400 cycles, crack length 240 μm. (d) 96,800 cycles, crack length 380 μm. (e) 97,400 cycles, crack length 570 μm. *(From Haworth, Singh, and Mueller.*[20]*)*

Fracture Mechanisms and Microfractographic Features   163

ing (3 to 4), then the reopening during the next tensile cycle (5 to 6). Also note the advancement of the crack tip from point 2 to point 6. It is further seen that slip bands have formed on the prepolished surface, showing that the crack opening is associated with slip.

The advancement of a fatigue crack can be seen in the micrographs in Fig. 3.54. Note that the crack advances about 140 μm from the location after 96,400 cycles to the location after 96,800 cycles, or a distance of about 0.3 μm per cycle (3000 Å, or about 1000 atom distances). Figure 3.55 shows schematically how the crack grows by slip.

As the fatigue crack propagates in stage II, it usually leaves on the fracture surface, behind the advancing crack front, regions of depression and elevation. An example is shown in Fig. 3.56. These *striations* are the most common characteristic of fatigue fractures seen on a microscopic scale. However, as shown later, fatigue cracks can propagate without forming striations; thus their absence does not mean the part did not fail by fatigue. Also their resolution will depend on the type of damage to the

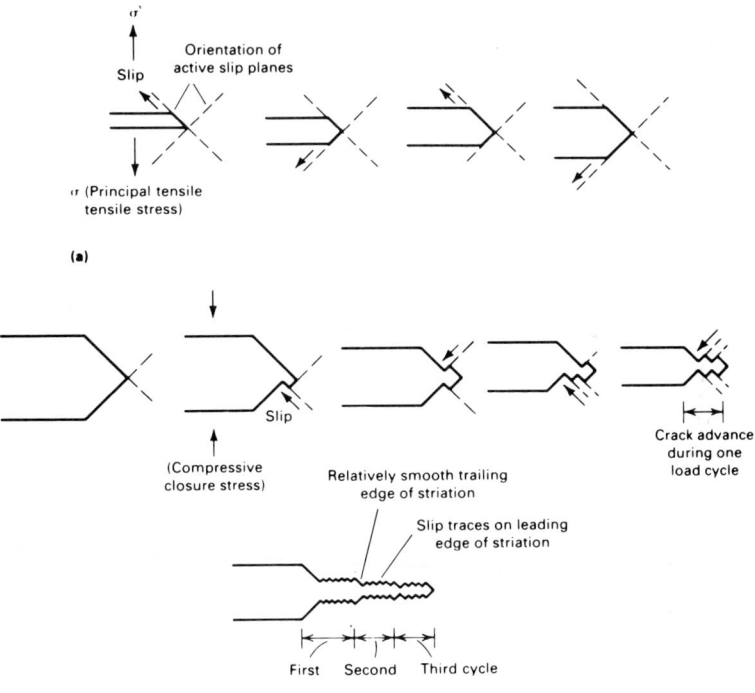

Figure 3.55 Schematic illustration of a mechanism of fatigue crack propagation in stage II by alternate slip at the crack tip. (*a*) Crack opening and crack tip blunting by slip on alternate slip planes with increasing tensile stress. (*b*) Crack closure and crack tip resharpening by partial slip reversal on alternate slip planes with increasing compressive stress. (*From Kerlins and Phillips.*[9])

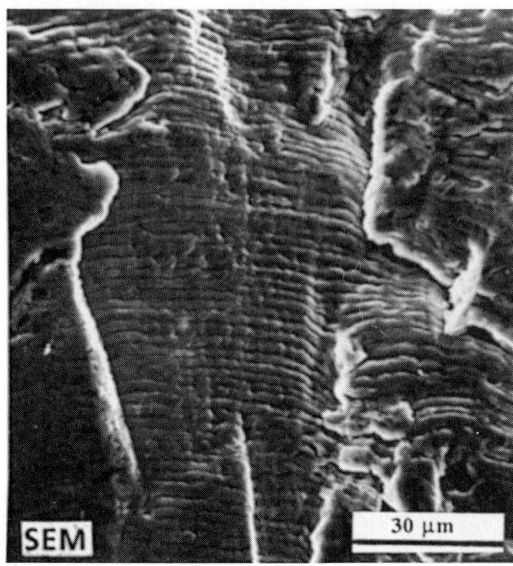

Figure 3.56 Fatigue fracture surface showing typical striations. Material was an 18:8 austenitic stainless-steel screw fastener. (*From Engel and Klingele.*[10])

surface when the crack closes (in the compression part of the loading cycle) and on the reaction of the fracture surface with the environment.

It has been clearly established that each striation is associated with the growth of the crack during one loading cycle. Although the fine-scale mechanism of the formation of the striations is not clear, Fig. 3.57 illus

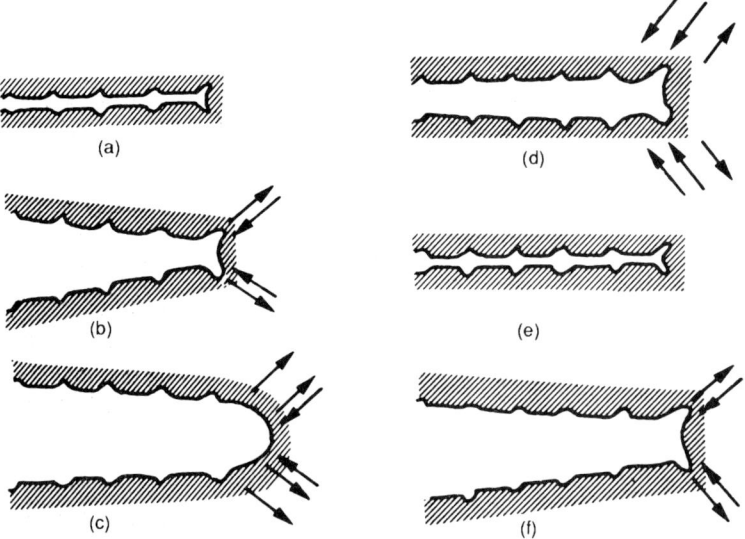

Figure 3.57 Possible mechanisms of the formation of fatigue striations. (*a*) Unstressed. (*b*) Small tensile stress. (*c*) Maximum tensile stress. (*d*) Small compressive stress. (*e*) Maximum compressive stress. (*f*) Small tensile stress. (*From Laird.*[18])

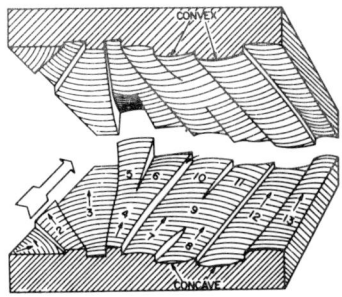

Figure 3.58 Local variation of the orientation of the fatigue striations due to microstructural features. (*From Beachem.*[8])

trates one way these striations could form. On a microscopic scale, the direction of crack propagation will depend on the microstructural features through which the crack is advancing, and this causes the striations to alter orientation locally, as depicted schematically in Fig. 3.58. This variation aspect of the orientation is seen in the fractograph of Fig. 3.59. The overall direction of crack movement, of course, will be that of the macroscopic crack propagation direction.

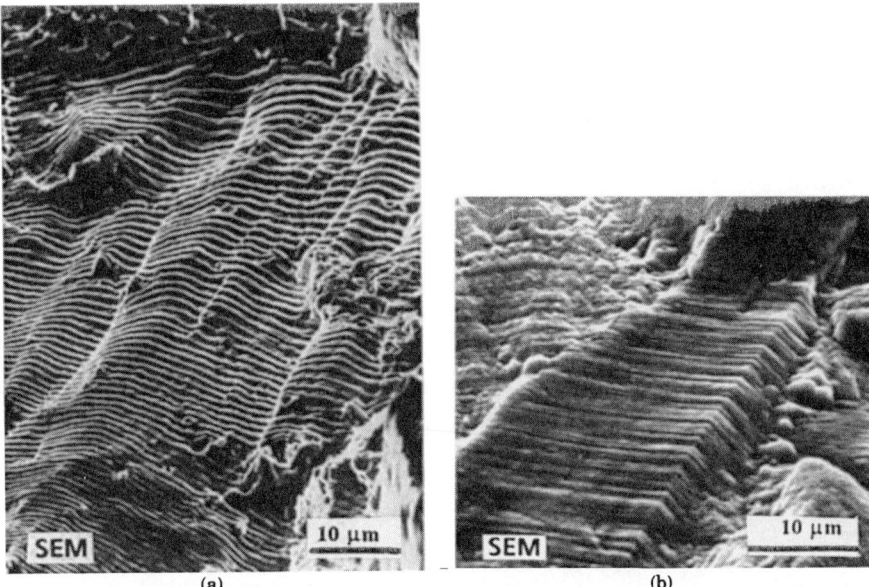

Figure 3.59 Fractographs of fatigue fractures showing typical striations. (*a*) Nickel alloy specimen (Nimonic) tested at 650°C under rotating-bending conditions. The propagation direction is from bottom to top. (*From Engel and Klingele.*[10]) (*b*) Crystallographically oriented fatigue striations on a fatigue fracture surface (stage II) from a laboratory tested nickel-based alloy (Inconel 718) tested at 600°C. (*From Engel and Klingele.*[10])

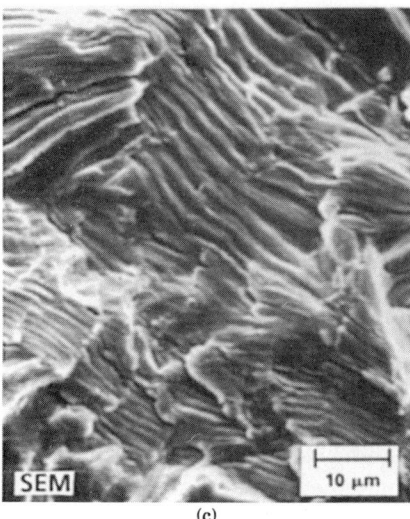

Figure 3.59 (*Continued*) Fractographs of fatigue fractures showing typical striations. (c) Fatigue fracture in commercially pure titanium. (*Courtesy of O. E. M. Pohler, Institut Straumann AG; from Kerlins and Phillips.*[9])

Several examples of fatigue fracture surfaces showing striations are given in Fig. 3.60. It is seen that the striation spacing is from approximately 1 to 10 μm, which is consistent with the distance the crack advances per cycle for the cases shown in Figs. 3.53 and 3.54. The correlation between the striation spacing and loading has been estab-

Figure 3.60 Examples of fatigue fracture surfaces showing the effect of loading on the fatigue striation spectrum. (a) One large striation at 65 N/mm² cycle, three striations closer to each other (75 N/mm²), and 20 striations very close to one another (60 N/mm²). (*From Henry and Horstmann.*[7])

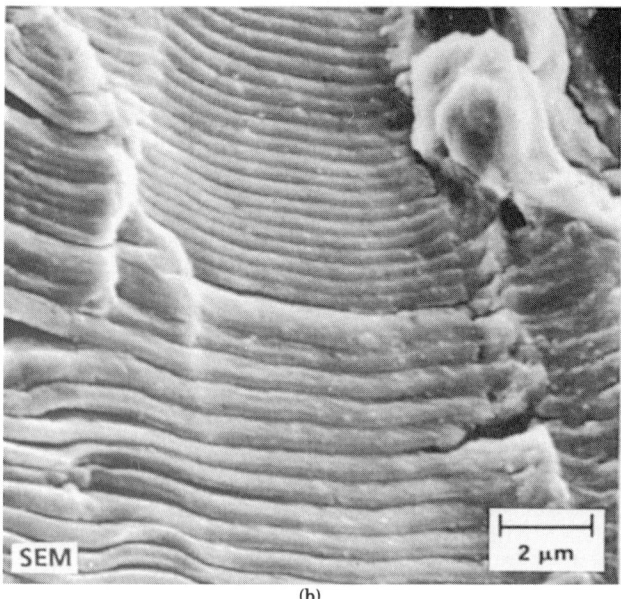

(b)

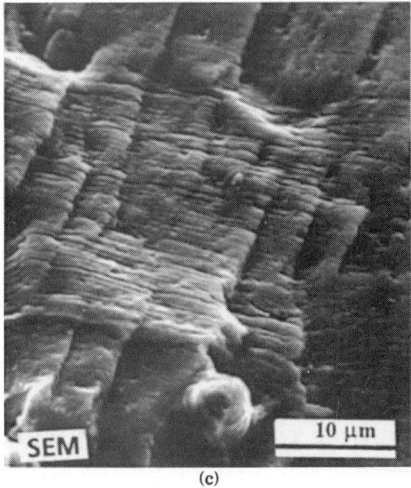

(c)

Figure 3.60 (*Continued*) Examples of fatigue fracture surfaces showing the effect of loading on the fatigue striation spectrum. (*b*) Spectrum-loaded fatigue fracture of a 7475-T7651 aluminum alloy test coupon showing an increase in the striation spacing due to higher alternating stress. (*From Kerlins and Phillips.*[9]) (*c*) Fatigue test of an aluminum alloy. In the center is a band of fatigue striations which originated from 45 cycles of higher stress amplitude. Above and below this band are fracture regions which originated from load cycles of lower amplitudes. (*From Engel and Klingele.*[10])

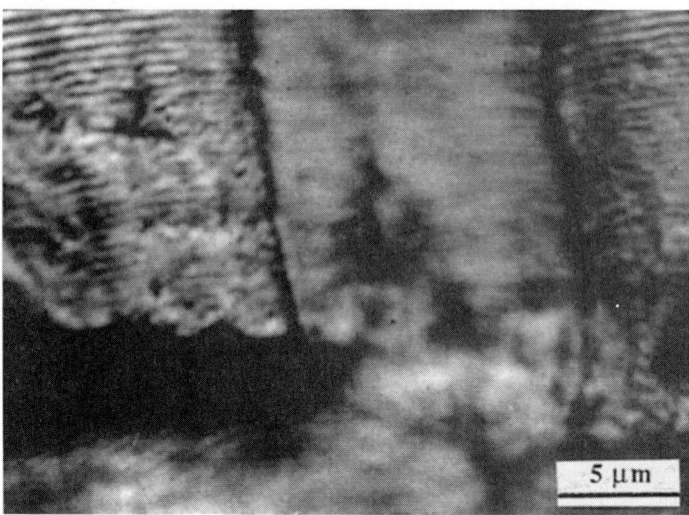

Figure 3.61 Fatigue striations as observed using an optical microscope. Material was an aluminum alloy tested in complete reversed bending, at a maximum stress of 172 MPa (25 ksi) at room temperature, to failure at 336×10^3 cycles. (*From Zapffe and Worden.*[21])

lished from controlled experiments. Examples showing the effect of the loading on striation spacing are given in Fig. 3.60. Thus examination of the striation spacing gives clues to the loading spectrum. The higher the loading frequency and the lower the stress magnitude, the finer the striations. Also as the crack propagates, the remaining material supporting the load decreases, thereby increasing the stress, and thus the crack propagation distance will eventually show an increase (if the maximum load in the cycle is constant).

It is important to note that the fatigue striation spacing may be at or below the limit of resolution of an optical microscope. In addition, the roughness of the surface makes it difficult to focus due to the low depth of field. An example of this effect is shown in Fig. 3.61. Thus examination must usually be made with an electron microscope, either of the surface directly with an SEM or of a replica with an SEM or a TEM. Note that this requirement is reflected in the relatively high magnification ($>2000\times$) used for most fatigue fractographs.

A common feature associated with fatigue striations is *secondary cracks*. Examples are shown in Fig. 3.62. The mechanism of their formation is not known.

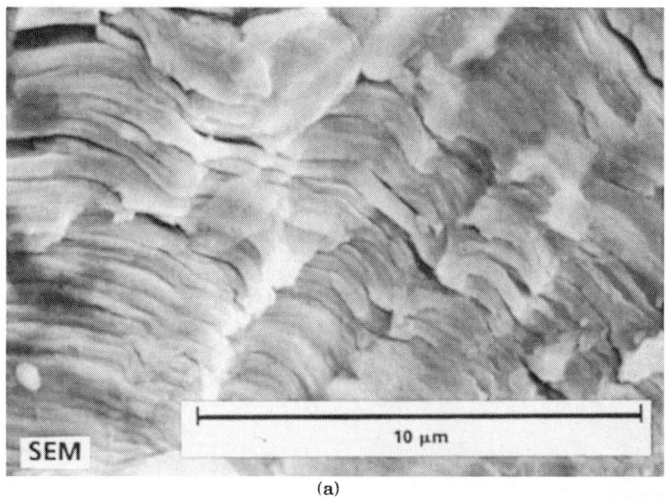
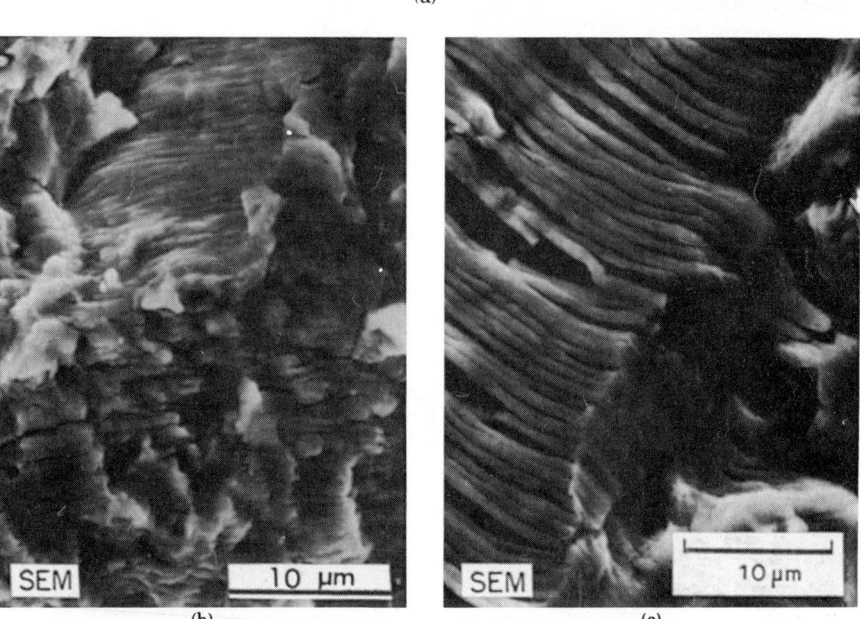

Figure 3.62 Examples of fatigue fractures showing secondary cracks associated with striations. (a) From a tantalum heat-exchanger tube. The rough surface appearance is due to secondary cracking caused by high-cycle low-amplitude fatigue. (*Courtesy of M. E. Blum, FMC Corporation; adapted from Kerlins and Phillips.*[9]) (b) From steel (0.38% C, 1.19% Si, 1.23% Mn, and 0.023% P) axle broken in service. (*From Henry and Horstmann.*[7]) (c) From steel containing 0.10% C, 0.27% Si, 0.43% Mn, 0.027% P, and 0.022% S. (*From Henry and Horstmann.*[7])

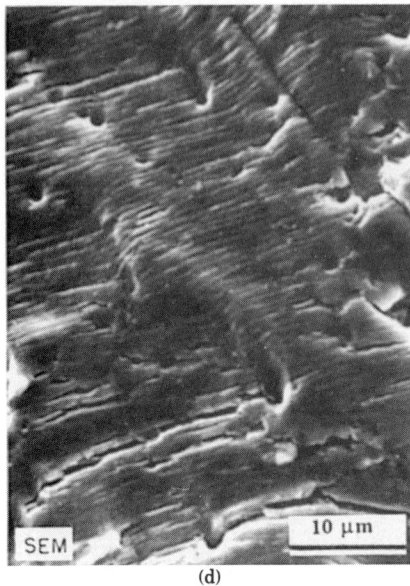

Figure 3.62 (*Continued*) Examples of fatigue fractures showing secondary cracks (arrows) associated with striations. (*d*) From Inconel 718 tested at room temperature. The crack propagation is from bottom to top. (*From Engel and Klingele.*[10])

Occasionally, fatigue cracks propagate by cleavage and thus without significant local plastic deformation. An example of the resulting fracture-surface topography is shown in Fig. 3.63. In such cases it is more difficult to correlate the fracture-surface features with the cyclic loading. Figure 3.64 shows a case where "striations" are found. However, these may be made prominent by the secondary cracking associated

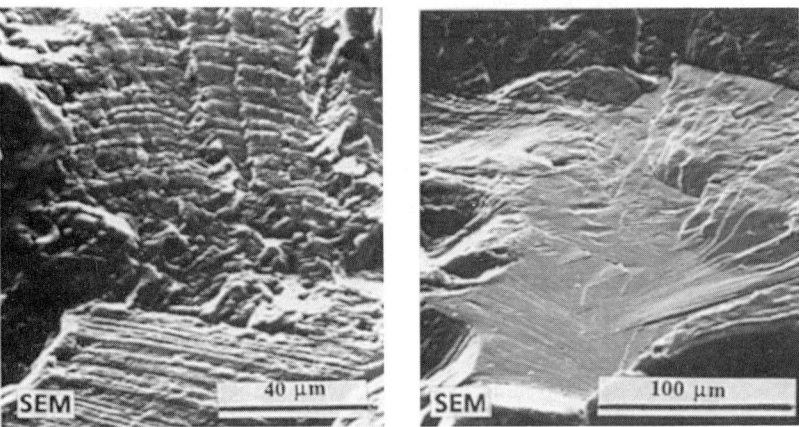

Figure 3.63 Fatigue fracture surface which fractured by brittle crack propagation on cleavage planes. This was produced by testing at room temperature a 1-mm-thick silicon sheet. Crack propagation is from bottom to top. (*From Engel and Klingele.*[10])

Figure 3.64 Striations on a fatigue fracture surface in which the crack propagated in a brittle manner. Material was a titanium casting alloy. (*From Engel and Klingele.*[10])

with each one. In rare cases the fatigue crack may propagate intergranularly, as illustrated in Fig. 3.65.

Another feature sometimes found on fatigue fracture surfaces is called *tire tracks*. Examples are given in Fig. 3.66. This feature is formed by the repeated impingement on one surface of a particle embedded in the other surface, as shown in Fig. 3.67.

Although in most cases fatigue cracks form striations, in many cases they do not, or they are extremely difficult to see or find. This is especially true in materials which have complex microstructures

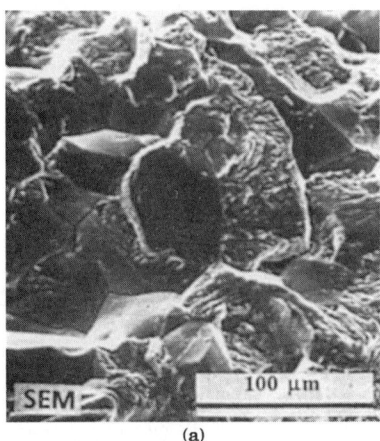
(a)

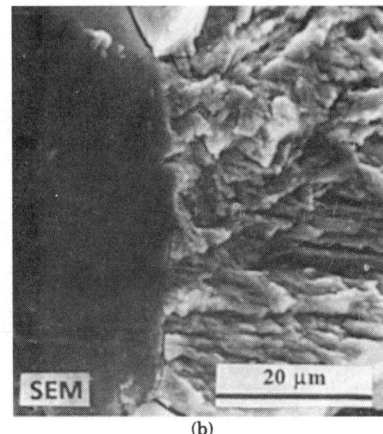
(b)

Figure 3.65 Fractographs showing an example where the fatigue crack propagated intergranularly. The fracture is from a gear tooth made from a case-hardening steel (0.16% C, 0.25% Si, 1.15% Mn, 0.95% Cr). The fracture in the core is on the right-hand side of (a) and propagated in a transgranular fashion, but in the 0.8-mm-deep case-hardened zone on the left, it propagated intergranularly. (b) Fractograph is a magnified view from the center of (a). (*From Engel and Klingele.*[10])

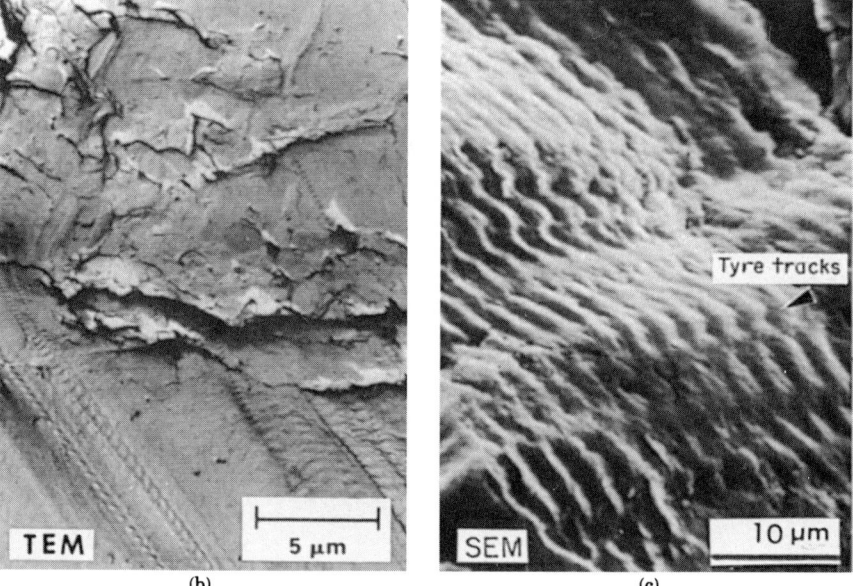

Figure 3.66 Examples of fatigue fracture surfaces showing tire tracks. (a) TEM replica of quenched and tempered 4140 steel. (*Courtesy of I. Le May, Metallurgical Services Ltd.; from Kerlins and Phillips.*[9]) (b) Alternate bending of a steel containing 0.09% C, 0.28% Si, 0.43% Mn, 0.025% P, and 0.024% S. (*From Henry and Horstmann.*[7]) (c) Threaded bar broken in service. Steel contained 0.61% C, 0.32% Si, 0.75% Mn, 0.027% P, and 0.028% S. (*From Henry and Horstmann.*[7])

Fracture Mechanisms and Microfractographic Features   173

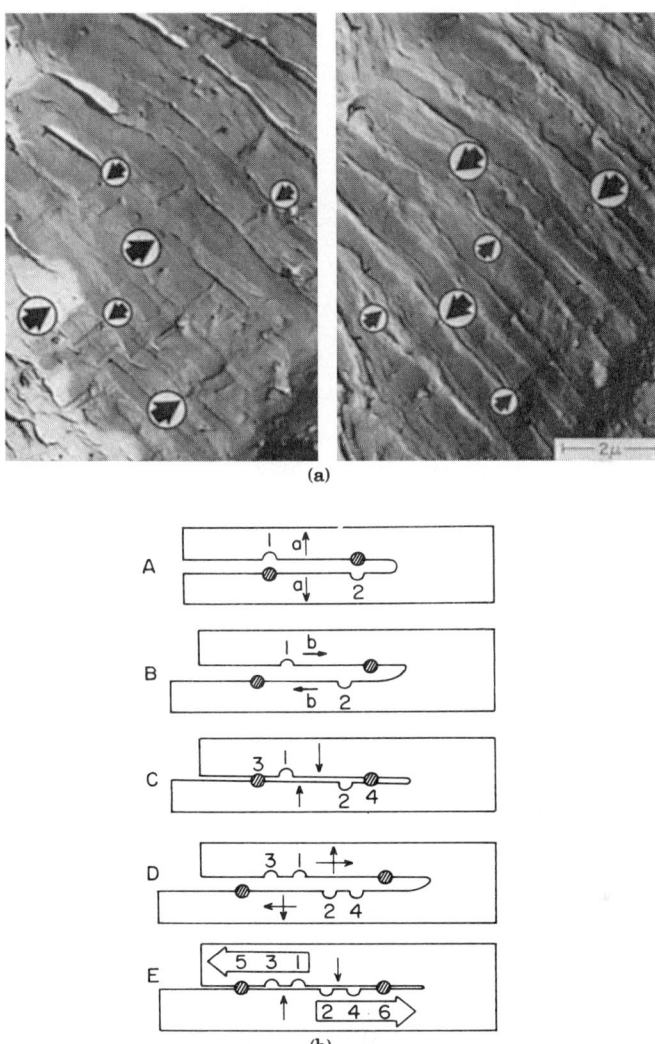

Figure 3.67 (a) Two-stage carbon TEM replica from same location of mating surfaces of a fatigue fracture, showing relationship between particles (large arrows) on one surface and tire tracks (small arrows) on the other. 2024-T3 aluminum alloy. (b) Schematic illustration of how tire tracks form, showing how the two surfaces offset and then reclose to form a series of impressions. (*From Beachem.*[8])

(such as tempered steels with a fine dispersion of carbides in ferrite). Figure 3.68 shows an example taken from a fatigue test. In Fig. 3.69 are examples of fracture surfaces in which the fatigue striations are either missing or ill-defined. Thus the absence of striations does not preclude fracture by fatigue loading.

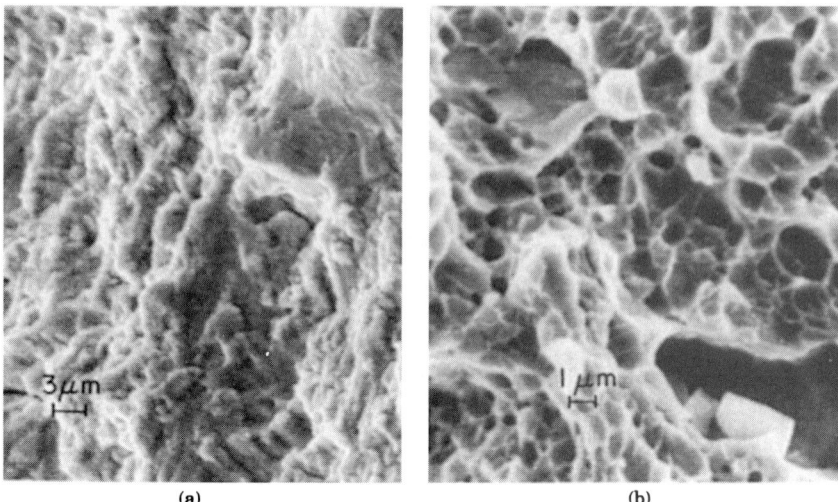

Figure 3.68 SEM fractographs taken from the fracture surface of a fatigue sample, showing that no striations are present. Arrows indicate crack propagation direction. This is the region of stage II propagation. The material was a 4140 steel tempered at 400°C. (a) $\Delta K = 21$ MN/m$^{3/2}$. (b) $\Delta K = 67$ MN/m$^{3/2}$. (*From Thielen and Fine.*[23])

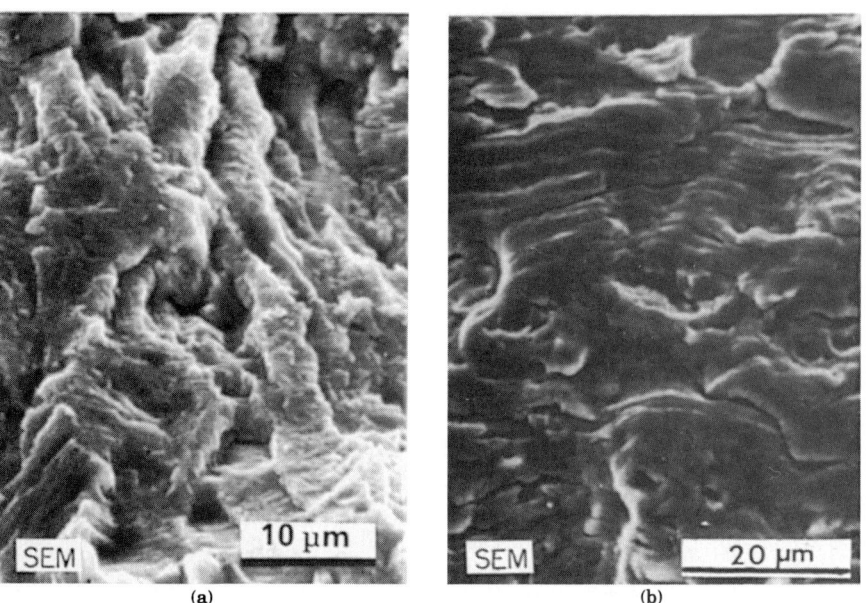

Figure 3.69 Examples of fatigue crack surfaces in which striations are absent or ill-defined. (a) Unalloyed carbon steel (0.45% C, 0.25% Si, 0.65% Mn). (*From Engel and Klingele.*[10]) (b) Drawn and galvanized annealed steel wire (0.46% C, 0.20% Si, 0.48% Mn, 0.027% P, 0.023% S). Rotative bending specimen. The striations are crushed and one mostly sees the secondary cracks. (*From Henry and Horstmann.*[7])

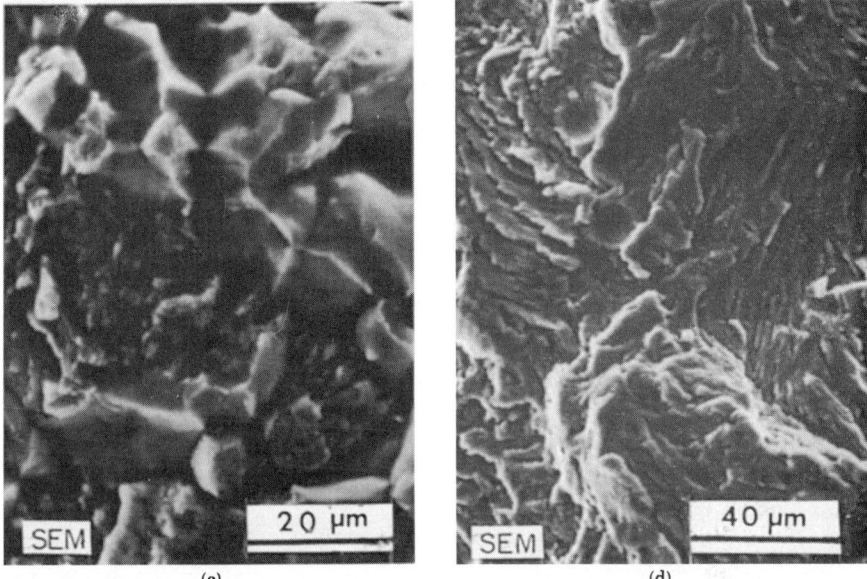

(c)   (d)

Figure 3.69 (*Continued*) Examples of fatigue crack surfaces in which striations are absent or ill-defined. (c) Fatigue rupture under tension-compression of a steel containing 0.51% C, 0.32% Si, 0.65% Mn, 0.013% P, 0.015% S, 1.12% Cr, and 0.28% V. Fatigue rupture along grain boundaries. Striations can only be seen in several parts of the rupture. (d) Threaded steel bar (0.61% C, 0.32% Si, 0.75% Mn, 0.027% P, and 0.021% S) broken in service. Pearlite structure. Here one can observe characteristic steps and striations besides pearlite. (*From Henry and Horstmann.*[7])

The final stage of fatigue crack propagation is stage III, which is the final overload fracture. The crack propagates in stage II to the point where the remaining supporting material is not sufficient to support the load on the next cycle (or the next few cycles), and fracture occurs usually catastrophically and usually with little gross plastic deformation. The fracture-surface topography depends on the fracture mechanism, which is usually by void coalescence or cleavage. It is usually easy to distinguish stage III from the other stages.

It might be expected that there would be a relationship between the fine fracture topography and the "macroscopic" conchoidal markings (see Sec. 4.5), which are characteristic of fatigue fracture. Figure 3.70a shows an SEM fractograph at low magnification in which the conchoidal marks are seen. (They were also quite visible in light microscopy.) However, upon imaging at increasing magnification the region containing one of these "lines," it is seen that the cause of the markings is less clear. There is a slight difference in coarseness of the surface on one side of the line compared to the other side (see Fig. 3.70e), which is what gives rise to the markings. (Note that in this

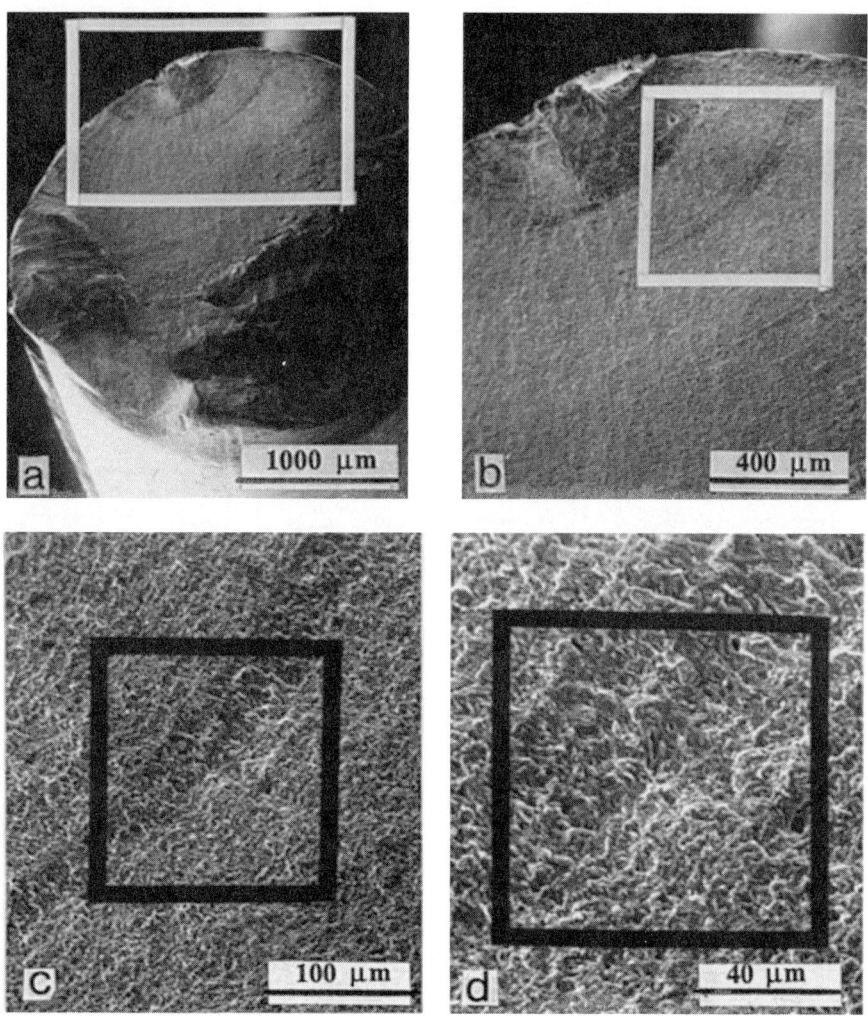

Figure 3.70 SEM fractographs showing the fine-surface topography associated with conchoidal markings. Each outlined area is shown at higher magnification in the following fractograph. The material was a 316 stainless-steel electrode wire from an electrostatic precipitator.

case only faint fatigue striations are observed.) In this example the loading spectrum appears to have been such that the fracture surface topography changed locally, which gave rise to the conchoidal markings.

The point of this example is that it is often not obvious exactly what causes the conchoidal markings to be visible. This can be caused by the intrinsic effect of the loading on the fine fatigue surface topography, by variations in rubbing of the fatigue fracture surface during service,

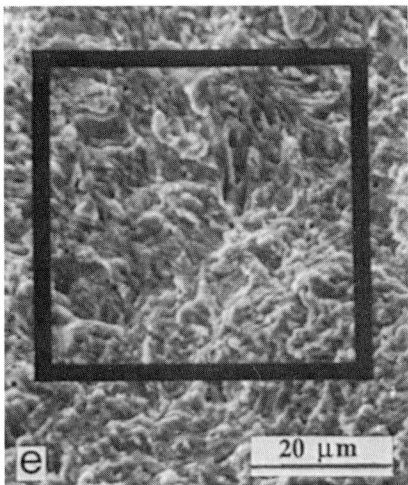 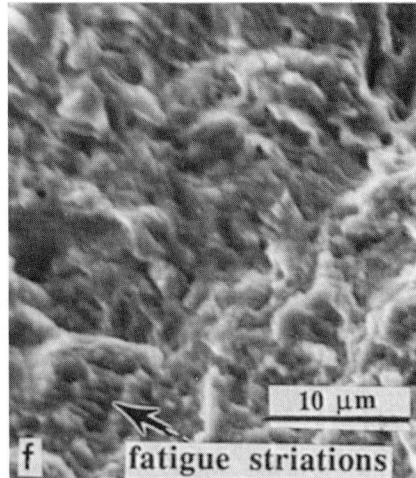

Figure 3.70 (*Continued*) SEM fractographs showing the fine-surface topology associated with conchoidal markings. Each outlined area is shown at higher magnification in the following fractograph. The material was a 316 stainless-steel electrode wire from an electrostatic precipitator.

and by different degrees of corrosion of the fracture surface as the crack advances.

3.10 High-Temperature Fracture Topography

In this section we examine the surface topography of fractures that occur at temperatures high enough that the mechanism may be basically different from those just described. Thus the term "high temperature" here does not necessarily mean elevated temperature, but refers to a temperature at or near the melting point, or to a temperature where creep-type processes control fracture. For example, these could occur in lead-based alloys at room temperature. From a practical standpoint, though, we are mainly concerned with fracture considerably above 25°C.

An important practical problem with fractography of high-temperature fractures is that after fracture the features of the fracture surface are usually affected by the environment. For example, the surface may be covered with a heavy oxide scale. This effect on the surface may make identification of the mechanism difficult. An example of this difficulty is illustrated in Figs. 3.71 and 3.72. The clear dimpled morphology of the fracture surface, allowing identification of the mechanism as void coalescence, is obliterated upon exposure of the fresh fracture surface to air. It also may be difficult to separate the effect of the environment on the fracture process, per se, from the effect it has on the fresh fracture surface after fracture or during fracture.

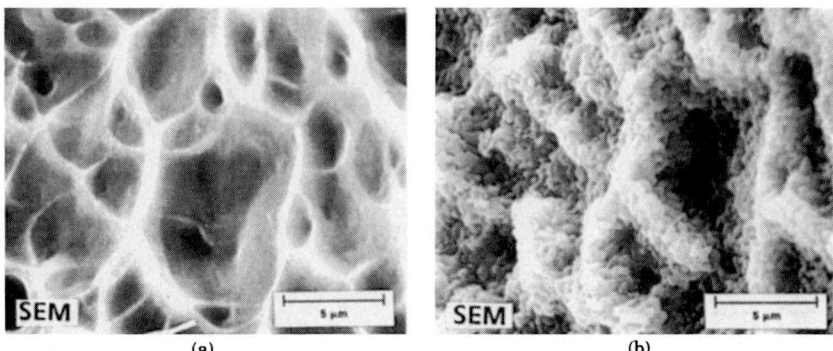

Figure 3.71 Fractographs illustrating the effect of oxidation on fracture-surface features. Material was an annealed Ti–6Al–6V–2Sn alloy. (a) Original fracture surface appearance. (b) Effect of a 15-min exposure in air at 800°C (1470°F). (*From Kerlins and Phillips.*[9])

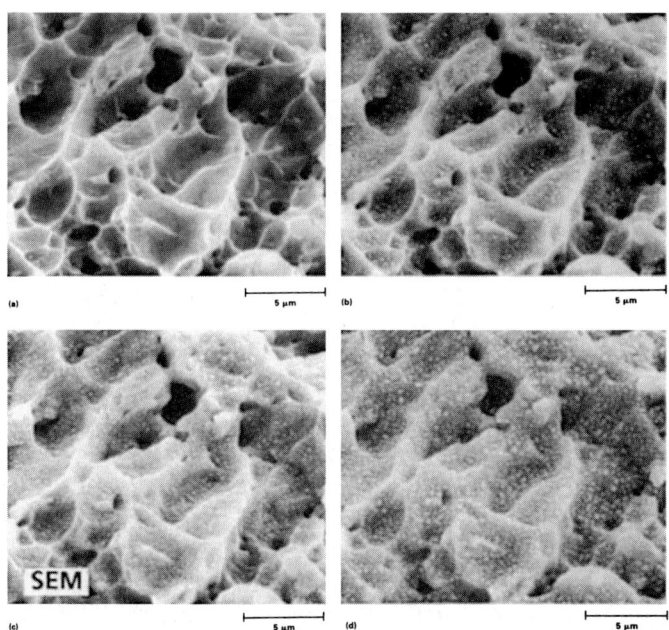

Figure 3.72 Fractographs illustrating the effect of oxidation on the fracture-surface features. Effect of a 700°C (1290°F) air exposure on an annealed Ti–6Al–2Sn–4Zr–6Mo alloy with an initial dimpled fracture surface. (a) As fractured. (b) Identical area after 3-min exposure. (c) After 10 min. (d) After 30 min. (*From Kerlins and Phillips.*[9])

Fracture Mechanisms and Microfractographic Features

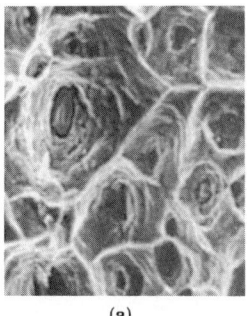 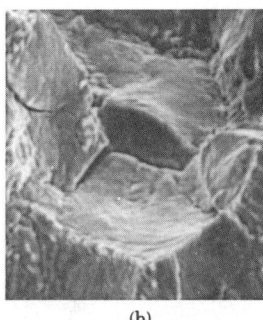 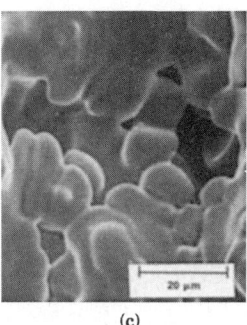

(a)   (b)   (c)

Figure 3.73 Effect of test temperature on fracture-surface topography. The sample was a Haynes 556 alloy and was tested at a strain rate of $1s^{-1}$. (a) Dimple rupture fracture at 1015°C (1860°F). At the bottom of many dimples are TaC inclusions. (b) Intergranular decohesion at 1523°C (2287°F). Secondary cracks are also visible. (c) Local eutectic melting of TaC + austenite at 1333°C (2431°F). (*Courtesy of J. J. Stephens, M. J. Cieslak, and R. J. Lujan, Sandia National Laboratories; from Kerlins and Phillips.*[9])

As the temperature of fracture increases, some alloys show a change in mechanisms from void coalescence to intergranular separation, especially if a liquid forms at the boundaries. An example is shown in Fig. 3.73. It is interesting to note that the tensile elongation of the sample tested at 25°C was about 70 percent, while that tested at 1253°C, which fractured intergranularly (Fig. 3.73b), only exhibited 8 percent elongation.

If fracture is associated with local melting, then a characteristic dendritic topography is found, as illustrated in Fig. 3.74. In this case the actual fracture occurred at the junction of a few dendrite arms (see ar-

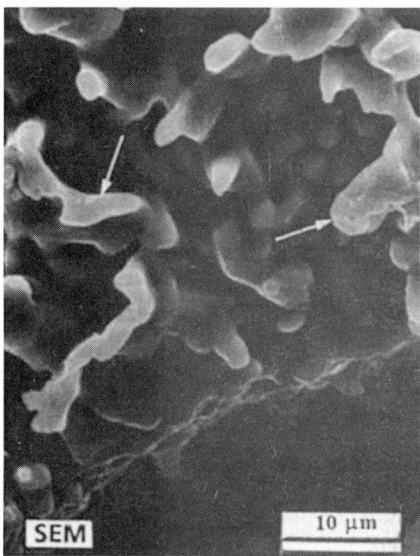

Figure 3.74 Fracture surface showing a dendritic appearance, associated with fracture of a liquid film between dendrites. This was a crack in a Waspaloy alloy immediately adjacent to an electron beam weld. The crack shows the characteristics of fracture in the pasty state, namely, the surface consisting of droplets and drawn-out peaks. (*From Engel and Klingele.*[10])

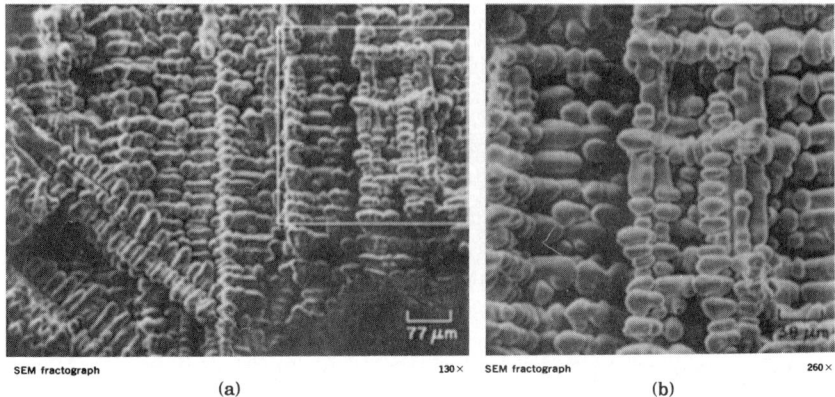

Figure 3.75 SEM fractographs of dendrites found on a fracture surface, but which are the surface of preexisting shrinkage cavities. The alloy was an 18% Ni, grade 300 maraging steel. Fractograph (b) shows at higher magnification the area outlined by the rectangle in (a). (*From* Metals Handbook.[11])

rows). A similar topography is found when the fracture exposes solidification voids, so that the dendritic surface is seen; an example is given in Fig. 3.75. However, in this case this is not a fracture surface but an exposed void surface. If incipient melting occurs along grain boundaries, the fracture will appear intergranular.

The fractograph in Fig. 3.76 illustrates the topography sometimes seen in fractures in which the alloy is partly liquid. The smooth regions

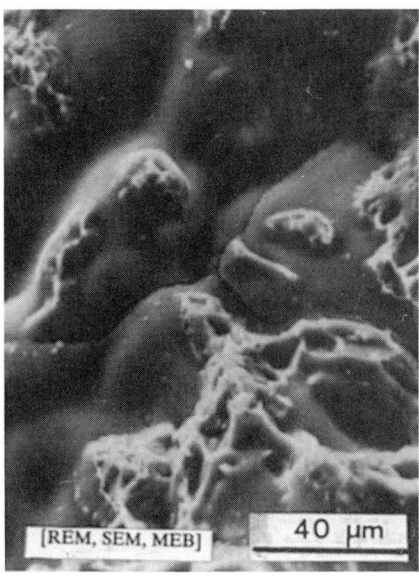

Figure 3.76 Fracture surface showing evidence of melting. This is from a crack in a brazed tube. The smooth regions are the intergranular zones that fractured at high temperature in the presence of liquid copper; the dimpled regions are where the material fractured during opening of the crack at room temperature to reveal the fracture surface. (*From Henry and Horstmann.*[7])

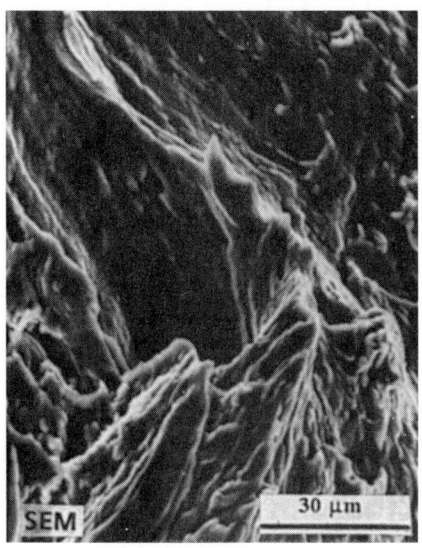

Figure 3.77 Fractograph showing extensive plastic deformation due to fracture at high temperature. (*From Engel and Klingele.*[10])

were molten at the temperature at which the part fractured. The rougher regions are the location of actual separation which occurred by void coalescence.

At high temperature, after the formation of microcracks and microvoids, the remaining connecting material can undergo extensive plastic deformation before final separation occurs. This creates highly elongated features on the fracture surface, as illustrated in Fig. 3.77. In some cases, very elongated "necks" are seen, as shown in Fig. 3.78.

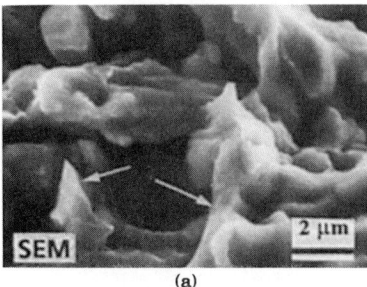
(a)

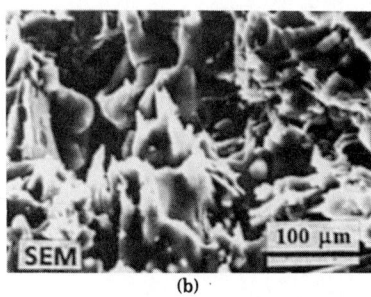
(b)

Figure 3.78 Fractographs showing regions exhibiting extensive plastic deformation (arrows). (*a*) Separation with the formation of drawn-out peaks in an electron-beam weld bead in a maraging steel (0.02% C, 0.1% Si, 0.1% Mn, 12% Co, 5% Mo, 18.5% Ni, 0.9% Ti). (*b*) Separation (hot tearing) produced in a nickel alloy (Al Si 12 Cu Mg) casting which involved sharp changes in cross section. The fracture surface shows rounded quench structures resembling shrinkage porosity, dimple formation and the characteristic, drawn-out peaks. (*From Engel and Klingele.*[10])

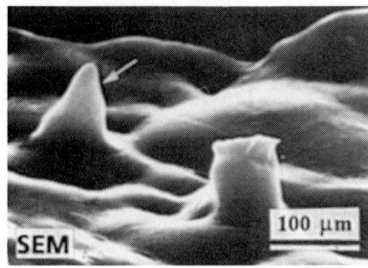

Figure 3.78 (*Continued*) Fractographs showing regions exhibiting extensive plastic deformation (arrows). (*c*) Intercrystalline separation, showing peaks formed at the softened grain boundaries in a steel forging. (*From Engel and Klingele.*[10])

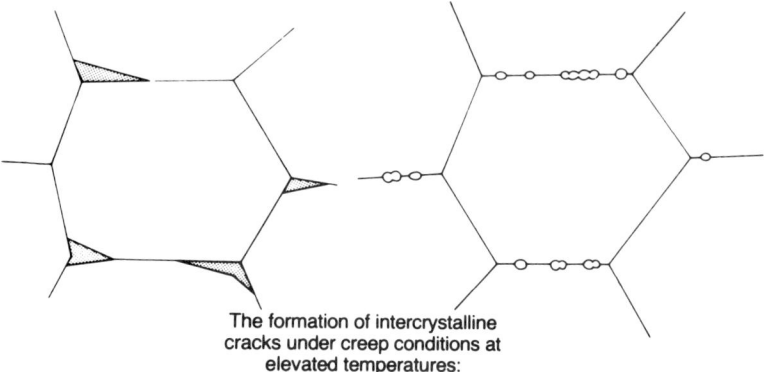

The formation of intercrystalline cracks under creep conditions at elevated temperatures:

(a) cracks formed at the intersections of several grain boundaries

(b) the formation of voids at grain boundaries aligned perpendicular to the principal tensile stress

After H. Böhm, Einführung in die Metallkunde (Introduction to Metallurgy), Bibliogr. Institut Mannheim, 1968

Figure 3.79 Schematic illustration of the formation of (*a*) wedge and (*b*) creep cavities. (*After Böhm*[24]; *from Engel and Klingele.*[10])

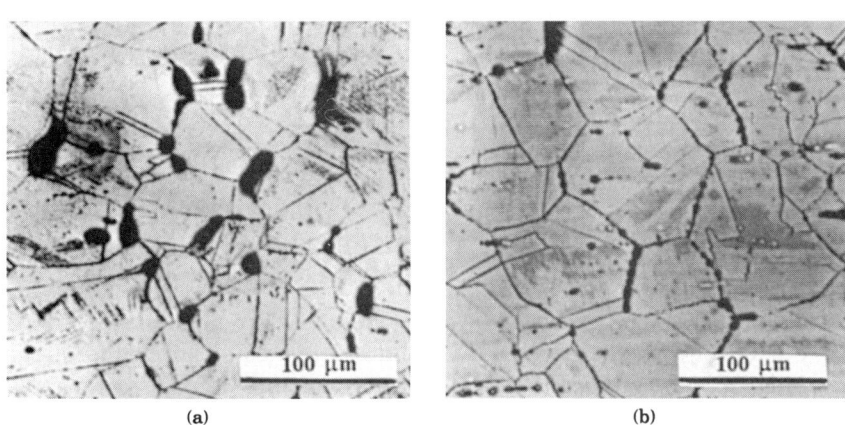

Figure 3.80 Microstructures from creep specimens showing creep cavities and wedge cracks. (*a*) Cracks initiated at triple boundaries. 17Cr–14Ni–Ti austenitic stainless steel creep-tested at 650°C (1202°F) under 80 N/mm² for 3383 h. (*b*) Beadlike cracks along grain boundaries. 17Cr–14Ni–Ti–B austenitic stainless steel creep-tested at 650°C (1202°F) under 130 N/mm² for 429 h. (*From Henry and Horstmann.*[7])

Creep fracture is a separate mechanism. Usually cracks are initiated at grain boundaries and at grain-boundary junctions, as shown schematically in Fig. 3.79. The former are called *creep voids* and the latter *wedge cracks*. Examples of these are shown in the microstructures in Fig. 3.80. Fractographs of creep voids and wedge cracks are given in Fig. 3.81. As pointed out, interaction of the fracture surface with the environment during and after the fracture may make it difficult to identify such features.

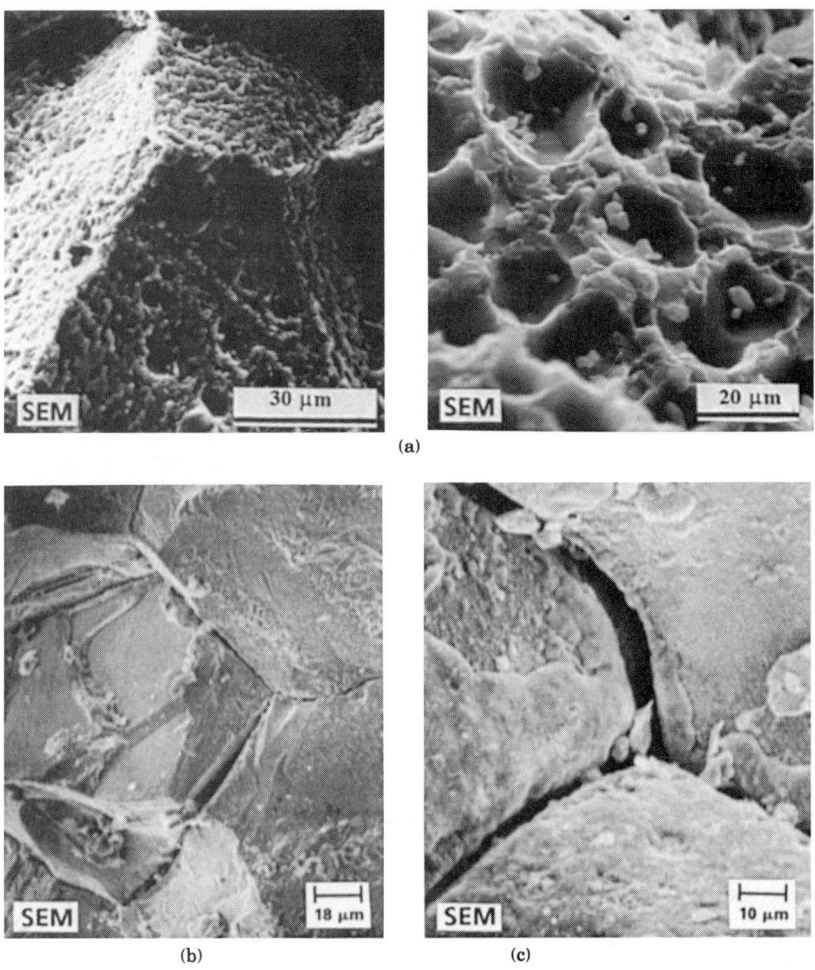

Figure 3.81 Fractographs showing creep cavities and wedge cracks. (a) Creep cavities (voids) which formed at 800°C (1472°F) at the grain boundaries of Nimonic 105. These were revealed by breaking the sample at −196°C (−321°F). (b) Wedge cracking in Inconel 625 and (c) wedge cracking in Incoloy 800. [(a) *From Engel and Klingele*,[10] (b) and (c) *from Kerlins and Phillips*.[9]]

3.11 Environmentally Assisted Fracture

The environment can have a potent effect on crack nucleation and propagation in alloys. In spite of extensive research on the environmental effect, the underlying mechanisms are not always understood since the processes have been found to be very complex. This is reflected in the wide variety of fracture-surface topographies found, showing that the type of fracture which occurs depends on the alloy and the environment, and therefore a single underlying fracture mechanism has not been found.

Thus the approach to recognizing that the environment may have influenced fracture cannot rely so clearly on the type of fracture topography. In addition, the "fresh" fracture surface after it forms may be al-

Figure 3.82 Fractographs illustrating a variety of stress corrosion cracking features. (a) Two-stage TEM replica of intergranular stress corrosion crack surface in AMS-6434 steel. (*From Beachem.*[8]) (b) Transcrystalline stress corrosion cracking in a ferritic welding metal (unalloyed, high-purity filler material) which initiated due to the presence of aqueous NaOH. This type of attack occurs only with high-purity unalloyed steels and under the influence of liquid NaOH, which formed as a result of the evaporation of water in a crevice formed due to bad manufacturing of a boiler. (c) and (d) Stress corrosion cracking in an austenitic stainless steel (0.1% C, 1% Si, 2% Mn, 18% Cr, 10% Ni, 2% Mo, 0.6% Ti) caused by $MgCl_2$ solution at over 100°C. Noticeable features are intergranular separation (c), and featherlike structures and large cleavagelike areas (d). (*From Engel and Klingele.*[10])

tered by subsequent chemical interaction with the environment, which makes interpretation difficult. In this section we present some examples of typical fracture topographies found to illustrate the influence of the environment.

An important type of fracture mechanism is *stress corrosion cracking*. It has been found that components loaded to below the yield strength may suddenly fracture in certain corrosive environments. Figure 3.82 shows examples of such fracture surfaces, which illustrate the diversity of the topography. Note that both intergranular and transgranular fractures are found. It may be difficult just from the fracture-surface appearance to conclude that the fracture was caused by corrosion assistance.

Hydrogen embrittlement is another environmentally assisted fracture mechanism commonly encountered in a number of alloys. The hydrogen can already be present in the material (such as by retention during casting), but the more common source is hydrogen picked up in service or during processing (such as plating). The hydrogen can be absorbed in sufficient quantities at quite low temperature, such as during chemical cleaning of the surface of the component or during plating.

The mechanism of the embrittlement is not clear. The type of crack which forms and propagates is sensitive to several factors, such as the alloy and the loading conditions. This may make it difficult to ascertain whether hydrogen embrittlement was involved from just the fracture surface appearance. Figure 3.83 shows examples of the fracture surfaces of hydrogen-embrittled alloys.

(a)

Figure 3.83 Fractographs illustrating a variety of hydrogen embrittlement fracture surfaces. (a) A gas-carburized steel screw (0.3% C, 0.25% Si, 0.75% Mn, 1% Cr) broke after being galvanized. Characteristic features of hydrogen-initiated fracture are the yawning grain boundaries, micropores, and hairline patterns (partly in the form of crows-feet markings). (*From Engel and Klingele.*[10])

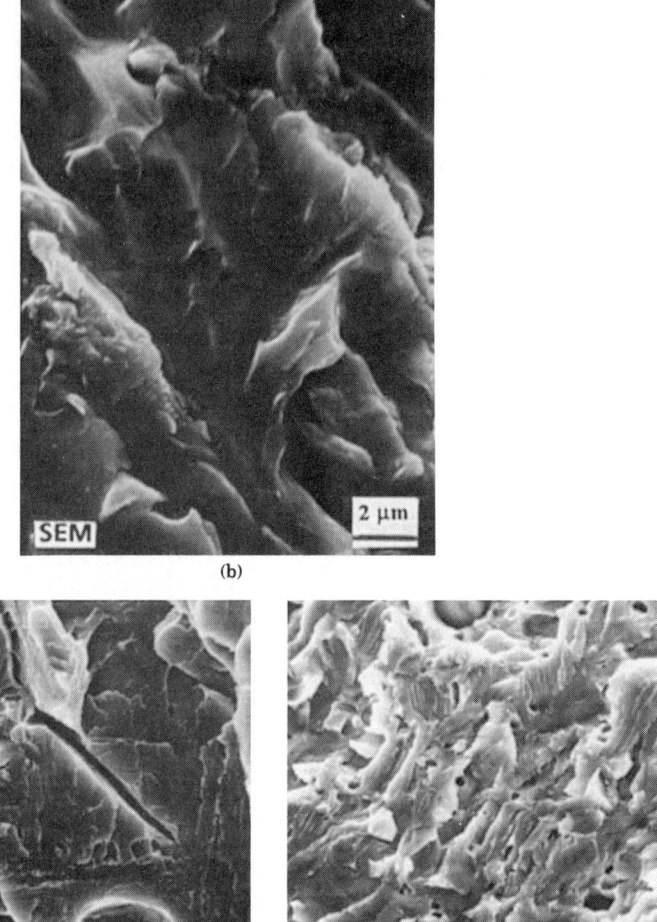

Figure 3.83 (*Continued*) Fractographs illustrating a variety of hydrogen embrittlement fracture surfaces. (*b*) A ferritic fine-grained steel (0.2% C, 0.5% Si, 1.5% Mn) was tested in the laboratory by straining it beyond the yield point in an aqueous solution of H_2S. Hydrogen, produced by the corrosion reaction $Fe + H_2S \rightarrow FeS + 2H$, caused the brittle fracture of the material. The fracture mechanism is mainly that of transcrystalline quasicleavage, but areas of pearlite can also be recognized (roughly in center of picture). (*c*) A precipitation hardening steel (17% Cr, 3% Ni, 2% Cu) became embrittled due to the influence of hydrogen. This steel is regarded as being extremely sensitive to the presence of hydrogen. The characteristic features indicating hydrogen embrittlement are: microquasicleavage facets, micropores, and ductile hairlines. (*d*) Flake cracking in unalloyed carbon steel. (*From Engel and Klingele.*[10])

Stress corrosion cracking and hydrogen embrittlement are usually found in statically loaded situations. If the load varies, then corrosion-assisted fatigue is encountered. Typical fracture surfaces are shown in Fig. 3.84. Usually the surfaces show signs of corrosion. Fatigue striations may be seen, but not always. They may have been obliterated by the corrosive attack of the fracture surface during crack propagation, or after fracture but before the part could be removed from the environment.

3.12 Flutes

A fractographic feature which has been found in alloys of complex microstructure (such as Ti alloys) is called *flutes*. Examples are shown in Fig. 3.85. In some alloys their formation is associated with

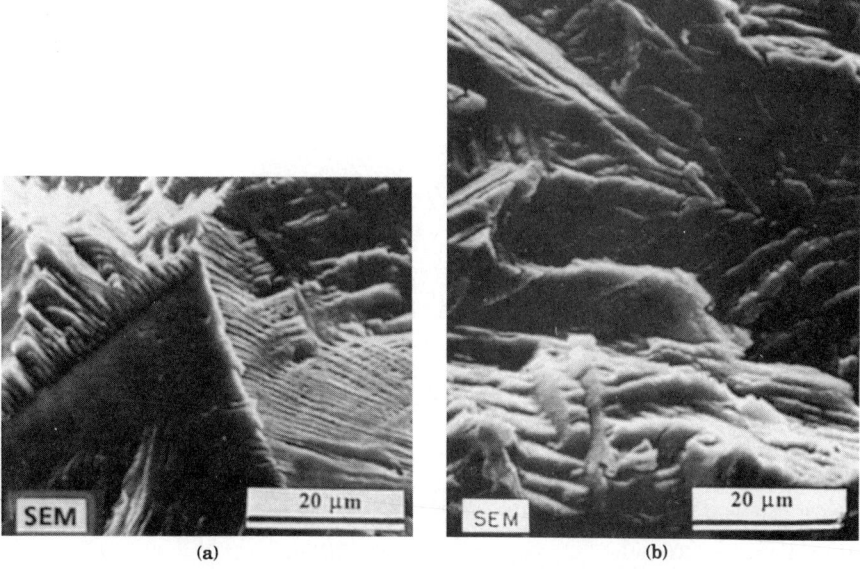

Figure 3.84 Fractographs of alloys which failed due to corrosion-assisted fatigue. (*a*) Corrosion fatigue in an austenitic stainless steel (17% Cr, 15% Ni, 4.3% Mo) caused by chloride-containing bleaching solution with a pH of 2. The fracture surface shows little evidence of plastic deformation and consists of smooth, crystallographically oriented surfaces. (*b*) Brass condenser tube (Cu Zn 28 Sn). The crystallographically oriented, featherlike structure on the fracture surface has resulted from transcrystalline fracture which has been accompanied by only a minimal amount of deformation. (*From Engel and Klingele.*[10])

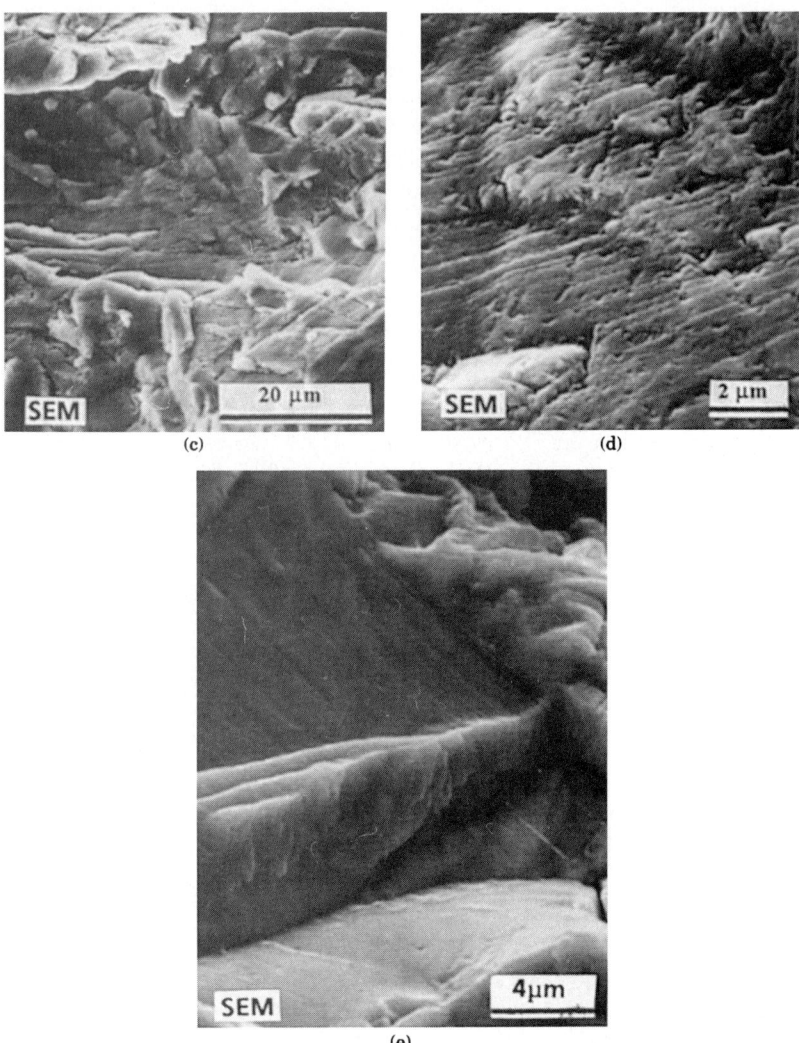

Figure 3.84 (*Continued*) Fractographs of alloys which failed due to corrosion-assisted fatigue. (*c*) Cu Zn 28 Sn. Fatigue striations on the smooth blocks are characteristic of this type of fracture. (*d*) (*e*) Cu Zn 20 Al. The smooth crack paths with fatigue striations are characteristic features of corrosion fatigue. [(*c*) and (*d*) *From Engel and Klingele.*[10] (*e*) *From Kerlins and Phillips.*[9]]

a variety of fracture modes, such as fatigue, overload, and stress corrosion cracking. They appear to form from planar slip as tear regions form between cleavage cracks. The elongated geometry is associated with the elongated grains which are present in the microstructure.

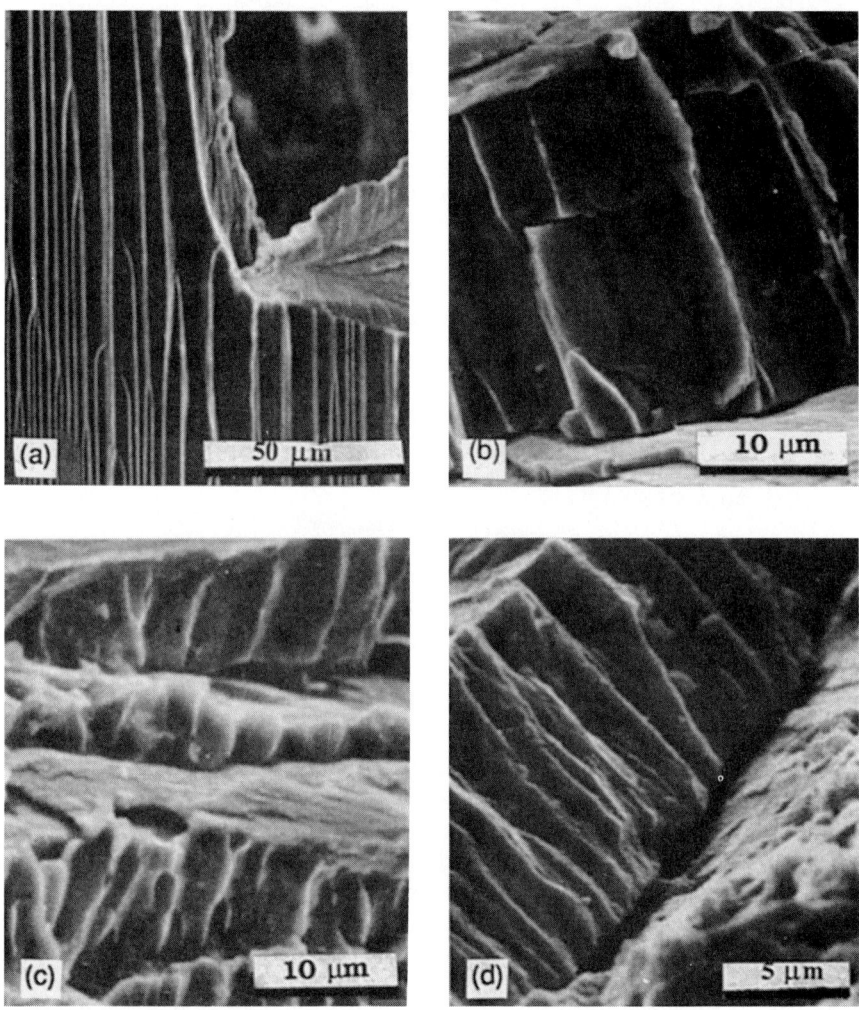

Figure 3.85 Examples of the fracture-surface feature called flutes. (a) Flutes and cleavage resulting from mechanical overload of a Ti–0.35O alloy. (b) Flutes and cleavage resulting from stress corrosion cracking of beta-annealed Ti–8Al–1Mo–1V alloy in methanol. (c) Flutes and cleavage in beta-annealed Ti–8Al–1Mo–1V resulting from sustained-load cracking in vacuum. (d) Flutes near the notch on the fracture surface of mill-annealed Ti–8Al–1Mo–1V alloy resulting from corrosion fatigue in salt water. (*Adapted from Meyn and Brooks.*[25])

3.13 Wear

The removal of small particles from surfaces by contact with other surfaces or by interaction with the moving surrounding environment (such as high-velocity water) is called *wear*. The basic characteristic of the process is fracture of material from the surfaces, and thus involves the fracture mechanisms described. However, it usually is difficult to characterize the wear process in a simple fashion due to factors such as melting and oxidation at the interface of the mating surfaces due to frictional heating, localized solid-state welding of surface asperities followed by fracture, embedding of debris in the surfaces, and so on. Because it is such a complicated process, a treatment of the subject is not included in this book. However, publications listed in the Bibliography (Phol, Tucker, and Wulpi in particular) contain treatments which will serve as starting points for understanding wear failures. Also, the glossary at the end of the book contains some of the common terms in wear failures.

3.14 Stereo Examination of Fracture Surfaces

Although electron micrographs of fracture surfaces (especially scanning electron micrographs) have a three-dimensional appearance, due to the great depth of field of the instruments, a much better feeling of the depth profile can be obtained by stereo imaging the surface (see Appendix 1A). Such pictures are easy to obtain. They require taking a picture of the region of interest, then tilting the specimen a few degrees and photographing the new image of the same area. The two pictures are then appropriately arranged and viewed with a stereo viewer to bring out the depth detail.

The added information obtained by stereo viewing cannot be overemphasized. To emphasize the advantage of stereo viewing, several stereo views of fracture surfaces, which should be examined using a stereo viewer, are presented here. Figure 3.86 shows optical stereo pairs, illustrating that this method can be used for viewing with an optical microscope. However, the image suffers from low resolution, so most stereo pictures are obtained using an electron microscope. Figures 3.87 and 3.88 present several stereo pairs for a variety of fracture surface topographies.

Stereo pairs can be used to obtain quantitative depth profiles. The procedure is outlined in detail in Appendix 1A.

3.15 Comparison of SEM and TEM Fractographs

Prior to the advent of the SEM, fractographs were obtained by imaging replicas of the fracture surface in a TEM, and thus the earlier doc-

Fracture Mechanisms and Microfractographic Features 191

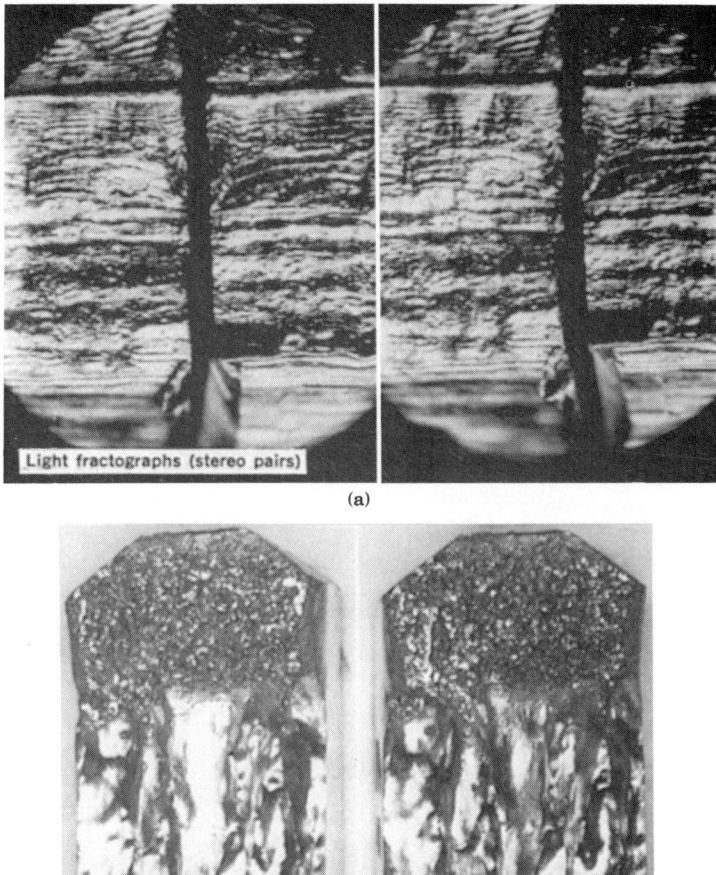

Figure 3.86 Optical stereo pairs of fracture surfaces. (a) Low-cycle fatigue fracture in aluminum alloy 1100. (b) Fracture surface of cast experimental low-carbon steel. (*From* Metals Handbook.[11])

umentation of the fracture surfaces is in this form. Now most observations are made with the SEM. However, there still are situations where a replica must be examined; for example, the broken component may be too large to place in an SEM. The appearance of the same fracture surface as observed in SEM and from a replica in TEM is somewhat different, and it is useful to recognize the appearance of common features (such as dimples) in the two types of images. Figure 3.89

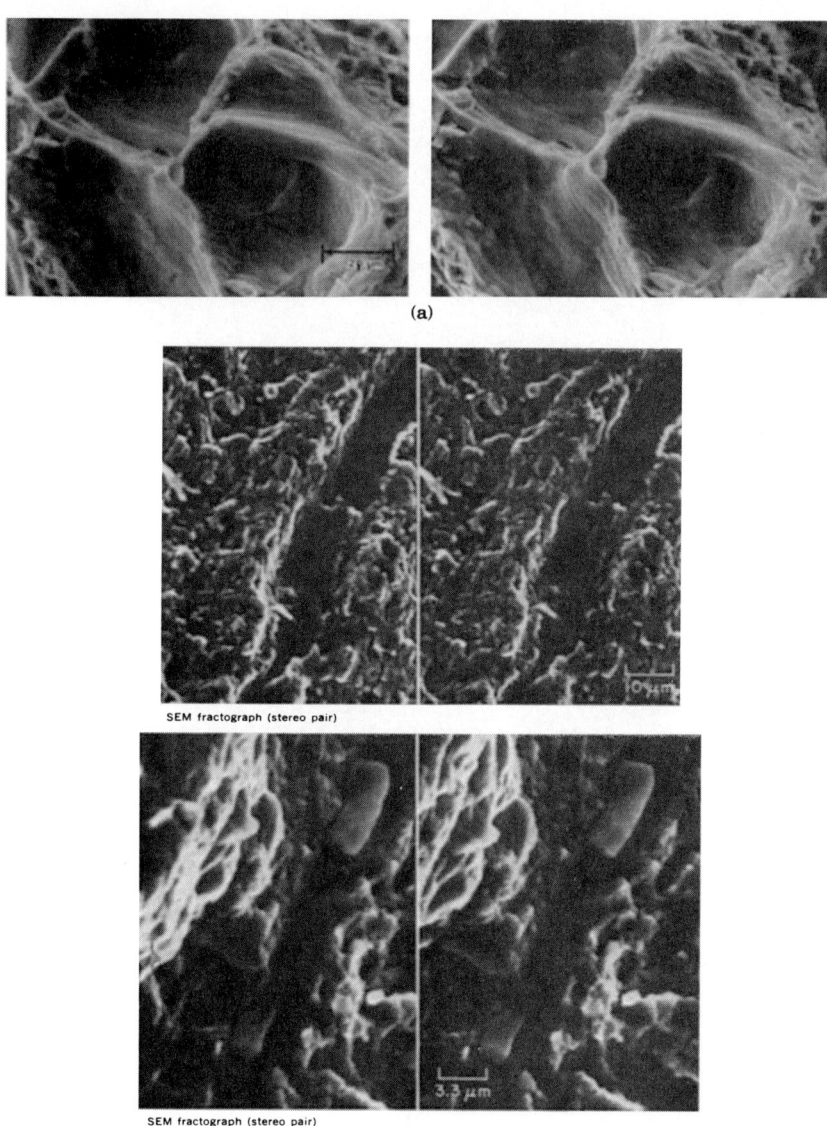

Figure 3.87 Stereo pairs of several types of fracture surface topographies. (*a*) Stereo pair showing deep dimples in the fracture surface of commercially pure titanium. (*Courtesy of M. Erickson-Natishan, University of Virginia; from Gabriel.*[14]) (*b*) Stringer and two stringer troughs on the fracture surface of 4340 steel heat-treated to a tensile strength of 1310–1450 MPa (190–210 ksi). Top—trough that earlier contained a stringer; bottom—trough that still contains two portions of a stringer. (*From Metals Handbook.*[11])

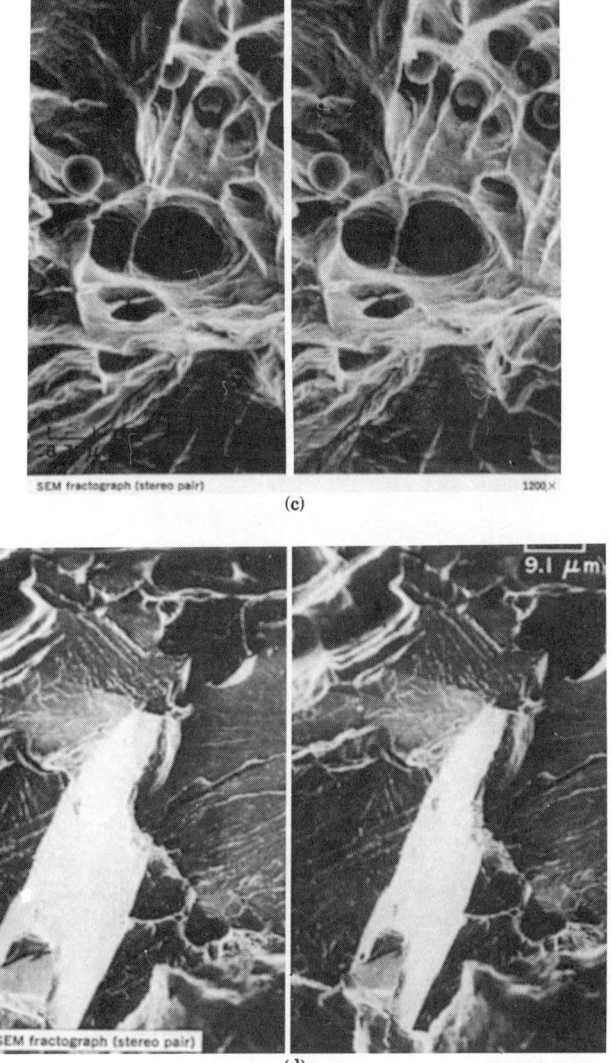

Figure 3.87 (*Continued*) Stereo pairs of several types of fracture surface topographies. (c) SEM fractographs (stereo pair) of the surface of a tensile-test fracture obtained at room temperature. The alloy was a low-carbon iron to which an appreciable amount of Fe_2O_3 had been added. (d) SEM fractograph (stereo pair) of the fracture surface of a Charpy impact test bar of high-purity iron. (*From* Metals Handbook.[11])

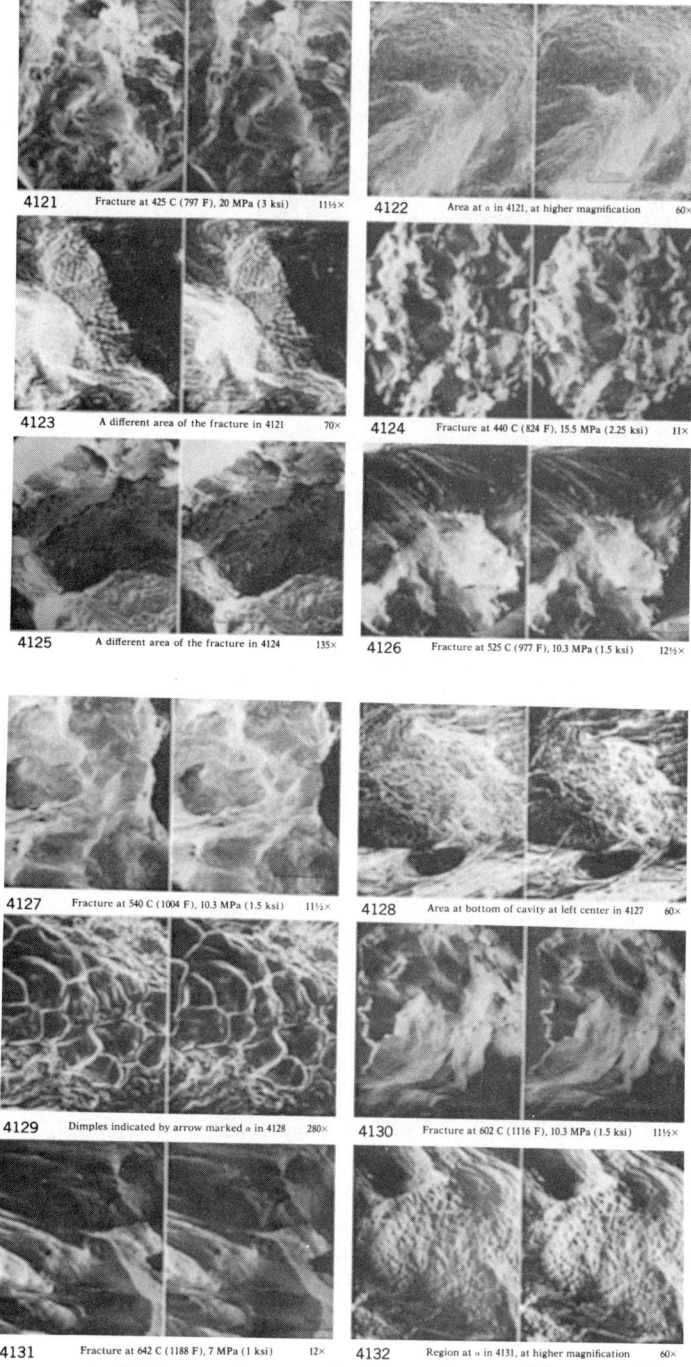

Figure 3.88 SEM stereo fractographs of several types of fractures. (*From Metals Handbook.*[11])

Fracture Mechanisms and Microfractographic Features 195

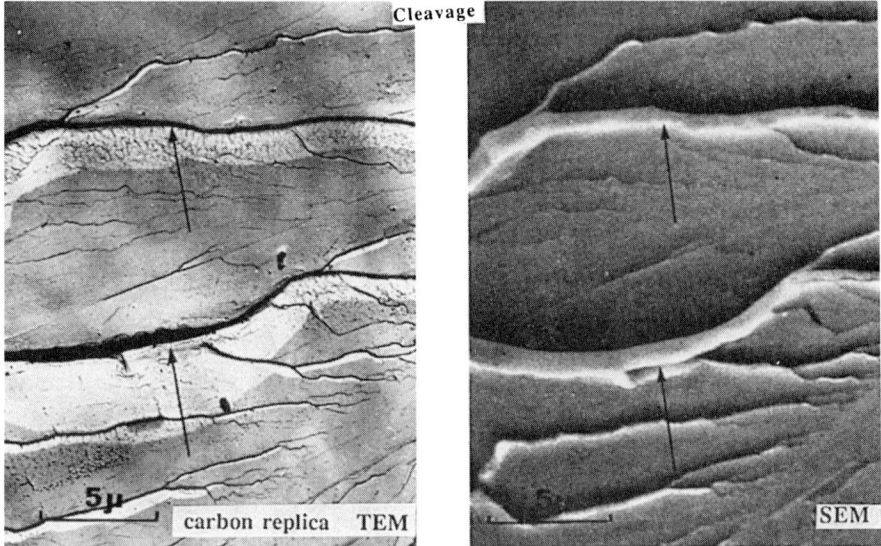

Acier C 0,05—Cr 2,25—Mo 1—Nb 1—cassé par choc à—196°C.

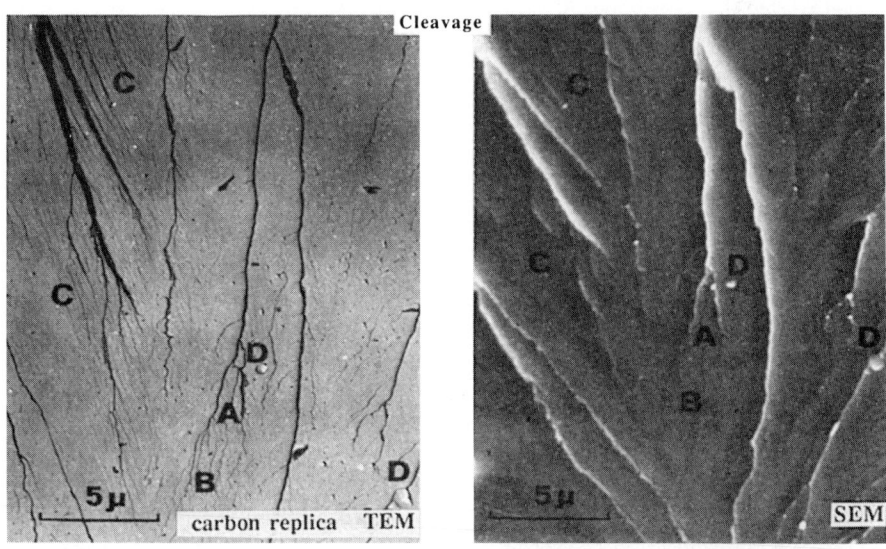

Acier C 0,05—Cr 2,25—Mo 1—Nb 1—cassé par choc à—196°C.

Figure 3.89 Fractographs comparing the exact same areas as observed on a carbon replica in TEM (left) and in SEM (right). (*From Maillard, Meny, and Champigny.*[27])

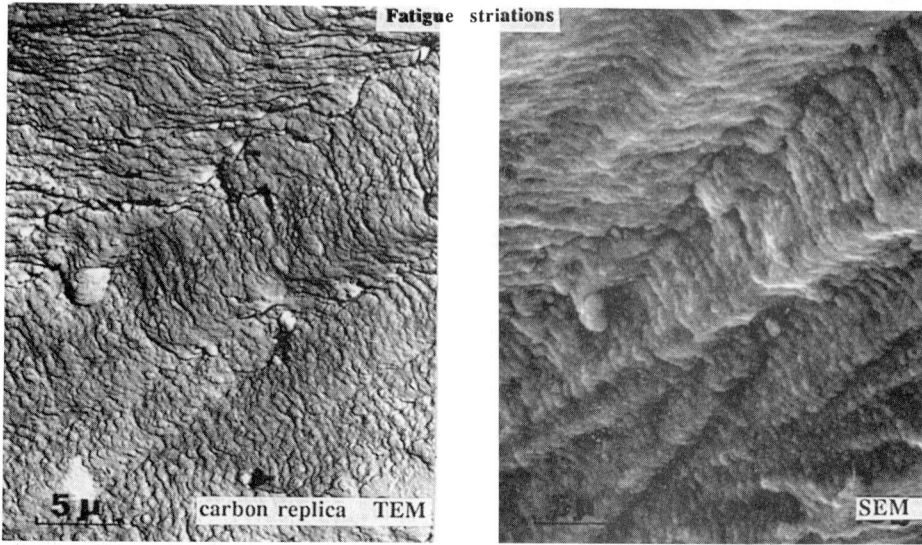

Acier C<0,03—Cr 16—Ni 12—Mo 2—Mn 1,2. Fatigue en traction compression à 580°C dans l'argon—nombre de cycles 25.535—ε0,5—fréquence 20 s.

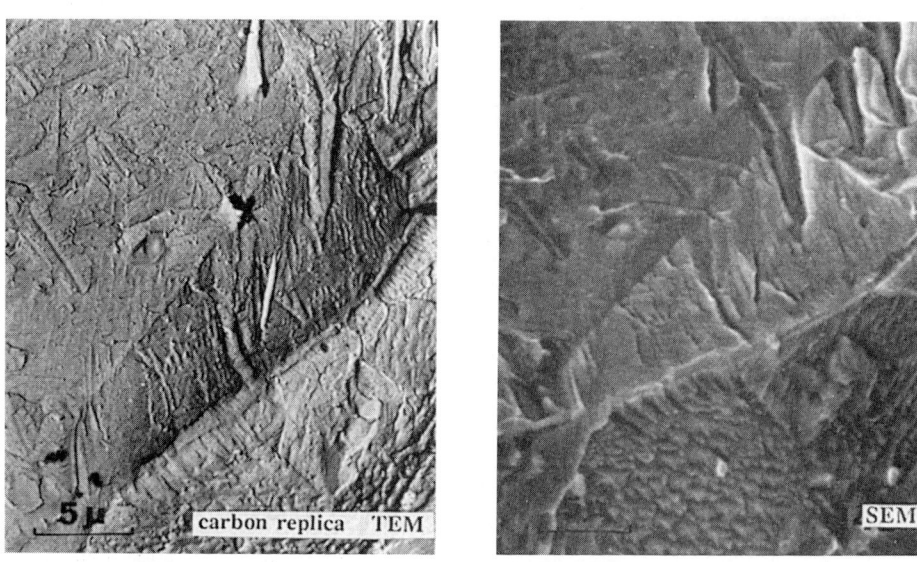

Acier C 0,15—Cr 16—Ni 2—cassé par choc à —196°C.

Figure 3.89 (*Continued*) Fractographs comparing the exact same areas as observed on a carbon replica in TEM (left) and in SEM (right). (*From Maillard, Meny, and Champigny.*[27])

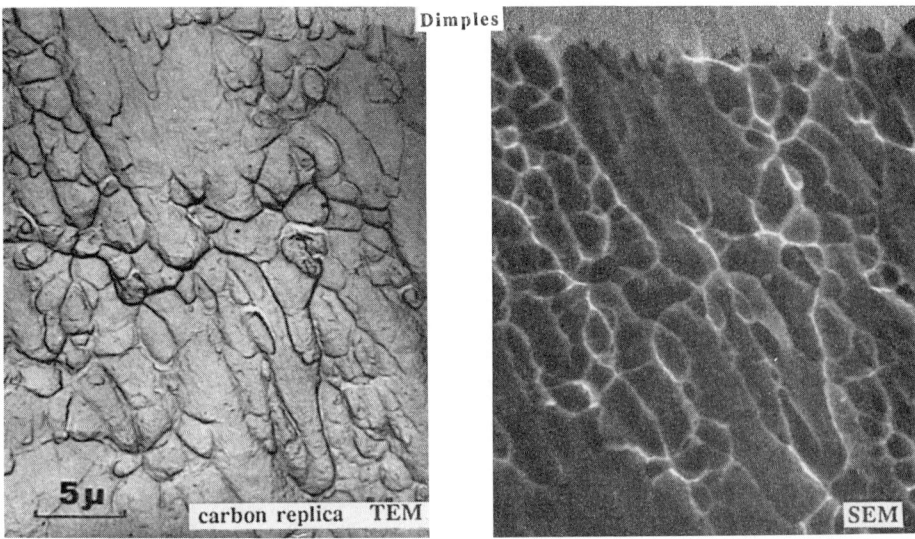

Acier C<0,08—Cr 17—Ni 12—traction à 20°C.

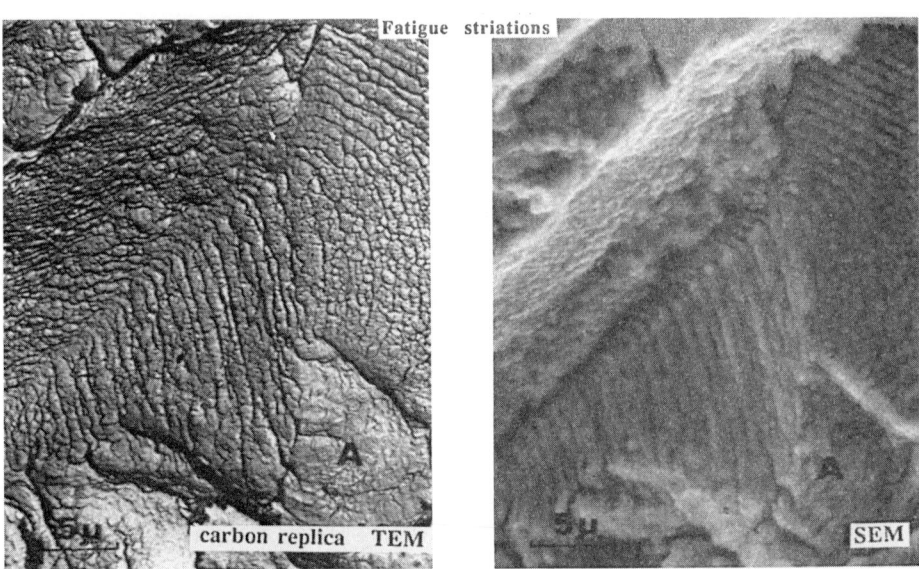

Acier C<0,03—Cr 16—Ni 12—Mo 2—Mn 1,2. Fatigue en traction—compression à 580°C dans l'argon—nombre de cycles 25.535—ε0,5—fréquence 20 s.

Figure 3.89 (*Continued*) Fractographs comparing the exact same areas as observed on a carbon replica in TEM (left) and in SEM (right). (*From Maillard, Meny, and Champigny.*[27])

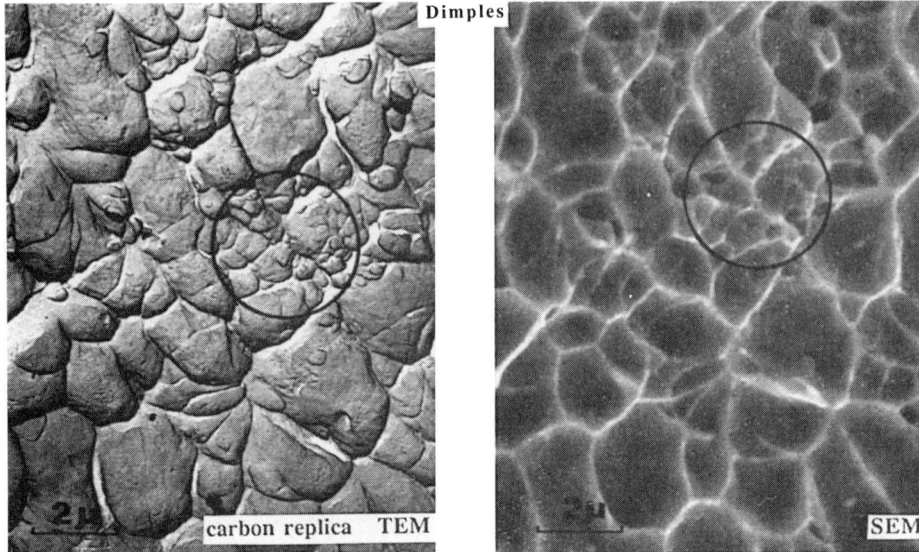

Acier C<0,08—Cr 17—Ni 12—traction à 20°C.

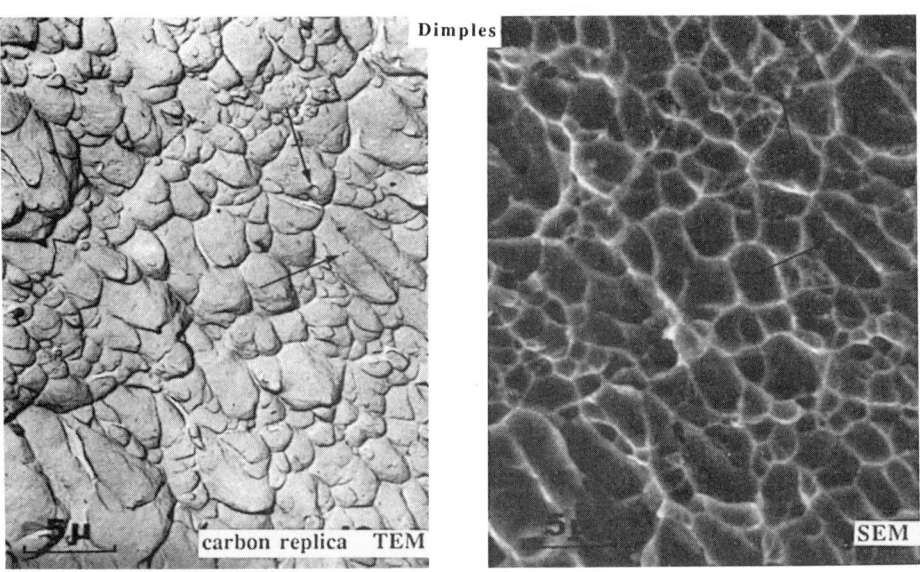

Acier C<0,08—Cr 17—Ni 12—traction à 20°C.

Figure 3.89 (*Continued*) Fractographs comparing the exact same areas as observed on a carbon replica in TEM (left) and in SEM (right). (*From Maillard, Meny, and Champigny.*[27])

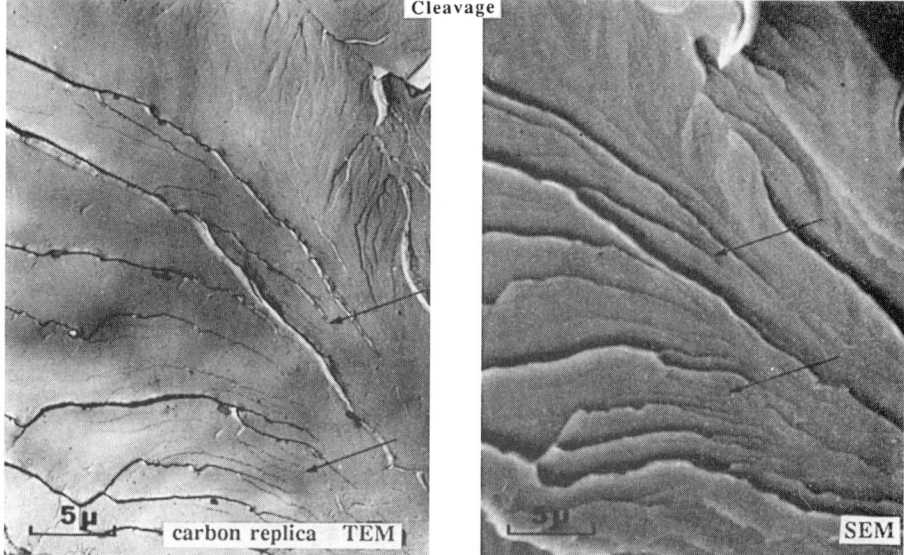

Fer pur cassé par choc à —196°C.

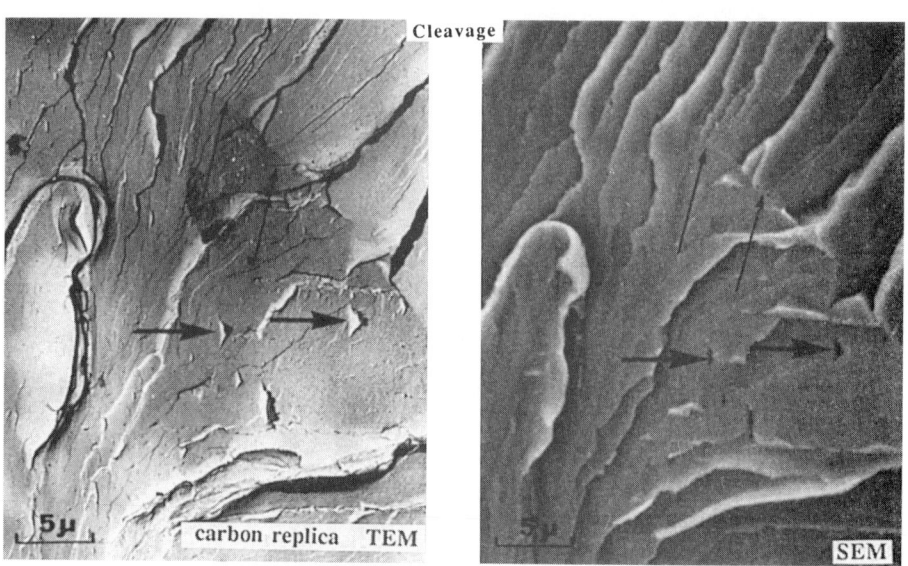

Fer pur cassé par choc à —196°C: origine des rivières ↑ ; Langrettes →.

Figure 3.89 (*Continued*) Fractographs comparing the exact same areas as observed on a carbon replica in TEM (left) and in SEM (right). (*From Maillard, Meny, and Champigny.*[27])

shows examples of the surface topography of the *same area* as imaged in SEM and TEM. Figure 3.90 contains stereo pictures of the same fracture surface (but not the exact same area) imaged with SEM and TEM.

3.16 Artifacts

There may be features on fracture surfaces which are artifacts introduced by sample preparation and which must be distinguished from true fracture features. Some of these are usually easily identifiable,

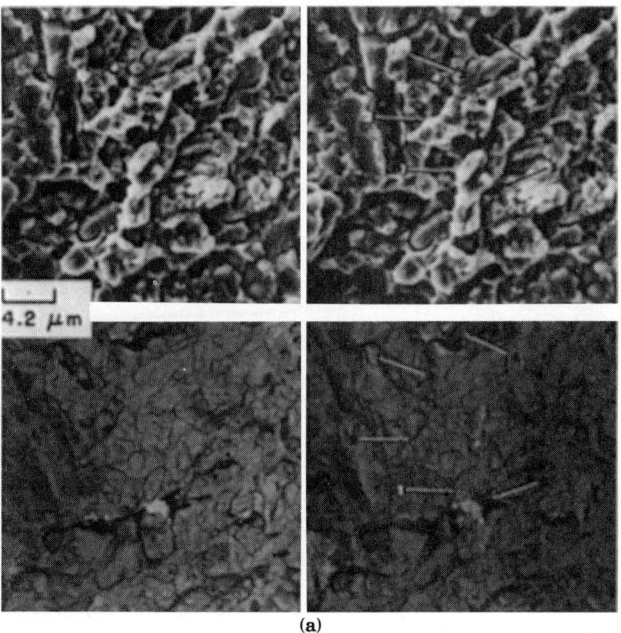

(a)

Figure 3.90 Stereo fractographs of various surfaces comparing the appearance of the identical area on the fracture surface as observed in TEM and in SEM. In each set of pictures, the upper pair are SEM micrographs, and the lower pair are TEM replica micrographs. (*a*) A room-temperature impact fracture in plain carbon steel after heat treatment at 800°C (1472°F) for 5 min and oil quenching. The specimen was fractured untempered, at a hardness of Rockwell C 65. The fracture surface exhibits dimples resulting from microvoid coalescence, and small local areas of quasicleavage facets. The roughly oval tip of a projection at arrow 1 in the SEM view can easily be identified in the TEM. A feature with a crudely triangular tip pointing to the left (arrow 2) is also readily recognized. The upper portion of this feature, however, appears much flatter in the TEM view than in the SEM, although local marks can be identified in each. Another site of conformance between the SEM and TEM stereofractographs is at arrow 3. Some details are lost at arrows 4 and 5 in the TEM pair, because of local tearing of the replica. (*From* Metals Handbook.[11])

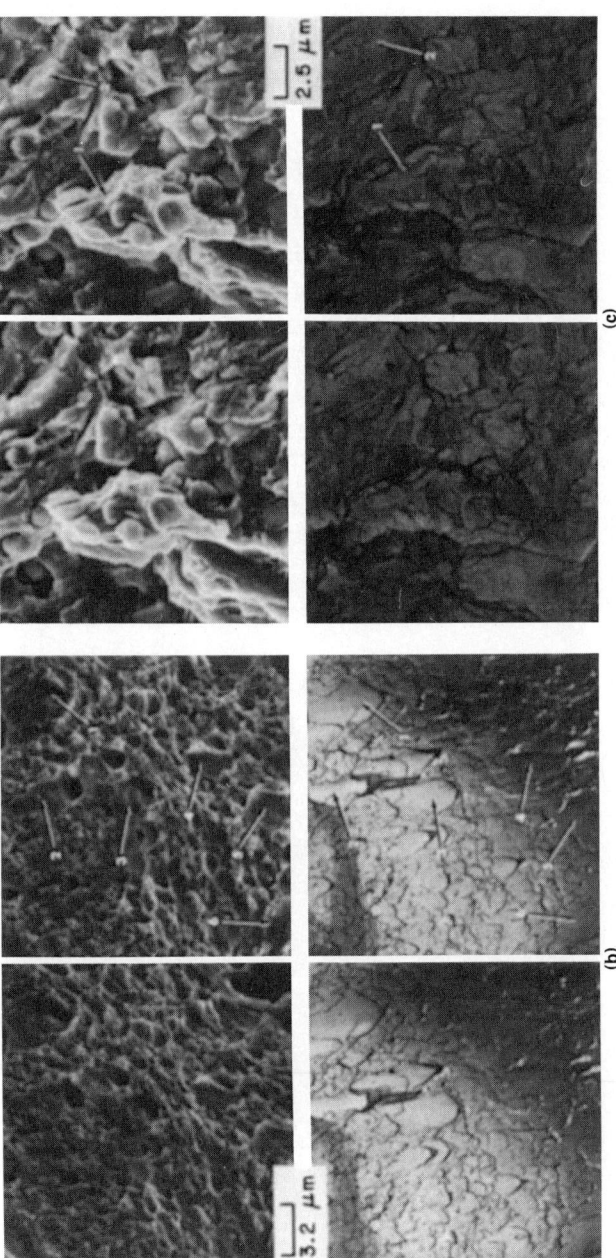

Figure 3.90 (*Continued*) (*b*) A room-temperature impact fracture in low-alloy manganese steel (0.60 to 0.70% C, 1.0 to 1.1% Mn) that was oil quenched from 900°C (1652°F) before being broken. The surface shows mainly shear dimples, some of which (at upper right in each stereo pair) are large enough to display areas of stretching. (It should be noted that a TEM replica of such large dimples may sag or collapse and, as in this comparison, introduce a very different appearance to the dimples from that displayed in an SEM view.) Specific dimples may be identified in each pair of views, and a number of these have been indicated by arrows 1 to 6. Observe, however, that the sense of depth, particularly in the larger dimples in the SEM pair, is quite lacking in the TEM. (*c*) This is a remarkably complex area, containing many small and varied quasicleavage facets and exhibiting a number of spheroidal inclusions. The most prominent feature is the high peak at arrow 1, which is reproduced in very faithful and identical detail in both stereo pairs. Note that the two inclusions between the twin tips of the peak are shown in both pairs, but that similar inclusions slightly nearer the top of the SEM pair are missing in the TEM because of damage to the replica. The top center area of the SEM view appears to record a greater number of surface complexities than does the same area in the TEM view; at arrow 2, however, the TEM view displays very fine fracture marks that cannot be *found* in the same location in the SEM view. (*From* Metals Handbook.[11])

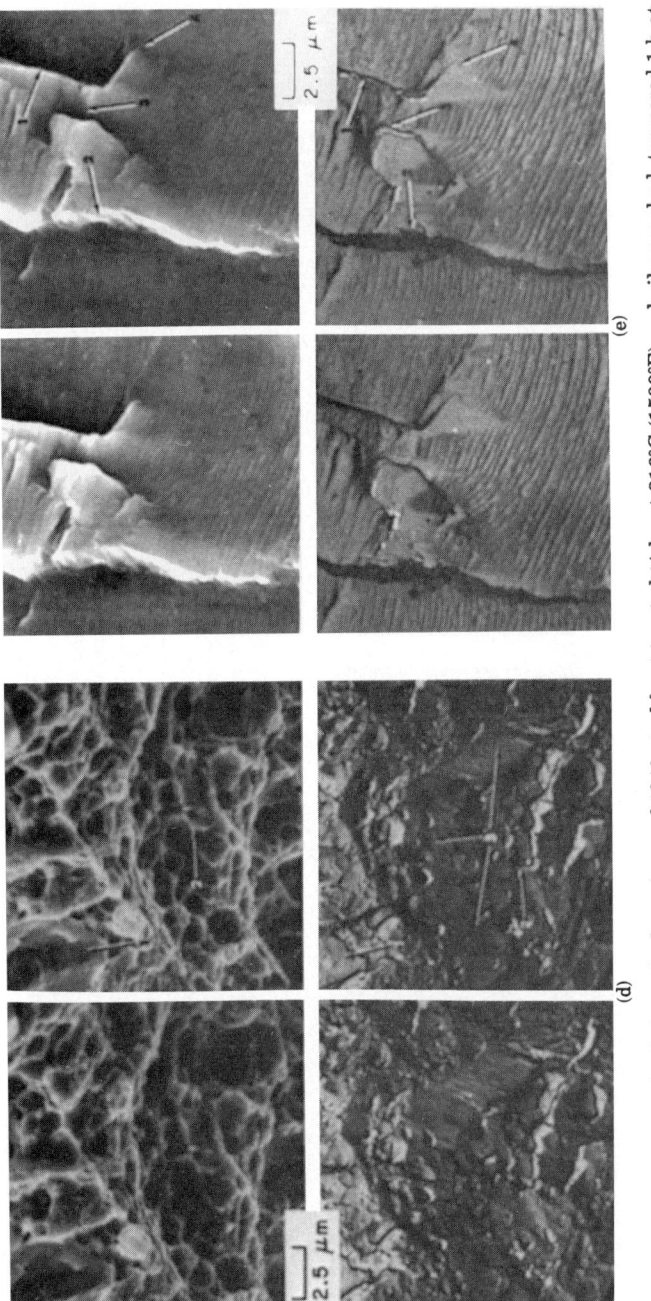

Figure 3.90 (*Continued*) (*d*) A notched specimen of 4340 steel heat treated ½ hr at 816°C (1500°F) and oil quenched, tempered 1 h at 260°C (500°F) and air cooled to give a hardness of Rockwell C 50, then broken by tension overload at room temperature. The surface is dimpled, with at least one area (at arrow 1) showing some evidence of stretching. At first glance, the TEM pair appears to bear little resemblance to the SEM, but careful scrutiny in 3-D of the sites at arrows 1 and 2 establishes that they are indeed identical. The major area of departure appears to lie in the regions marked 3 in the TEM pair. (*e*) A low-cycle fatigue-test fracture in aluminum alloy 7075-T6. The fracture surface shows well-defined fatigue striations that are clearly visible in both the SEM and the TEM pairs but with more detail and contrast in the TEM. Note the excellent match of the reproductions made by the two techniques. The vertical surface at arrow 1 is quite accurately registered in the TEM fractograph, but at the other side (at arrow 2) the replica has been torn. The detail in the crevices marked by arrows 1 and 3 is more clearly displayed in the TEM pair but could be matched in the SEM if the exposure were adjusted for this purpose. The sharp line at arrow 4 in the TEM fractograph bears some resemblance to a replica tear, but it is also visible in the SEM and actually is a secondary crack in the aluminum. (*From* Metals Handbook.[11])

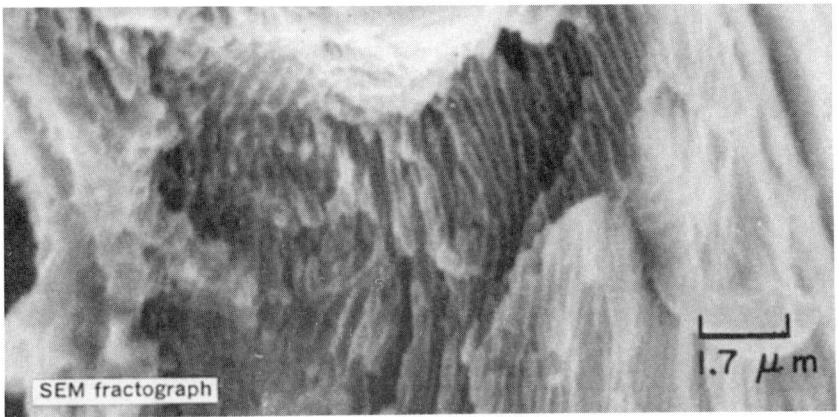

Figure 3.91 Fractograph showing what appear to be fatigue striations, but these are actually caused by fracture through the alternate lamellar layers of ferrite and iron carbide in pearlite. Note that in several local areas the pearlite lamellae have been separated by secondary cracks. (*From* Metals Handbook.[11])

such as debris and corrosion products. Others can cause false interpretation of the fracture process.

Figure 3.91 shows a fracture surface which includes regions containing parallel lines. These are similar to striations, but in this case they are caused by fracture through the parallel lamellae of ferrite and iron carbide in pearlite. Other features which can be mistaken for fatigue striations are *Wallner lines* (Fig. 3.92) (sometimes found on

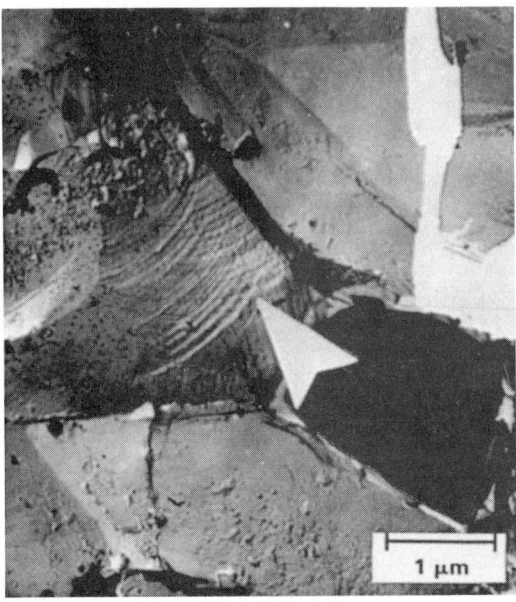

Figure 3.92 TEM replica of Wallner lines, a feature which sometimes occurs on the fracture surface of very brittle materials (such as glass). The specimen was WC–Co. Etched with 5% HCl. (*Courtesy of S. B. Luyckx, University of Witwatersrand; from Kerlins and Phillips.*[9])

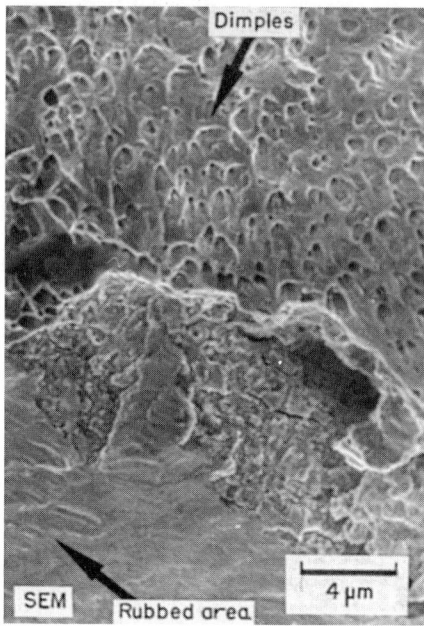

Figure 3.93 Fractograph showing smooth regions which are rub marks, due to contact of the two surfaces after fracture occurred.

fracture surfaces of brittle materials and phases, but their origin is not well established), slip traces (Figs. 3.39 and 3.23), and tire tracks (Fig. 3.67).

Figure 3.93 shows a fractograph in which smooth regions appear. This is a true feature of the fractured surface; however, it is not associated with the fracture process but was formed by rubbing of the two surfaces together (rub marks) after the fracture surfaces formed. This is a common feature in fatigue fractures. Figure 3.94 illustrates a feature called *mud cracks*. This is a region of deposited corrosion products and not an actual feature of the fracture process. Examination of the fracture surface in between repeated cleaning (for example, with replication tape; see Sec. 1.4.2) will allow detection of such artifacts.

In the preparation of TEM replicas, a number of artifacts occur which have been recognized. One problem is tearing, stretching, or scraping of the plastic replica on removal from the fracture surface. Examples of the effect of these on fractographs are shown in Fig. 3.95. Since the replica is usually shadowed with a heavy metal to improve contrast (see Sec. 1.4.2), it is possible to overheat the plastic replica in this step. Testing of the consistency of the appearance of successive replicas will usually isolate this effect. Excessive heating of the replica by the electron beam may cause the heavy metals

(a)

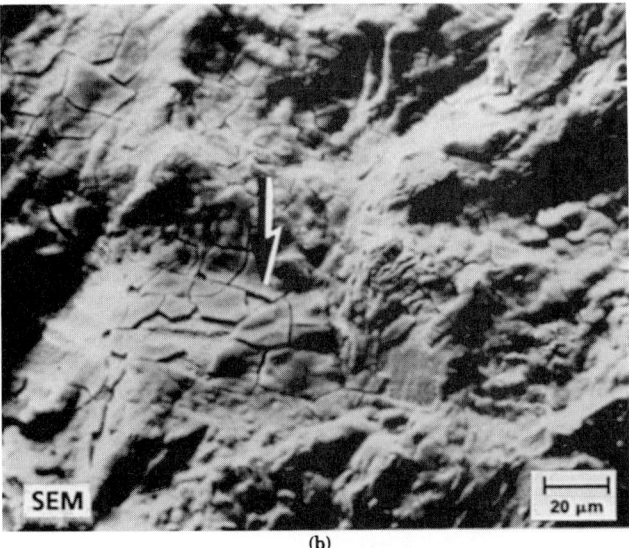

(b)

Figure 3.94 Fractographs showing mud cracks, a feature associated with corrosion products on the fracture surface. (a) Surface of an intergranular fracture in aluminum alloy 7079-T651 caused by stress-corrosion cracking in a 3½% NaCl solution. (*From Meyn.*[28]) (b) Fracture surface of 316L austenitic stainless-steel surgical implant, broken in service in body fluid. (*From Metals Handbook.*[26])

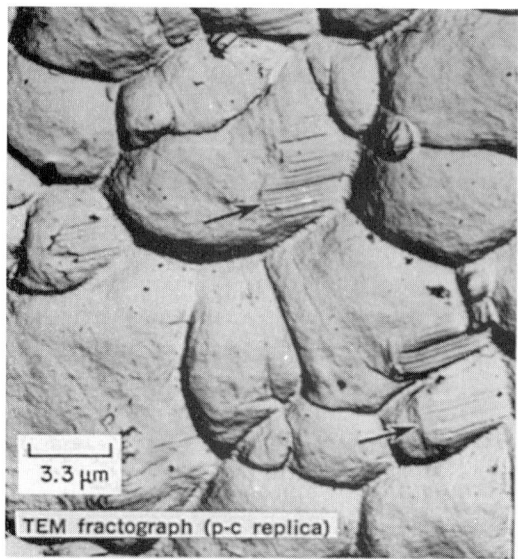

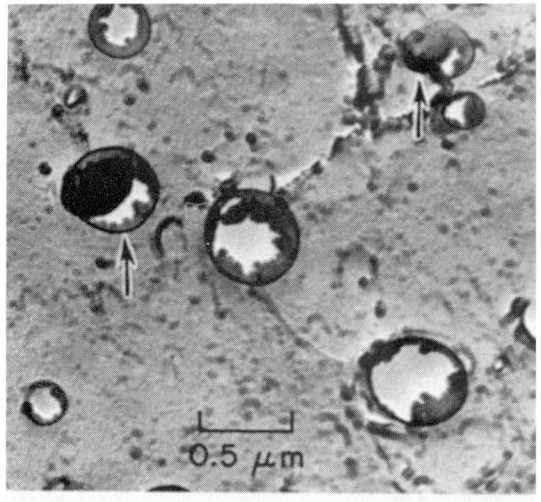

Figure 3.95 Examples of artifacts in TEM replicas. (*a*) Scraping artifact frequently encountered in the two-stage process. (*b*) The thin film of plastic between the bubble and the replicated surface is seen to be partly removed from theplastic replica at the bubbles indicated by arrows.

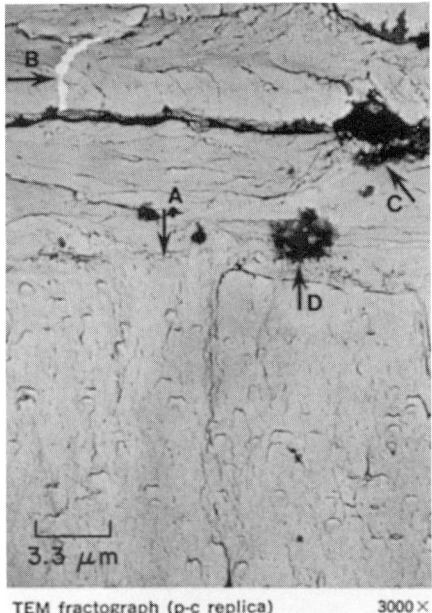

TEM fractograph (p-c replica)　　3000×
(c)

Figure 3.95 (*Continued*) Examples of artifacts in TEM replicas. (c) The appearance of the bottom half of the fractograph (below arrow A) is the result of tears in the first-stage plastic replica, and is not at all representative of the fracture surface. A crack in the carbon second-stage replica is indicated by arrow B; a particle from the improperly cleaned fracture surface is indicated by arrow C, and residual plastic from incomplete dissolution of the first stage is indicated by arrow D. (*From* Metals Handbook.[11, 26])

used for shadowing (see Sec. 1.42) to become reticulated (patterned), as shown in Fig. 3.96. Again, whether such an appearance is an artifact can be determined by the preparation of successive replicas. A problem in two-stage plastic replica preparation (see Sec. 1.42) is that the plastic may not be completely removed from the carbon film. The appearance of the residual plastic film in TEM is shown in Fig. 3.97. Even if the replica contains no artifacts after removal from the surface to be replicated, the carbon film, which is very thin (such as 200 Å), may tear and will give an appearance similar to that illustrated in Fig. 3.98; however, this artifact is usually easy to recognize.

Figure 3.96 Example of reticulation of a replica caused by excessive heating in the electron beam of the TEM. This was a Pd-shadowed plastic-carbon replica of a fracture in nickel showing dimples. Reticulation was caused by melting of the shadowing metal in the microscope and the formation of globules. (*From* Metals Handbook.[26])

TEM fractograph (p-c replica)   22,500×

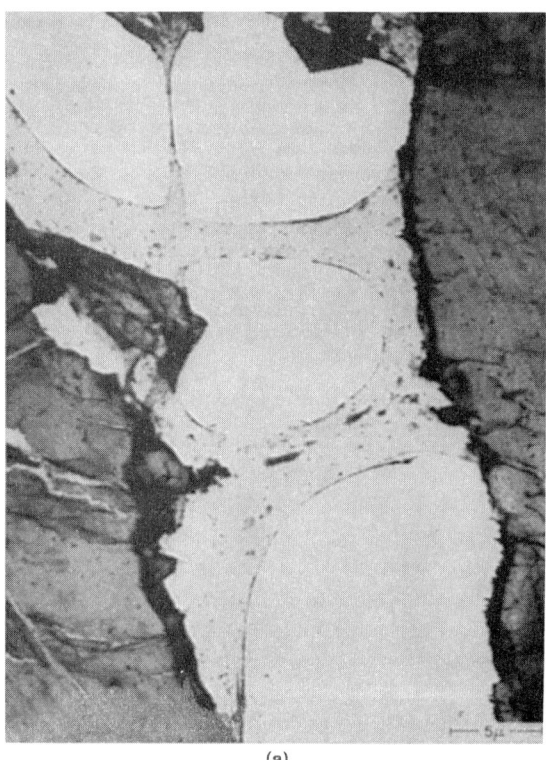

(a)

Figure 3.97 TEM replicas showing incompletely removed plastic from a carbon film. (*From Beachem.*[8])

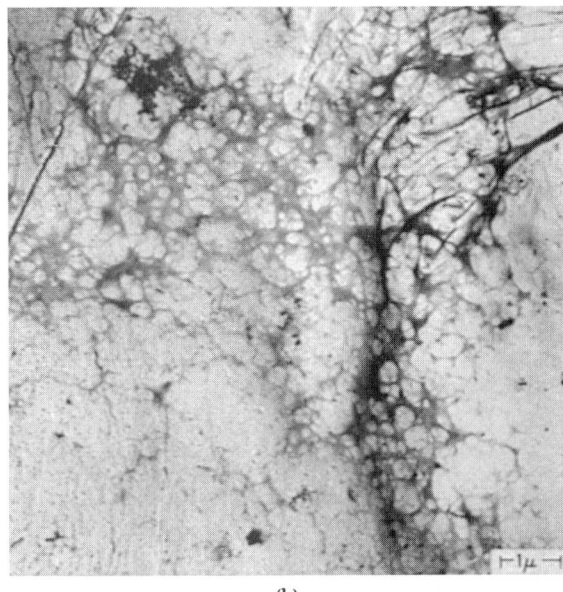

(b)

Figure 3.97 (*Continued*) TEM replicas showing incompletely removed plastic from a carbon film. (*From Beachem.*[8])

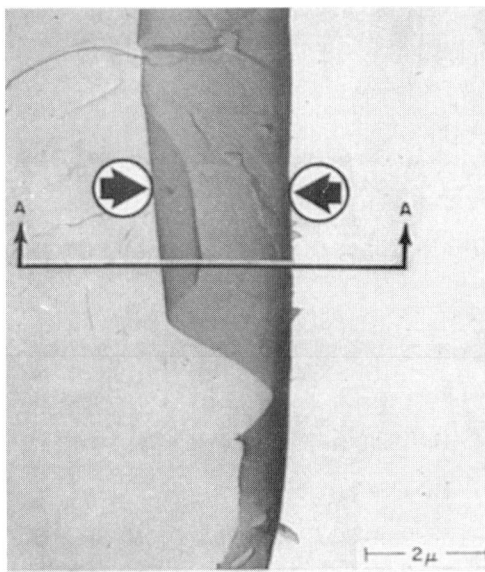

Figure 3.98 TEM replica showing the effect of curling of the carbon film. This was a direct carbon replica of a flat fracture surface (iron cleaved at dry-ice temperature). A region of gradually changing density is shown between the arrows. (*From Beachem.*[8])

References

1. W. Boas and E. Schmid, "Über die Temperaturabhängigkeit der Kristallplastizität," *Z. Physik*, vol. 61, p. 767 (1930). (Reprinted with permission of Springer-Verlag, Heidelberg, Germany.)
2. A. Guy, *Introduction to Materials Science*, McGraw-Hill, New York, 1972.
3. J. D. Verhoeven, *Fundamentals of Physical Metallurgy*, Wiley, New York, 1975. (Reprinted with permission of John C. Wiley & Sons, Inc.)
4. C. E. Birchenall, *Physical Metallurgy*, McGraw-Hill, New York, 1959.
5. C. S. Roberts, *Magnesium and Its Alloys*, Wiley, New York, 1960.
6. C. R. Brooks, *Heat Treatment, Structure and Properties of Non-Ferrous Alloys*, American Society for Metals, Metals Park, Ohio, 1982.
7. G. Henry and D. Horstmann, *De Ferri Metallographia*, vol. V: *Fractography and Microfractography*, Verlag Stahleisen m.b.H., Düsseldorf, Germany, 1979.
8. C. D. Beachem, "Microscopic Fracture Processes," in H. Liebowitz (ed.), *Fracture*, vol. 1: *Microscopic and Macroscopic Fundamentals*, Academic Press, New York, 1968.
9. V. Kerlins and A. Phillips, "Modes of Fracture," in *Metals Handbook*, 9th ed., vol. 12: *Fractography*, American Society for Metals, Metals Park, Ohio, 1987, p. 12.
10. L. Engel and H. Klingele, *An Atlas of Metal Damage*, Carl Hanser Verlag, Munich, Germany, 1981.
11. *Metals Handbook*, 8th ed., vol. 9: *Fractography and Atlas of Fractographs*, American Society for Metals, Metals Park, Ohio, 1974.
12. D. Broek, "Some Contributions of Electron Fractography to the Theory of Fracture," *Int. Met. Rev.*, vol. 19, p. 135 (1974).
13. S. P. Lynch, "Ductile and Brittle Crack Growth: Fractography, Mechanisms and Criteria," *Mat. Forum*, vol. 11, p. 268 (1988).
14. B. L. Gabriel, "Scanning Electron Microscopy," in *Metals Handbook*, 9th ed., vol. 12: *Fractography*, American Society for Metals, Metals Park, Ohio, 1987, p. 166.
15. W. D. Syniuta and C. J. Corrow, "Scanning Electron Microscopic Studies of Fracture Mechanisms of SAE 52100 Bearing Steel," *Wear*, vol. 15, p. 171 (1970).
16. A. W. Thompson and J. C. Chesnutt, "Identification of a Fracture Mode: The Tearing Topography Surface," *Met. Trans.*, vol. 10A, p. 1193 (1979).
17. J. C. Grosskreutz, "Fatigue Mechanisms in the Sub-Creep Range," in S. S. Manson (ed.), *Metal Fatigue Damage—Mechanism, Detection, Avoidance, and Repair*, American Society for Testing and Materials, Philadelphia, Pa., 1971.
18. C. Laird, "The Influence of Metallurgical Structure on the Mechanisms of Fatigue Crack Propagation," in *Fatigue Crack Propagation*, American Society for Testing and Materials, Philadelphia, Pa., 1967.
19. H. Verhoff and P. Newmann, "In situ SEM Experiments Concerning the Mechanism of Ductile Crack Growth," *Acta Met.*, vol. 27, p. 915 (1979).
20. W. L. Haworth, V. K. Singh, and R. K. Mueller, "Holographic Detection of Fatigue-Induced Surface Deformation and Crack Growth in a High-Strength Aluminum Alloy," *Met. Trans.*, vol. 11A, p. 219 (1980).
21. C. A. Zapffe and C. O. Worden, "Fractographic Registrations of Fatigue," *Trans. ASM*, vol. 43, p. 958 (1951).
22. C. D. Beachem, "Microscopic Fatigue Fracture Surface Features in 2024-T3 Aluminum and the Influence of Crack Propagation Angle Upon Their Formation," *Trans. ASM*, vol. 60, p. 324 (1967).
23. P. N. Thielen and M. E. Fine, "Fatigue Crack Propagation in 4140 Steel," *Met. Trans.*, vol. 6A, p. 2133 (1975).
24. H. Böhm, *Einführung in die Metallkunde*, Bibliogr. Institut, Mannheim, Germany, 1968.
25. D. A. Meyn and E. J. Brooks, "Microstructural Origin of Flutes and Their Use in Distinguishing Striationless Fatigue Cleavage from Stress-Corrosion Cracking in Titanium Alloys," in L. N. Gilbertson and R. D. Zipp (eds.), *Fractography and Materials Science*, American Society for Testing and Materials, Philadelphia, Pa., 1981. (Copyright © ASTM; reprinted with permission.)

26. *Metals Handbook,* 9th ed., vol. 12: *Fractography,* ASM International, Metals Park, Ohio, 1987.
27. A. Maillard, L. Meny, and M. Champigny, "Comparaison de Microfractographies Types Obtenues par Microscopie à Balayage et par Microscopie Conventionnelle," *Micron,* vol. 2, p. 290 (1971).
28. D. A. Meyn, "Fractographic Diagnosis of Stress Corrosion Cracking of Al–Zn–Mg Alloys," *Corrosion,* vol. 26, p. 427 (1970).

Bibliography

Among the books and articles that cover microfractography, the following have been found to be especially useful. From these, and from the references cited in this chapter, other references can be traced.

Bhattacharyya, E. S., V. E. Johnson, S. Agarwal, and M. A. H. Howes (eds.), *IITRI Fracture Handbook, Failure Analysis of Metallic Materials by Scanning Electron Microscopy,* IIT Research Institute, Chicago, Ill., 1979.

Gilbertson, L. N., and R. D. Zipp (eds.), *Fractography and Materials Science,* STP 733, American Society for Testing and Materials, Philadelphia, Pa., 1981.

Hertzberg, R. W., "Fracture Surface Micromorphology in Engineering Solids," in J. E. Masters and J. J. Au (eds.), *Fractography of Modern Engineering Materials: Composites and Metals,* STP 948, American Society for Testing and Materials, Philadelphia, Pa., 1987, pp. 5–36.

Jacoby, G., "Application of Microfractography to the Study of Crack Propagation under Fatigue Stresses," AGARD Rep. 541, North Atlantic Treaty Organization, London, 1966.

Lange, G. A. (ed.), *Systematic Analysis of Technical Failures,* DGM Informationsgesellschaft-Verlag, Braunschweig, Germany, 1986.

Louthan, M. R., Jr., and T. A. Place (eds.), *Microscopy, Fractography and Failure Analysis,* Failure Analysis and Prevention Lab., Virginia Polytechic Institute, Blacksburg, Va., 1986.

Phol, M., "Material Failure through Wear," in G. A. Lange (ed.), *Systematic Analysis of Technical Failures,* DGM Informationsgesellschaft-Verlag, Braunschweig, Germany, 1986.

Rice, R. C. (ed.), *Fatigue Design Handbook,* 2d ed., Society of Automotive Engineers, Warrendale, Pa., 1988.

Ryder, D. A., "The Elements of Fractography," AGARD Rep. 155, North Atlantic Treaty Organization, London, 1971.

Tucker, R. C., Jr., "Wear Failures," in *Metals Handbook,* 9th ed., vol. 11: *Failure Analysis and Prevention,* American Society for Metals, Metals Park, Ohio, 1986.

Whiteson, R. V., A. Phillips, and V. Kerlins, "Electron Fractographic Techniques," in R. F. Bunshah (ed.), *Techniques for the Direct Observation of Structure and Imperfections,* Interscience, New York, 1968, pp. 445–497.

Wulpi, D. J., *How Components Fail,* American Society for Metals, Metals Park, Ohio, 1985.

Chapter

4

Fracture Modes and Macrofractographic Features

Persian SIGLOS, fifth century B.C., showing impression of fractured bronze (?) punch × 4.
C. S. SMITH
A History of Metallography,
University of Chicago Press,
Chicago, 1960

4.1 Introduction

When the fracture surface of a broken component is examined with the unaided eye or at low magnification (such as <20×), distinct topographical features are usually seen. Deduction of the mode of fracture based on this macroscopic examination relies on the correlation of characteristic features with known fracture conditions. Fortunately there exists a wealth of experimental observations of the fracture surfaces of test samples fractured under known loading conditions, and in most cases the topographical features are quite unique for each type of loading condition. The orientation of the fracture surface of a broken component often reveals unambiguously the loading condition which led to fracture, and this was examined in Chap. 2. The microscopic process whereby metallic materials fracture was called mechanism (or micromechanism), and the associated fracture-surface fine-scale topography was the subject of Chap. 3.

In Chap. 4 the types of fracture mode and the associated macroscopic fracture-surface topographical features are examined. The term *fracture mode* is used to indicate the type of external loading that causes fracture; this could be called fracture macromechanism. The simplest modes are described first (such as simple tensile loading), then more complex cases are examined. We conclude the chapter by illustrating the microfractography of test samples of known fracture modes which have specific macrofractographic appearances.

4.2 Tensile Overload

In the tensile overload mode it is assumed that the external load is uniaxial, and that it is monotonically increasing until the material begins to neck. For samples tested in a conventional tensile tester, the test is conducted at approximately a constant extension rate, so that after necking the load actually decreases until fracture (see Sec. 2.2).

The fracture-surface appearance depends on the type of material, its microstructure, and the testing conditions (such as test temperature and strain rate). Under tensile loading, the material may show extensive plastic deformation prior to fracture, in which case it is considered to be ductile (see Sec. 2.11). If little plastic deformation occurs, the fracture is called brittle. However, the amount of plastic deformation which separates these two categories is somewhat arbitrary, although less than 5 percent elongation at fracture in a tensile test is a convenient value to consider a material brittle (see Sec. 2.11).

First we consider a ductile overload mode. In this case there is extensive elongation at fracture, and the cross section will be decreased in the vicinity of the fracture. In a cylindrical test sample, this usually results in the fracture appearance shown in Fig. 4.1. There is a central rough region whose overall plane is normal to the loading axis. On the

Fracture Modes and Macrofractographic Features   215

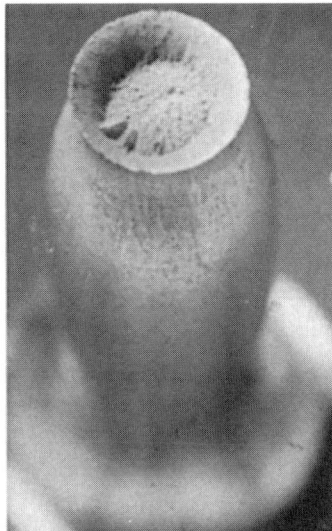
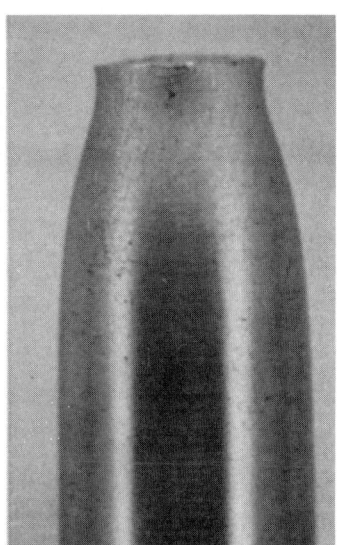

Figure 4.1 Broken tensile sample showing ductile fracture with the common cup-and-cone appearance. (*After Vander Voort.*[1])

periphery is a surface inclined at 45° to this plane, called a *shear lip*. This type of fracture is called *cup-and-cone fracture*. The origin of such a macroscopic geometry was discussed in Sec. 2.8. It was pointed out there that in the tensile test, once necking begins, the triaxial stress which develops favors plastic deformation by shear in the center of the plane of the minimum cross section (at the neck location). Thus the examination of cross sections on the axial plane of such samples loaded to near fracture shows that fracture initiates in the center of the cylinder (Fig. 4.2). When the cracks develop sufficiently, the intense triaxiality at the periphery of the internal crack causes the development of large shear stresses at 45° in the outer ring of connecting material, and this region then fractures along a general plane at this angle, forming the shear-lip region. Tensile test samples of rectangular cross section which fracture in a ductile manner also show similar characteristics (Fig. 4.3).

The fracture surface of such necked samples usually looks similar to that in Fig. 4.4. In the center there is a *fibrous zone*, so called because its appearance is similar to that of a broken fiber structure. Emanating from the outer periphery of this zone are somewhat coarser markings, called *radial marks*. These terminate at the outer circumferential zone containing the shear lip. The extent of each zone depends on the material, its structure, and the test conditions. Examples are

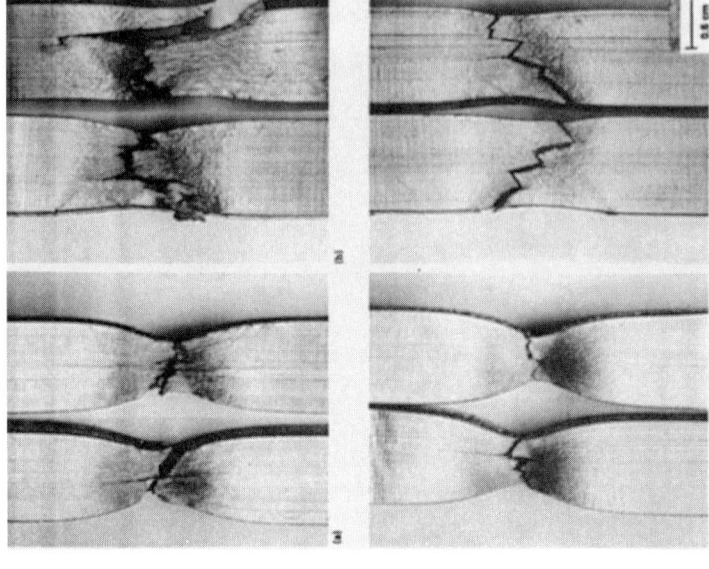

Figure 4.3 Broken rectangular steel tensile samples showing necking. (a) Longitudinal specimen finished rolled at 315°C (600°F). (b) Transverse specimen. (c) Longitudinal specimen finished rolled at 150°C (300°F). (d) Transverse specimen. (*From B. L. Bramfitt and A. R. Marder, Met. Trans., vol. 8A, p. 1262, 1977; after Vander Voort.*[1])

Figure 4.2 Cross section through a tensile sample just prior to fracture. Note that the fracture crack has initiated in the center of the sample. (*After Vander Voort.*[1])

216

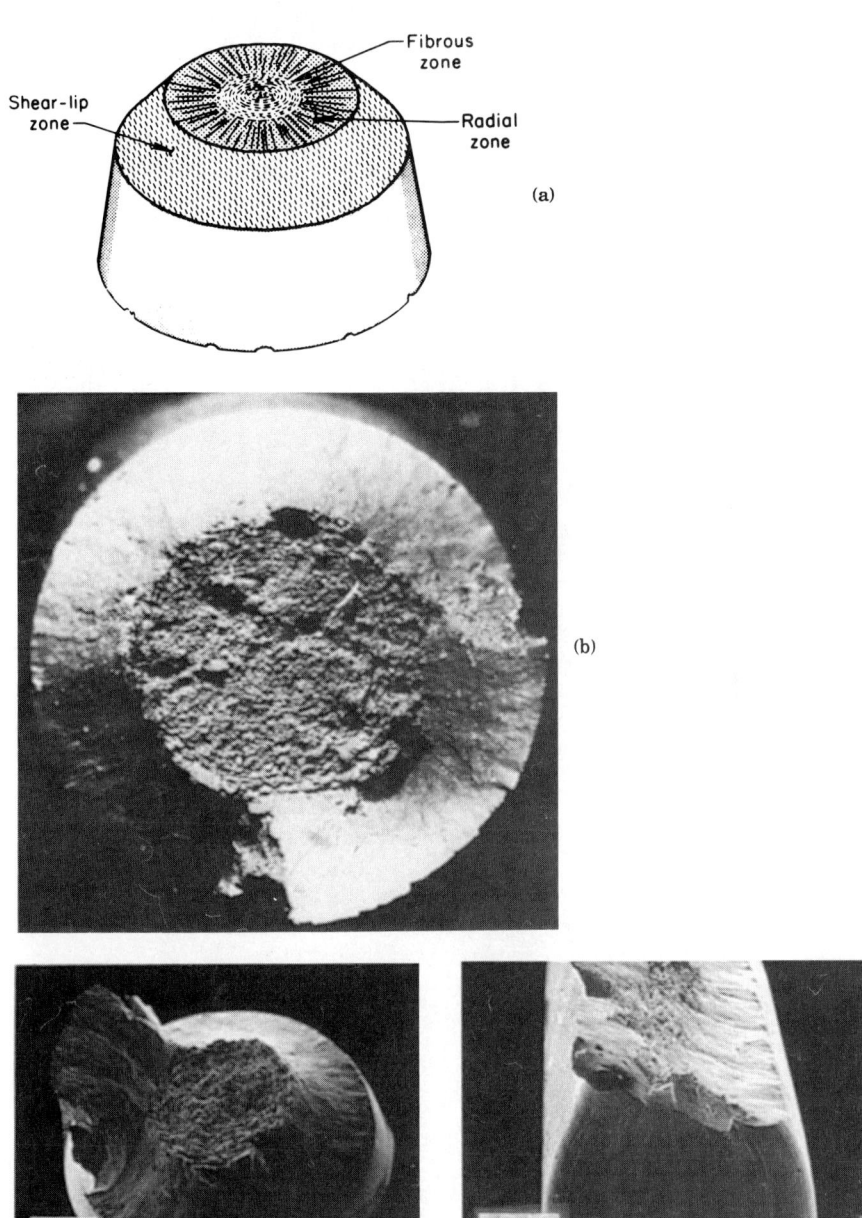

Figure 4.4 (*a*) Features of the fracture surface of a ductile tensile sample. The surfaces of the fibrous and radial zones are usually normal to the tensile axis; the shear-lip surface is always at about 45° to the tensile axis. (*From* Metals Handbook.[2]) (*b*) Fracture surface of a broken ductile tensile sample of 4340 steel tested at 120°C (248°F). The structure was tempered martensite, and the hardness was 46 Rockwell C. (*From* Metals Handbook.[2]) (*c*) Ductile cup-and-cone fracture of a tensile sample of a maraging steel (0.02% C, 0.1% Si, 0.1% Mn, 13% Cr, 8% Ni, 5% Co, 0.8% Ti). (*From Engel and Klingele.*[3]) (*d*)Necking and shear fracture in a rectangular tensile test sample made from a maraging NiCoMo-containing steel. (*From Engel and Klingele.*[3])

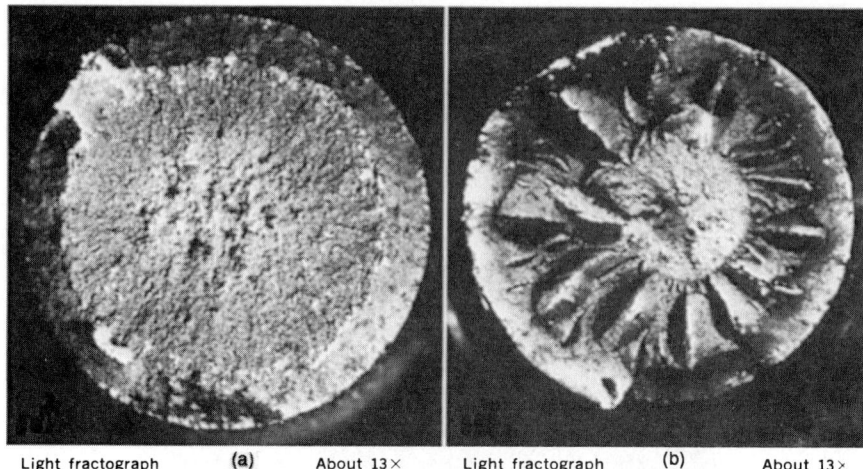

Light fractograph   (a)   About 13×   Light fractograph   (b)   About 13×

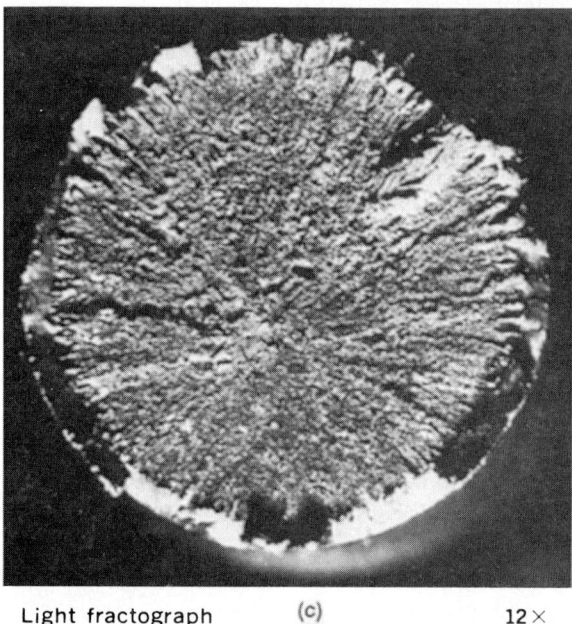

Light fractograph   (c)   12×

Figure 4.5 Fracture surfaces of broken 4340 steel tensile samples, showing the variations in topographical features indicated in Fig. 4.4a. (a) Pearlitic structure, hardness 15 Rockwell C. Inner zone is randomly fibrous; outer surrounding zone (inside the outer shear lip) has radial marks. (b) Tempered martensite structure, hardness Rockwell C 28. Inner, fibrous zone is circumferentially ridged; intermediate zone has coarse radial marks; outer ring is shear-lip zone. (c) Tempered martensite, tested at −196°C (−321°F). No fibrous zone, only radial marks. (*From* Metals Handbook[2])

Fracture Modes and Macrofractographic Features   219

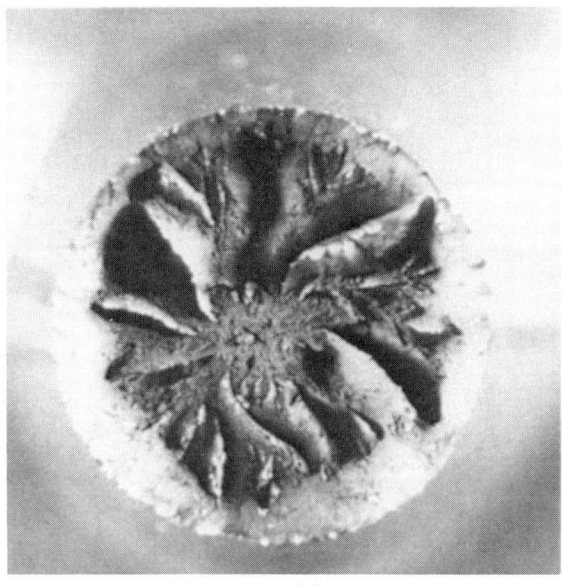

Light fractograph   (d)   14 ×

Figure 4.5 (*Continued*) Fracture surfaces of broken 4340 steel tensile samples, showing the variations in topological features indicated in Fig. 4.4a. (*d*) Tempered martensite, hardness Rockwell C 28. Curved coarse radial shear marks. (*From* Metals Handbook[2])

shown in Fig. 4.5. In general, as the strength increases, the amount of necking is less, and there is a reduction in the extent of the shear-lip zone. Also, the central fibrous zone covers less area and the radial marks are less coarse. These effects are illustrated in Fig. 4.6. Similar effects are seen as the test temperature decreases, as shown in Fig. 4.7. The general features for rectangular and square cross-sectioned tensile samples are presented in Fig. 4.8; they are seen to be similar to those of cylindrical samples.

If the sample is brittle, there is little macroscopic plastic deformation and fracture usually occurs on a plane which is macroscopically normal to the tensile load axis. The appearance of such tensile samples is shown in Fig. 4.9. On the fracture surface there is little shear lip and the fibrous zone is small. The radial marks are finer, but their direction still indicates the origin of fracture.

The macroscopic fracture topography is affected by constraints on the sample. As the width-to-thickness ratio increases, the relative amount of the fracture surface containing the shear-lip zone increases also

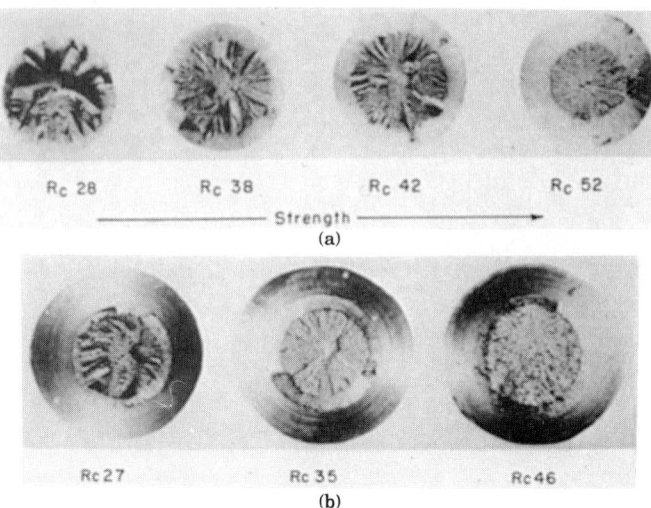

Figure 4.6 Fracture surfaces of broken tensile samples of the same steel heat-treated to different strengths. (*a*) Unnotched specimens. (*b*) 0.100-in notch-radius specimens. (Fracture originates at the center for these specimens as the notch is not sharp enough to initiate fracture at the notched surface.) (*After Nunes, Carr, and Larson.*[4] Copyright © 1968. Reprinted by permission of John Wiley & Sons, Inc.)

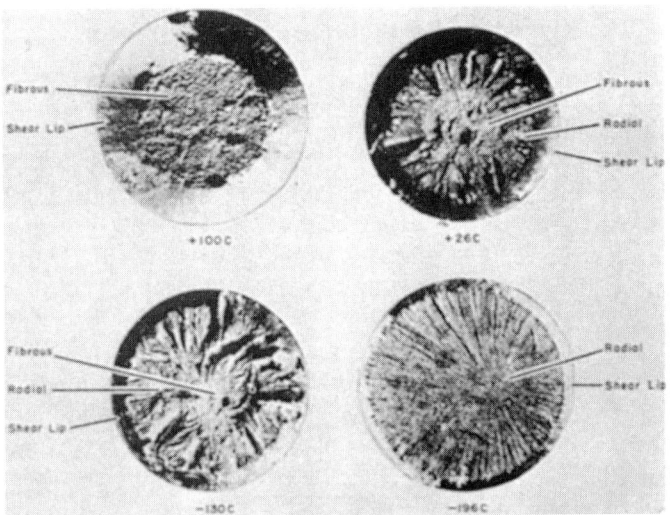

Figure 4.7 Fracture surfaces of tensile samples of the same material broken at different temperatures. (*From Nunes, Carr, and Larson.*[4] Copyright © 1968. Reprinted by permission of John Wiley & Sons, Inc.)

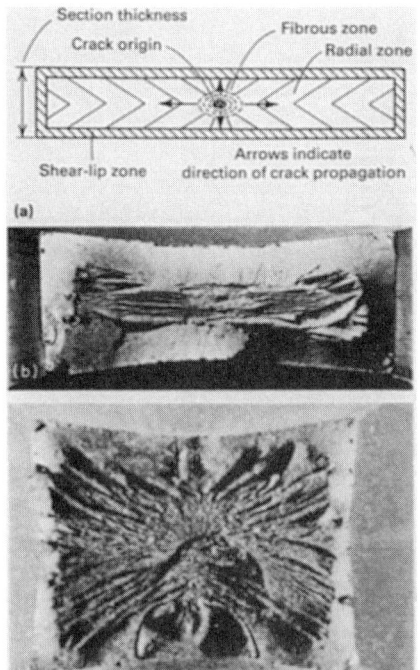

Figure 4.8 Fracture surfaces of broken rectangular tensile samples. (a) Schematic representation of tensile fracture features. (b) and (c) show that the features are affected by the width-to-length ratio; (c) shows a much smaller shear lip zone. (*From Nunes, Carr, and Larson*[4]; *adapted in* Metals Handbook.[5] Copyright © 1968. Reprinted by permission of John Wiley & Sons, Inc.)

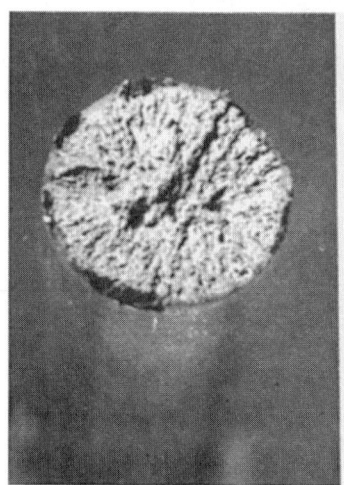

Figure 4.9 Broken tensile sample showing brittle fracture. (*After Vander Voort.*[1])

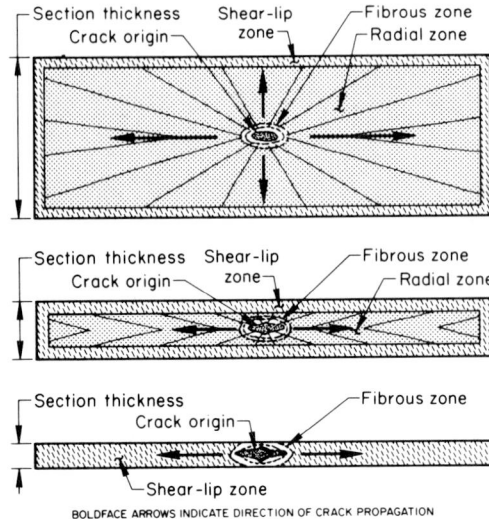

Figure 4.10 Schematic illustration of the fracture-surface features of plate samples of different thicknesses broken in tensile loading. (*From Nunes, Carr, and Larsen*[4]; *adapted in* Metals Handbook.[5] Copyright © 1968. Reprinted by permission of John Wiley & Sons, Inc.)

(Fig. 4.10). Of special importance is that the radial marks develop into a pattern called *chevron marks* (Fig. 4.8), which point back toward the origin of the fracture. Thus they are a very important feature in reconstructing the fracture path and locating the fracture origin. An example of the fracture surface of a plate showing the chevron pattern is given in Fig. 4.11.

Constraints induced by notching the tensile sample affect the fracture topography. Due to the intense triaxiality at the root of the notch, the cracks will initiate in this region and not in the center, and then progress inward. The fracture features usually observed are indicated schematically in Fig. 4.12a. Note that the fibrous zone is on the outside, compared to the case of an unnotched sample (Fig. 4.4), and there is no shear lip. The radial marks emanate from the periphery of the fibrous zone, in toward the final fracture region at the center. How-

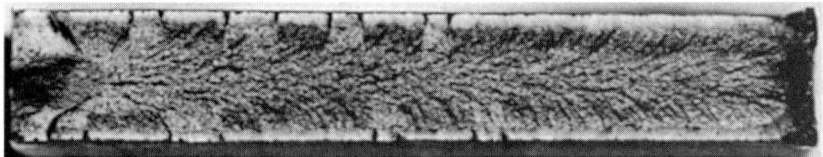

Figure 4.11 Fracture surface of a broken plate which shows radial marks in a chevron pattern, pointing back toward the origin of the fracture, which is at the left end of the specimen. (*From* Metals Handbook.[2])

Fracture Modes and Macrofractographic Features 223

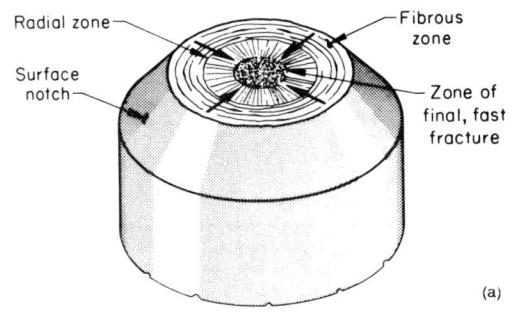

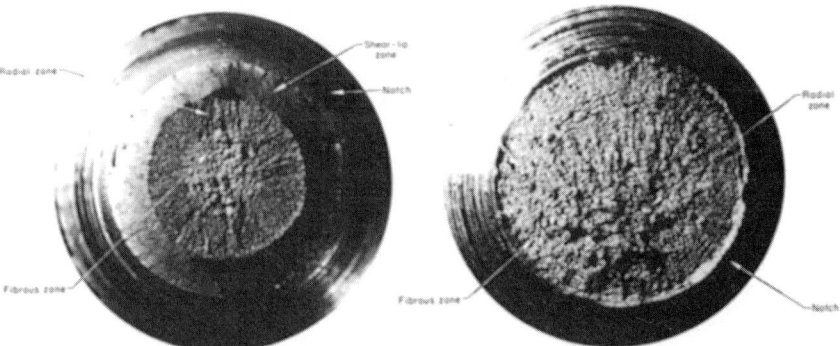

Figure 4.12 (a) Schematic illustration of the fracture surface zones of a notched tensile specimen. There is no shear lip, and the final fracture zone is in the center. (Compare to Fig. 4.4a.) (b) Fracture surfaces of notched tensile samples showing the effect of notch radius on the fracture appearance. The material is 4340 steel, tested at −40°C (−40°F). Left: notch-root radius 0.1 in., tensile strength 1544 MPa (224 ksi). Right: notch-root radius 0.01 in., tensile strength 1758 MPa (255 ksi). (*From* Metals Handbook.[2])

ever, these radial marks are similar in appearance to those for unnotched samples (Fig. 4.4). Similar features are found in tensile tests of notched plates (Fig. 4.13). Figure 4.14 shows the effect of the test temperature on the appearance of the fracture surface of notched tensile samples.

Figure 4.15 illustrates two sets of bolts broken by tensile overload. The bolt on the right in Fig. 4.15a is ductile, showing necking and considerable elongation; the one on the left broke in a brittle manner with little elongation. In Fig. 4.15b, the bolt on the left has a cup-and-cone fracture surface, and the distance between the threads increases as the fracture surface is approached, showing increased axial elongation. The bolt on the right in Fig. 4.15b broke in a brittle manner with no obvious macroscopic dimensional change, and the fracture plane is relatively smooth and normal to the tensile load axis.

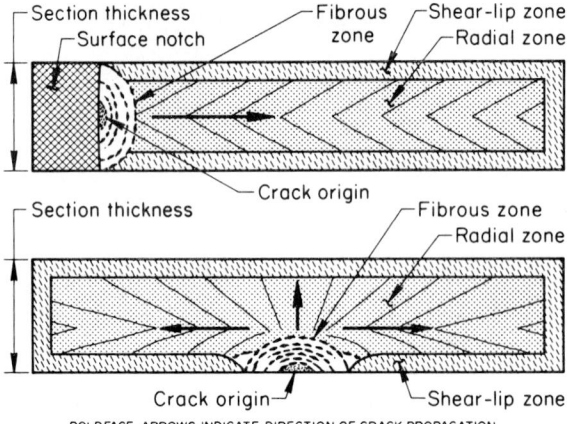

Figure 4.13 Schematic illustration of the fracture-surface features for notched rectangular tensile samples. (Compare to Fig. 4.10.) A fibrous zone is generated from the notch and extends for a short distance. A radial zone then supplants the fibrous zone, and a shear-lip zone is formed wherever the radial zone approaches the edge of the specimen. (*From Nunes, Carr, and Larson*[4]; *adapted in* Metals Handbook.[5] Copyright © 1968. Reprinted by permission of John Wiley & Sons, Inc.)

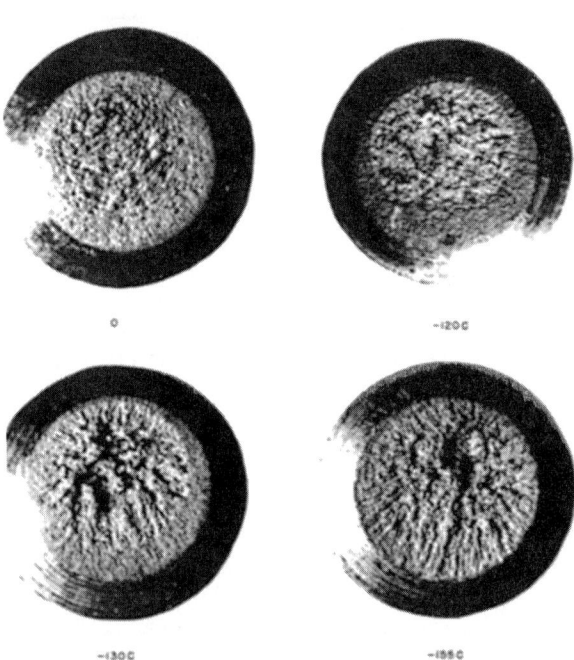

Figure 4.14 Fracture surfaces of notched tensile samples of the same material broken at different temperatures. (*From Nunes, Carr, and Larson.*[4] Copyright © 1968. Reprinted by permission of John Wiley & Sons, Inc.)

Fracture Modes and Macrofractographic Features 225

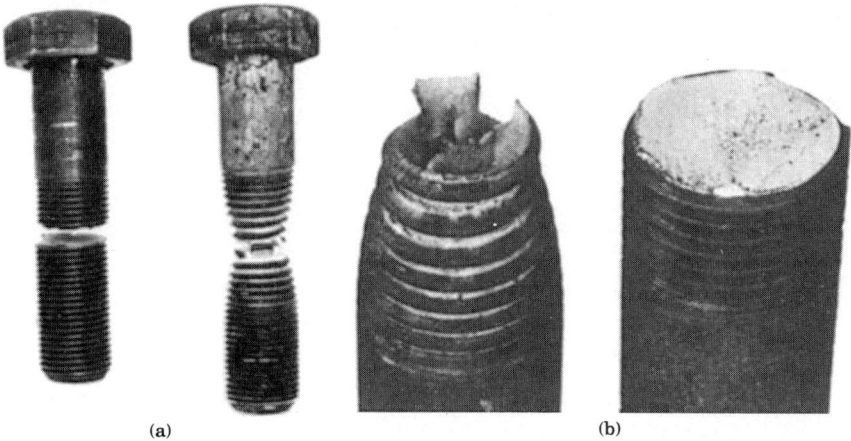

(a)   (b)

Figure 4.15 Steel bolts broken in tensile overload. (*a*) Brittle versus ductile fracture. The bolt on the left was water-quenched with a hardness of 47 Rockwell C and has no obvious deformation. The bolt on the right was annealed to a hardness of 95 Rockwell B (equivalent of 15 Rockwell C) and shows extensive permanent deformation. (*b*) Two identical steel bolts given different heat treatments. The annealed bolt on the left shows extensive necking and a cup-and-cone fracture. The austenitized then brine-quenched bolt on the right shows a relatively flat fracture surface, normal to the loading axis, and no obvious plastic deformation. (*From Wulpi.*[6])

4.3 Torsion Overload

The stress state associated with pure torsion was described in Chap. 2. In cylinders the maximum normal stress is at 45° to the axis, and the maximum shear stress is at 90°. Fracture can occur in a ductile or in a brittle manner. If the material fails in a ductile fashion, then obvious plastic deformation will be detected. For example, if the cylinder has axial marks on the surface (such as drawing lines), then these will be spiraled around the surface. A clear example is shown in Fig. 4.16. A case showing twisted splines which reveal torsion fracture is illustrated in Fig. 4.17. If this effect is not revealed by such surface markings, it sometimes can be revealed by surface etching, as shown in Fig. 4.18.

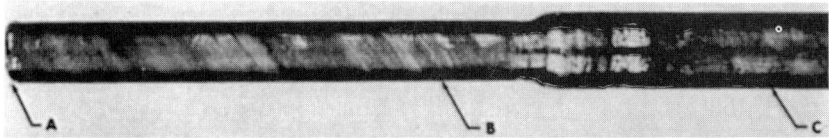

Figure 4.16 Surface of a shaft which broke in torsion overload. The fracture surface at point *A* is perpendicular to the page. Between *A* and *B* helical lines are seen, which are torsion deformation bands. (*From* Metals Handbook.[7])

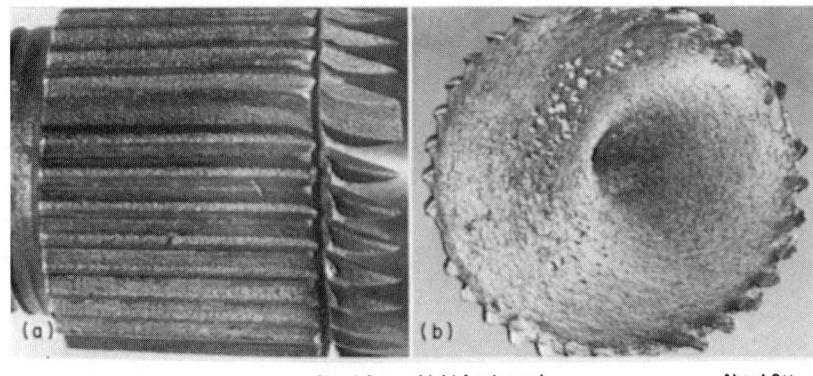

Photograph   About 2×   Light fractograph   About 2×

Figure 4.17 Splined shaft which broke in torsion overload. The 6118 steel shaft had a hardness of 23 Rockwell C. (*a*) The splines on the left were constrained in the holder, whereas the part on the right was free to twist, showing that the fracture was ductile. (*b*) The fracture surface shows a swirl pattern, characteristic of torsion, and is normal to the shaft axis. If there is combined bending, the region of final fast fracture will be off center. (*From* Metals Handbook.[2])

In ductile torsion overload fracture, the fracture surface is at 90° to the axis, as shown in Figs. 4.17 and 4.18. The surface appearance usually shows a "swirl" pattern (Fig. 4.17), but such a pattern can be made by the surfaces rubbing after fracture, and hence is not necessarily indicative of torsional overload fracture.

If the material is brittle, then it is expected that the fracture will be along planes sustaining the maximum normal stress, which is at 45° to the cylinder axis (see Fig. 2.12). Figure 4.19 shows a sketch of the fracture surface of a rod of chalk which has been broken in pure torsion. The fracture surface is a spiral at 45° to the axis. Figure 4.20 shows the fracture surface of a hardened steel rod which fractured in a brittle manner in torsion overload. The cracks on the surface of a shaft which fractured in a brittle manner are shown in Fig. 4.21; the cracks are at 45° to the axis and hence normal to the maximum normal stress.

In brittle fracture, the fracture-surface appearance is that associated with fast fracture. A very common type of brittle torsion fracture is found in spiral springs made of high-strength steel. An example which typifies the surface appearance is given in Fig. 4.22. In this case, initially a small fatigue crack had formed and served as the origin of the torsion overload fracture. Note that the fracture surface is helical and at 45° to the wire axis. The fracture surface shows only radial marks, characteristic of fast fracture, and they point back to the location of the original fatigue crack.

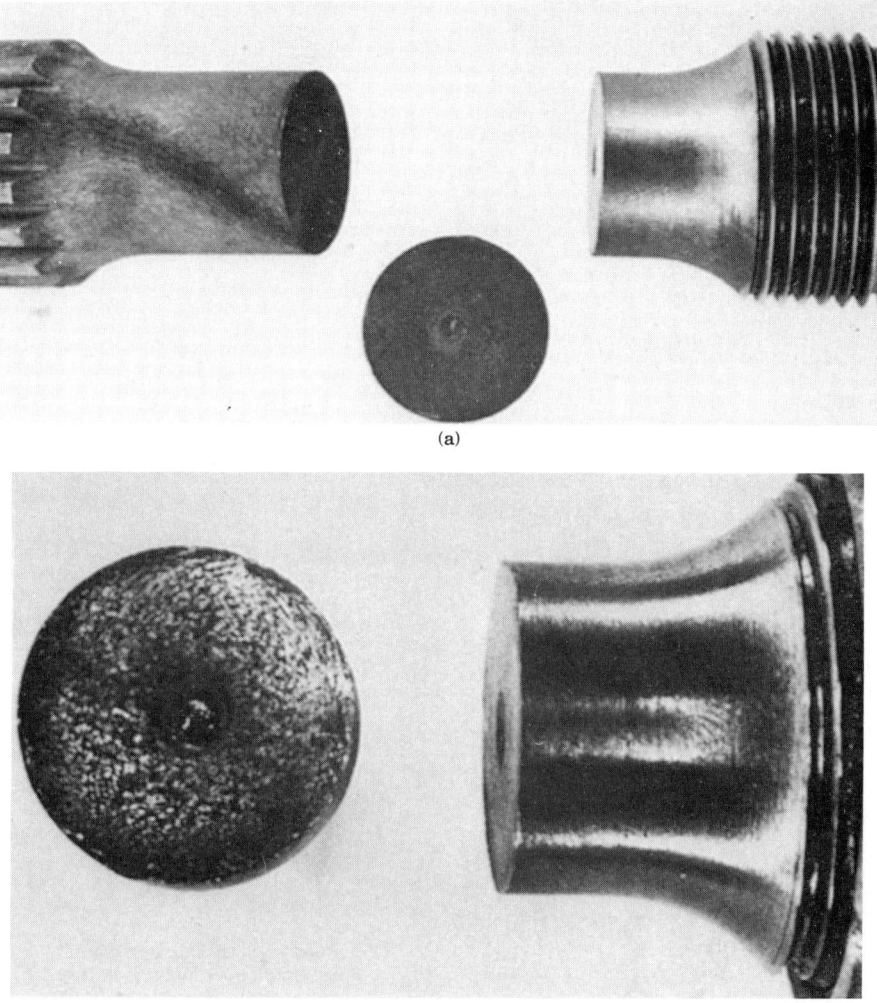

Figure 4.18 (a) Shaft that broke in torsion overload. (b) Higher magnification view of (a). 1035 steel drive shaft. Hardness of Rockwell C 34. Note that the fracture surface is flat and perpendicular to the shaft axis and that it shows a swirl pattern. The dark spiral on the shaft in (a) was revealed by etching, and shows that the shaft was subjected to plastic deformation by torsion. (*From* Metals Handbook.[2])

4.4 Bending Overload

In bending fracture, generally the fracture surface is similar to that produced by tensile overload. The difference arises due to the fact that one side of the component is in tension and the other in compression. Thus the crack forms on the tension side and propagates across until

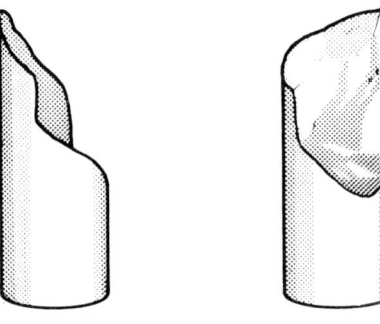

Figure 4.19 Fracture appearance of a piece of brittle chalk after breaking it in torsion overload. (*From* Metals Handbook.[2])

Figure 4.20 Shaft broken in torsion overload, showing brittle fracture. The steel contained 0.8% C, 0.15% Si, 0.60% Mn, 0.021% P, 0.032% S, 0.03% Ni, and 0.03% Cr and was quenched and tempered. (*From Henry and Horstmann.*[8])

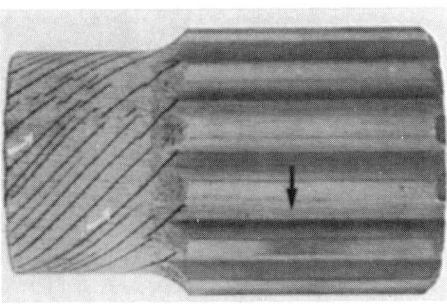

Figure 4.21 Case-hardened steel shaft which failed in torsion in a brittle manner. The arrow shows the direction that this end was twisted. The black lines at 45° are cracks, which show that the fracture is brittle (*see* Fig. 2.12). (*From Wulpi.*[6])

separation occurs. Figure 4.23 shows a broken bending sample that was ductile. The sample was notched, which introduced stress triaxiality, but the loading rate was low. Thus the fracture proceeded from the tension side to the opposite side with considerable plastic deformation and the formation of shear lips on the sides at 90° to the notch.

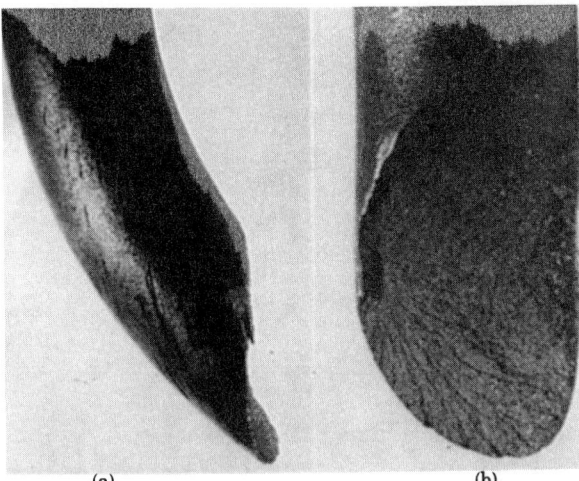

(a) (b)

Figure 4.22 Spiral spring which broke in a brittle manner. (a) Fractured 13-mm-diameter spring of AISI 10B62 steel wire with a hardness of 477 Brinell. Note the spiral gouges, which are screw marks that were generated during coiling. (b) Fracture surface shows a small fatigue crack which originated at a screw mark, but only grew slightly before torsion overload fracture occurred. (*From Metals Handbook.*[5])

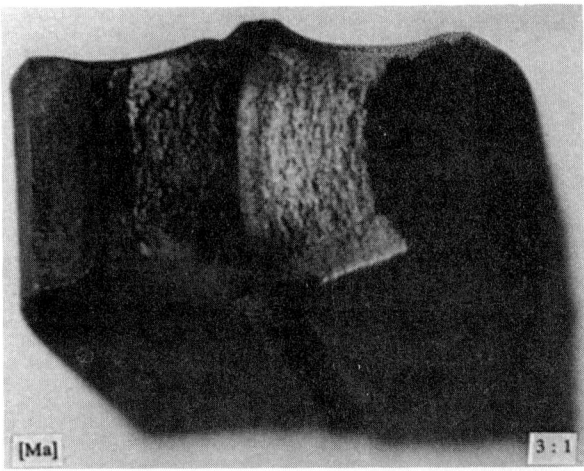

Figure 4.23 Ductile bending overload fracture. This was a notched bending specimen made of a steel containing 0.10% C, 0.02% Si, 0.30% Mn, 0.032% P, 0.024% S, and 0.003% N. (*From Henry and Horstmann.*[8])

Figure 4.24 Brittle bending overload fracture. This steel sample contained 0.09% C, 0.02% Si, 0.34% Mn, 0.042% P, 0.028% S, and 0.010% N and was in the aged state. The brittle cleavage rupture first propagated in the zone under tension perpendicular to the principal stress direction, then branched out. (*From Henry and Horstmann.*[8])

Figure 4.24 shows the surface of a brittle sample broken in bending. No gross plastic deformation occurred, and a secondary crack formed. Note that there are no clear distinguishing surface features to help locate the origin of the fracture. This is more clearly seen on the fracture surface of Fig. 4.25. However, in some cases radial marks do form and point back to the origin of the fracture. An example is shown in Fig. 4.26, the case of a broken crane hook. (In this case the fracture began below the surface.)

Shafts frequently break by overload bending, and a typical fracture appearance is shown in Fig. 4.27. The direction of bending is obvious from the slant of the fracture. The location of the origin of the fracture

Figure 4.25 Brittle bending overload fracture of a sample notched on both sides. The crack propagated from notch to notch. (Material as in Fig. 4.24.) (*From Henry and Horstmann.*[8])

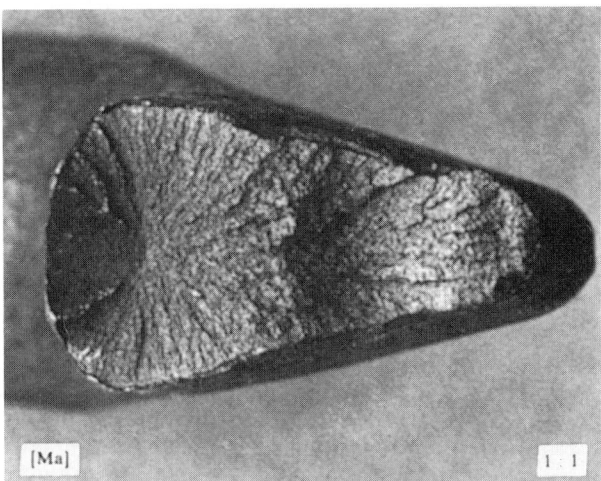

Figure 4.26 Fracture surface of a steel crane hook broken by bending overload. The steel contained 0.24% C, 0.40% Si, 0.39% Mn, 0.095% P, and 0.024% S. (*From Henry and Horstmann.*[8])

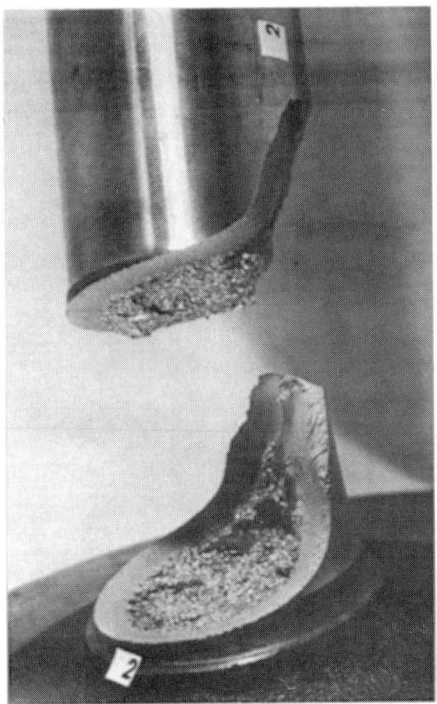

Figure 4.27 Steel axle shaft broken in a laboratory by a single bending impact overload. The steel was modified SAE 1050 containing 0.50% C, 0.95% Mn, 0.25% Si, 0.01% S, and 0.01% P. The hot-rolled and upset shaft had an induction-hardened case (60 Rockwell C) and a pearlite–primary ferrite core (20 Rockwell C). (*From* Metals Handbook.[5])

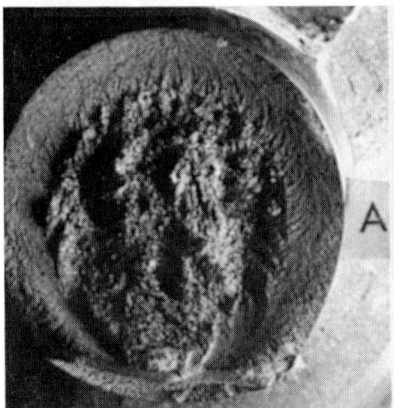 

Figure 4.28 Fracture of an axle shaft broken in a laboratory by a bending overload. Note that the chevron marks (in the induction-hardened surface) and radial marks (in the softer core) point to the origin of the fracture. The material was modified AISI 1050 steel, induction-hardened and tempered to 60 Rockwell C at the bearing surface. (*Courtesy of Z. Flanders, Packer Engineering Associates; from Metals Handbook.*[5])

is deduced from the clear chevron marks on the periphery of the fracture surface (Fig. 4.28). In this case the distinct difference in appearance of the periphery and the center of the fracture surface is caused by the different microstructural characteristics in the two locations. The outer layer was induction-hardened, and hence here the crack ran through a brittle material. In the center the material was tougher, and coarse radial marks were formed. Note that these also point toward the origin of the fracture, and they merge into finer radial marks, which in turn merge with the chevron marks in the vicinity of the origin.

4.5 Fatigue Fracture

A fatigue fracture surface usually displays macroscopically two distinct regions: the fatigue crack propagation region and the final overload region. The latter has characteristics similar to those described in previous sections, since this surface was created when the supporting material was too slight to sustain the next (and final) load cycle. The final fracture is nearly always a catastrophic event and occurs with almost no plastic deformation. The lack of gross plastic deformation is also a characteristic of the propagation of the fatigue crack. Thus a very distinguishing characteristic of a component which broke due to fatigue loading is that the parts exhibit almost no gross plastic deformation.

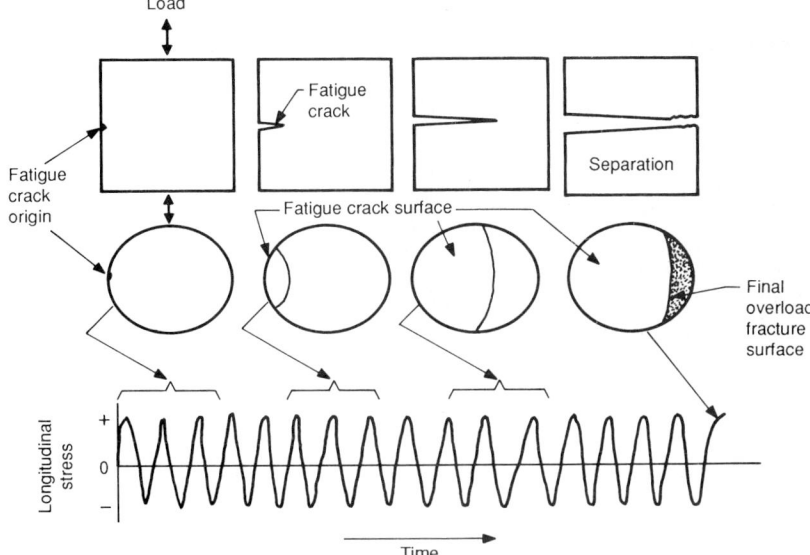

Figure 4.29 Schematic illustration of the location of a fatigue crack and number of cycles.

The main interest in this section is the fractographic features which fatigue fractures exhibit upon visual examination or at low (such as 20 ×) magnification. We first present the appearance of test samples broken under a loading cycle of constant amplitude. Then the effect of loading conditions and specimen geometry are examined. Finally, a few examples of the fracture surface of machine components which fractured in fatigue follow.

Figure 4.29 shows schematically the growth of a fatigue crack and its relation to the number of loading cycles. It is emphasized that the stress generated by the load would be calculated to be below that required for yielding. However, due to the intense stress concentration at the root of the crack, local plastic deformation will occur upon each cycle. As discussed in Chap. 3, frequently striations are associated with each load cycle. But the propagation distance per cycle is quite small and the associated fatigue striations are too small to be resolved, and hence are not macrofractographic features. Figure 4.30 shows fracture surfaces of test samples broken in fatigue loading. It is important to realize that these samples were tested continuously until fracture, using a cyclic load of constant frequency and of a constant maximum load. The distinction between the area of fatigue crack growth and that of final overload separation is not necessarily very clear.

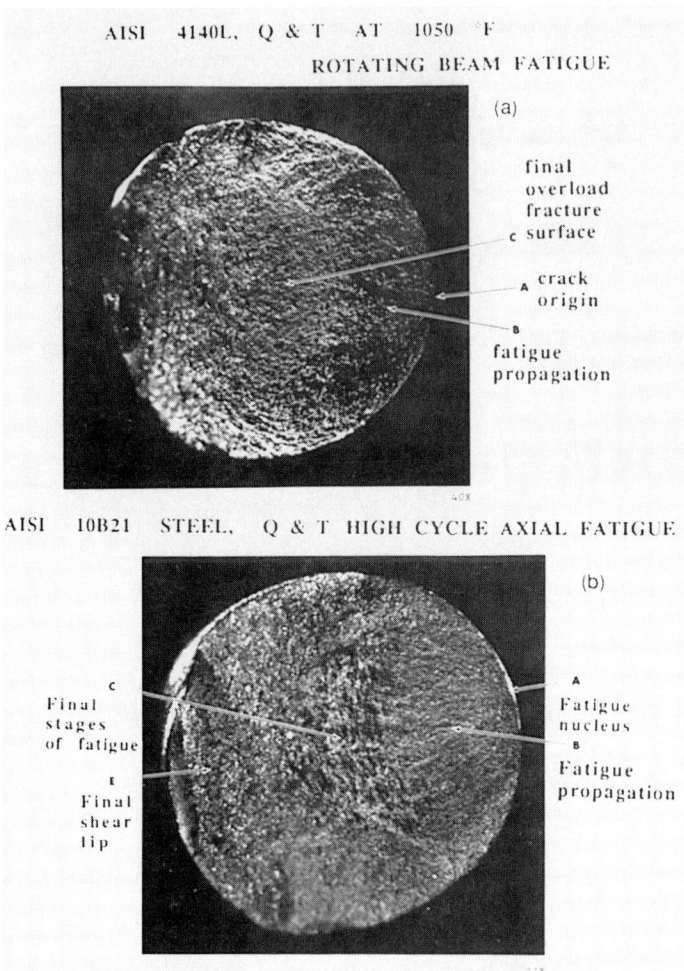

Figure 4.30 Fracture surfaces of samples tested in fatigue with a constant maximum load and frequency. (a) AISI 4140L steel, quenched and tempered at 1050°F (566°C), rotating beam fatigue. (b) AISI 10B21 steel, quenched and tempered, high-cycle axial fatigue. (*From Bhattacharyya et al.*[9])

However, in most (but not all) fatigue fractures a surface feature is found which is usually an indicator that fracture was by fatigue. This is illustrated in Fig. 4.31. The markings are called by several names based on their appearance: *oyster-shell marks, tide marks, ripple marks, beach marks, stop marks, arrest marks, clamshell marks, or conchoidal marks*. We will use the last term for them. (This term is also used in mineralogy to describe the fracture appearance of shells.) These markings are associated with surface-topography changes that

Fracture Modes and Macrofractographic Features   235

Light fractograph   About 0.95×

Figure 4.31 Fracture surface of a sample which broke in fatigue loading, illustrating the conchoidal marks and the final overload region typical of fatigue fractures. This was an axle of 8640 steel with a hardness of approximately 30 Rockwell C. The loading was nonrotational. The arrow indicates the location of the origin of crack initiation, which was at a discontinuity in a cut thread. (*From* Metals Handbook.[2])

cause the surface appearance to vary locally. They are caused by such factors as variation in the crack growth velocity or surface oxidation. We now examine how these markings develop, assuming that such factors allow them to form, and we see what information they convey about the loading conditions.

To illustrate how these markings form, consider a cylindrical test bar loaded cyclically in tension-compression-tension, as shown in Fig. 4.32. Let the crack form only on one side and progress across toward the other side. After a certain advancement, let the loading cease and the surface oxidize. Now resume the loading, with the crack advancing, and then let the loading cease again. The new surface now oxidizes, and the previous surface oxidizes further. Thus the initial fracture surface has a thicker oxide layer on it. This process continues until fracture occurs, as shown in Fig. 4.32. The final overload fracture surface will show features associated with overload fracture, and the part of the surface along which the fatigue crack grew will show conchoidal marks because of the different thicknesses of oxide. Thus

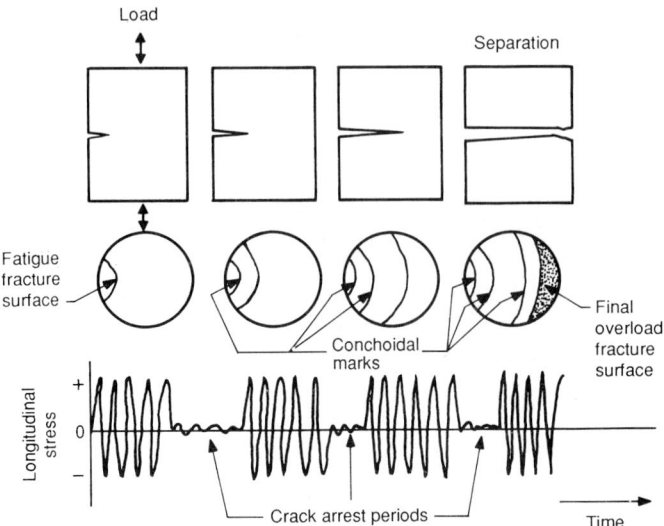

Figure 4.32 Schematic illustration of the formation of conchoidal marks in a fatigue test.

in this example the appearance of these marks relies on the crack not moving continuously until fracture, but growing intermittently. Other mechanisms also cause these conchoidal marks to form, such as rubbing of the mating surfaces in different fashions if the load amplitude varies. It is important to note that fatigue fracture propagation without any arrests will probably not produce conchoidal marks. This was the case in Fig. 4.30, where fracture surfaces of fatigue specimens were tested uninterruptedly to fracture.

The example in Fig. 4.32 can be used to illustrate the effect of the load level (stress level) on the fractographic features. If the load is relatively low, then it takes many cycles for the fatigue crack to propagate to a location where the remaining material will fracture catastrophically on the next (and final) load application. Thus the relative amount of the fracture surface covered by the conchoidal marks will be large. If the stress level is high, then the crack will not propagate far before final fracture occurs. This is illustrated in Fig. 4.33a. A similar situation exists if the sample is loaded in alternating unidirectional bending, as shown in Fig. 4.33b. If the sample is loaded in reversed bending, then the crack will probably initiate simultaneously on opposite sides, as shown in Fig. 4.33c. Again, a low nominal stress creates a fracture surface with a large amount of it covered with conchoidal marks.

In simple rotating bending, every point on the surface is placed in longitudinal tension sometime during a rotation. If only one fatigue

Fracture Modes and Macrofractographic Features   237

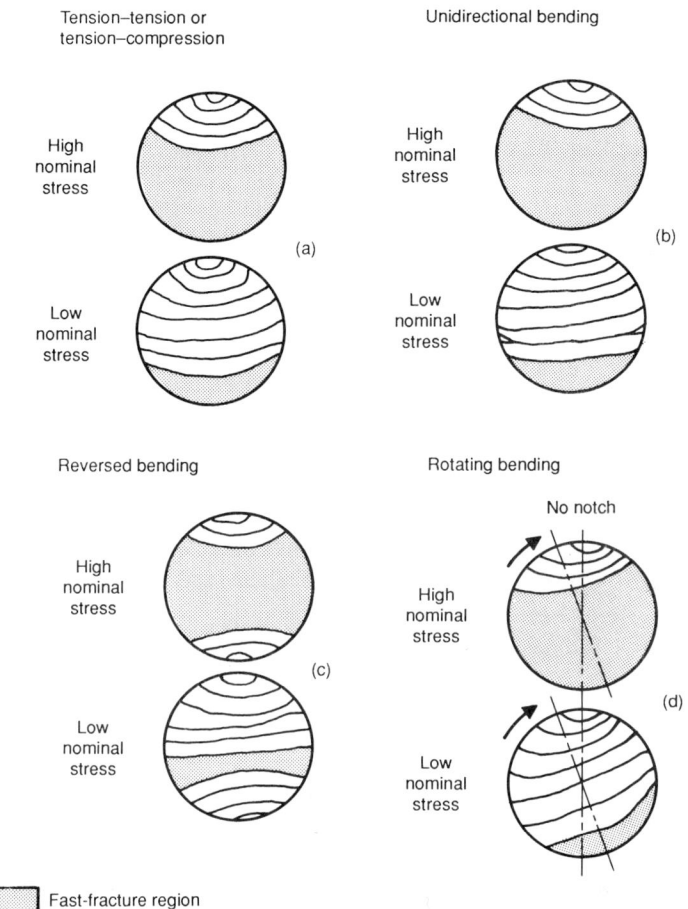

Fast-fracture region

Figure 4.33 Schematic illustration of a fatigue fracture surface showing the effect of loading conditions and stress level on the relative area covered by the fatigue crack. (a) Tension-tension or tension-compression. (b) Unidirectional bending. (c) Reversed bending. (d) Rotating bending. (*Adapted from Vander Voort.*[1])

crack initiates, then the appearance of the fracture surface will be similar to that shown schematically in Fig. 4.33d. However, the fatigue crack may initiate at many locations simultaneously and advance from the periphery. This is shown schematically in Fig. 4.34a. If the fatigue cracks originate at several locations but at different locations axially, then when the various advancing fatigue cracks meet at the different levels, accommodation occurs to join these into one fatigue crack. This leaves offsets on the surface called *ratchet marks*, as illustrated in Fig. 4.34b. Figure 4.35 shows a fatigue fracture surface with many faint ratchet marks along the periphery. The fracture sur-

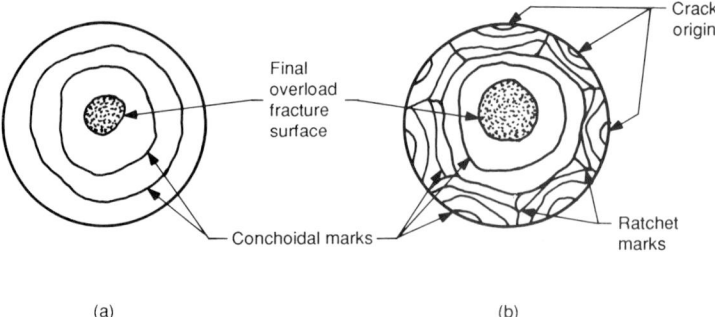

Figure 4.34 Schematic illustration of the fracture-surface appearance of a fatigue fracture which occurred in rotating bending. The fatigue crack originated simultaneously (a) at different locations on the same circumferential plane, (b) at different locations but not on the same circumferential plane, forming ratchet marks.

Figure 4.35 Fracture surface of a fatigue failure showing fine ratchet marks on the periphery, which formed along a sharp snap-ring groove. The eccentric pattern of oval conchoidal marks indicates that the load on the shaft was not balanced. This was a 1050 steel shaft with a hardness of about 35 Rockwell C. It was subjected to rotating bending. (*From Wulpi.*[6])

face in Fig. 4.36 depicts an extreme case, where the individual fatigue crack growth regions did not join before fracture occurred.

The fatigue fracture surface appearance is affected by the stress state, which can be illustrated by considering the effect of a notch. The more acute the notch is, the more likely that a fatigue crack

Fracture Modes and Macrofractographic Features   239

Figure 4.36 Fatigue fracture in which the fatigue cracks formed at six locations around the periphery, but at distinctly different levels. The locations were near the runouts of six grooves. Grinding damage in the fillet was the cause of the fracture. The component was a 4817 steel shaft, carburized and hardened to a surface hardness of 60 Rockwell C. It was loaded in rotating bending. (*From Wulpi.*[6])

will initiate at more than one location on the periphery. However, the subsequent growth of the crack is not greatly affected. This is due to the fact that once the crack forms (enters stage II, see Chap. 3), the root of the fatigue crack is sharp, and this will be true whether the initial crack formed at a notch or not. However, if the nominal stress is relatively low, then it is more likely that the crack will move along the periphery faster than it moves across the diameter, giving a different appearance to the fracture surface. These features are illustrated schematically in Fig. 4.37, and Fig. 4.38 shows examples of this effect in terms of stress concentration at the surface.

The effects of the type of loading, stress level, and notch severity on the fracture-surface topography of fatigue fractures in cylinders are summarized schematically in Fig. 4.39. In rotating bending under low nominal stress, the pattern of the conchoidal marks and the location of the final overload region have been correlated with the direction of rotation of the shaft. This is indicated by the arrows in Fig. 4.39. Thus in actual component fatigue failures, the direction of rotation prior to failure might be deduced. Also, the shape of the conchoidal marks and the location of the final overload fracture give information about the loading of shafts. Three examples illustrating these points are shown in Fig. 4.40.

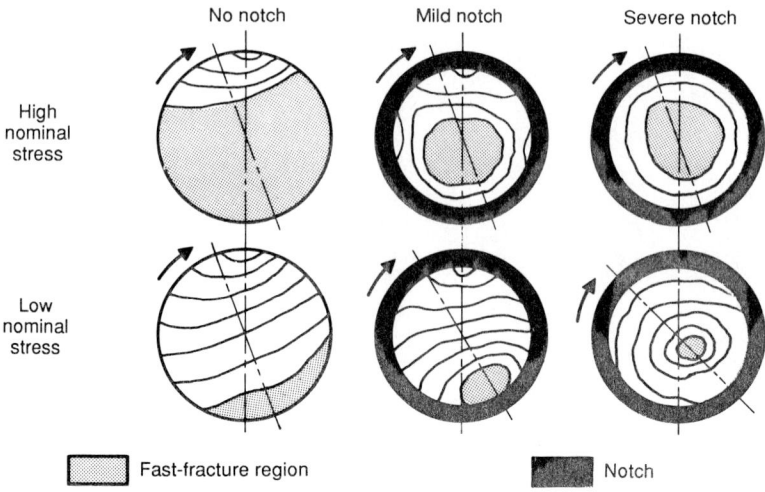

Figure 4.37 Schematic illustration of the effect of notch severity and stress level on the fatigue fracture surface appearance in rotating bending of cylinders. (*Adapted from Vander Voort.*[1])

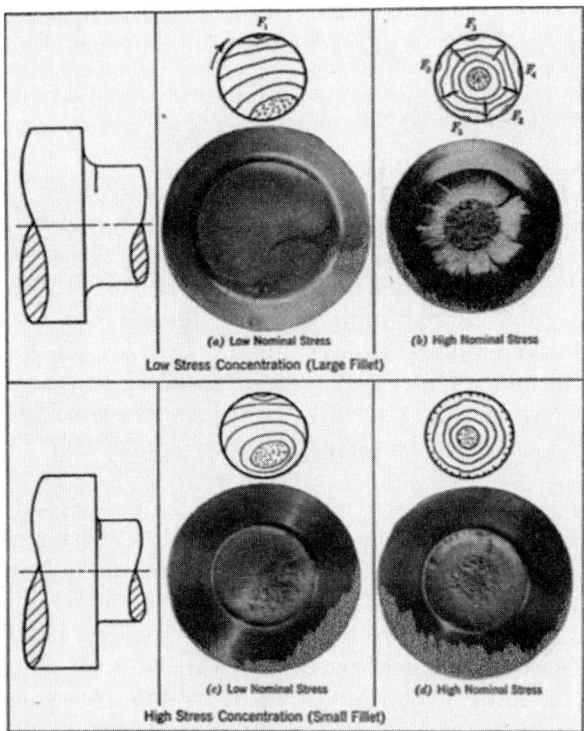

Figure 4.38 Fatigue fracture surfaces illustrating the effect of stress concentration on the fracture-surface appearance. A rotating shaft was subjected to bending in a fixed plane. (*From Peterson.*[10])

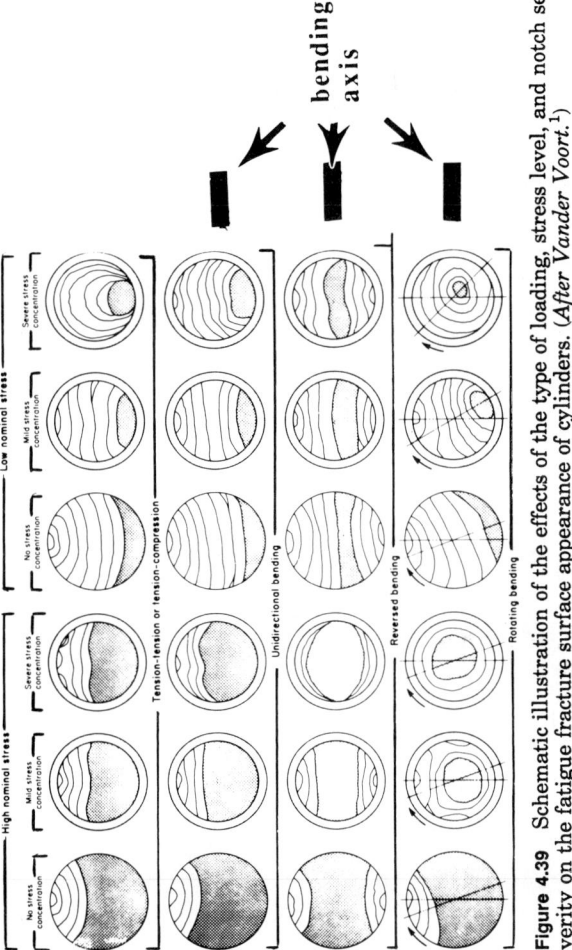

Figure 4.39 Schematic illustration of the effects of the type of loading, stress level, and notch severity on the fatigue fracture surface appearance of cylinders. (*After Vander Voort.*[1])

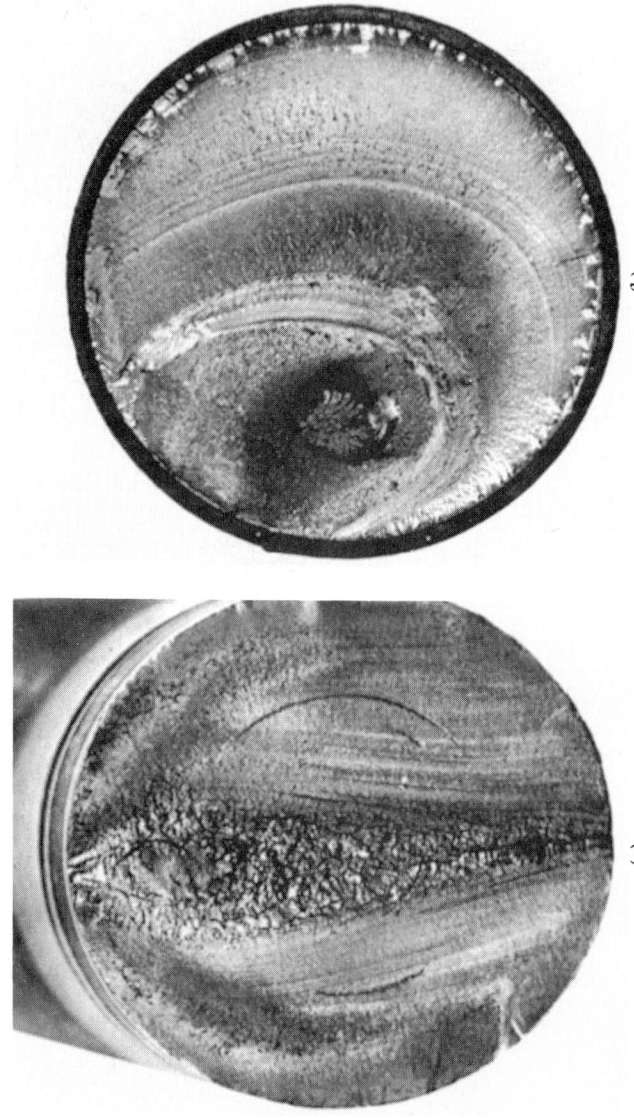

Figure 4.40 Fatigue fracture surfaces revealing information about the loading conditions. (a) Reversed bending fatigue of a 1.6-in-diameter shaft of 1046 steel with a hardness of approximately 30 Rockwell C. The symmetrical conchoidal mark pattern indicates that each side of the shaft was subject to the same maximum stress and the same number of load applications. (*From Wulpi.*[6]) (b) Rotating bending of a 1050 steel shaft with a hardness of about 35 Rockwell C. The numerous ratchet marks indicate that fatigue cracks initiated at many locations along a sharp snap-ring groove. The eccentric pattern of oval conchoidal marks indicates that the load was not balanced. (*From Wulpi.*[6])

Fracture Modes and Macrofractographic Features   243

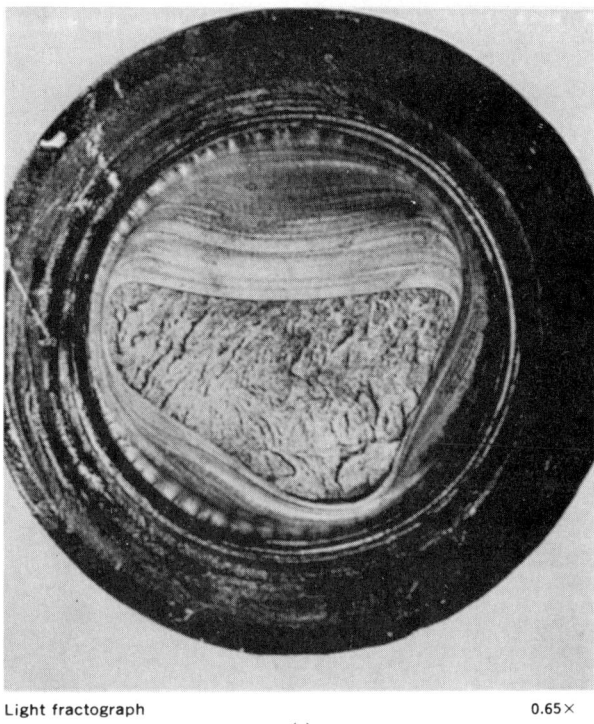

Light fractograph   0.65×
(c)

Figure 4.40 (*Continued*) Fatigue fracture surfaces revealing information about the loading conditions. (c) 4-in-diameter axle of 8640 steel with a hardness of about 30 Rockwell C. The shaft was contained within a shrink-fitted collar, which is the outer circular area. This nonrotating axle was subjected to bending stresses in three directions, which produced this unusual pattern of conchoidal marks. (*From Metals Handbook.*[2])

In alternating torsion loading, the situation is somewhat more complex than for the type of loading just described. Since fatigue cracks propagate only under the tensile component induced by the load, in torsion it is expected that in cylinders the cracks will be at 45°. This is shown schematically in Fig. 4.41a, and an example of such a fatigue failure is given in Fig. 4.42. If the cracks originate at several circum-

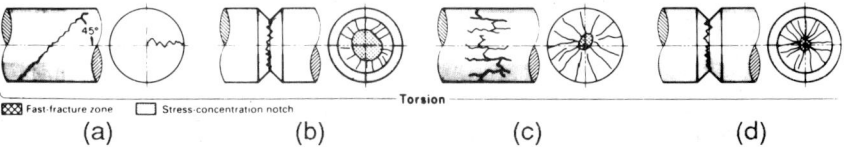

Figure 4.41 Schematic illustration of the effects of stress level and notch severity on the fatigue fracture of cylinders in reversed torsion loading. (*After Vander Voort.*[1])

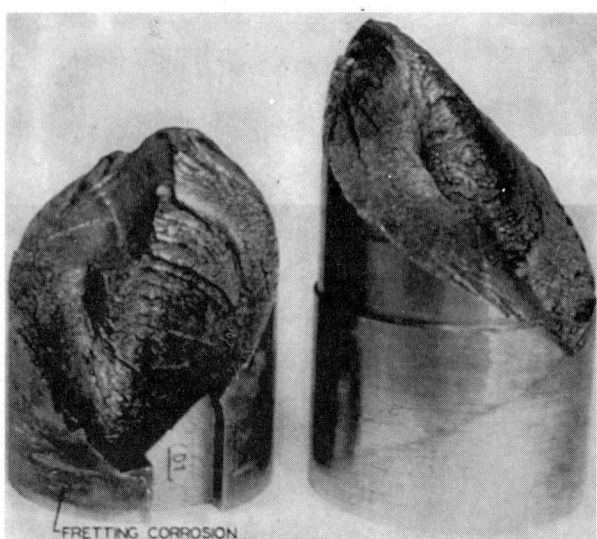

Figure 4.42 Fatigue fracture of a shaft which broke due to reversed torsion loading. (*From Peterson.*[10])

ferential locations, then the type of fractures shown in Figs. 4.43 and 4.44 may occur.

If the nominal stress is low, the cracks may follow a more devious path, growing both axially and diametrically, as depicted in Fig. 4.41c. An example is given in Fig. 4.45. The effect of notches on torsion fatigue fracture is shown schematically in Fig. 4.41b and d. The effect of stress concentration on the crack growth direction for a splined shaft is illustrated in Fig. 4.46. In Fig. 4.46b the contour lines

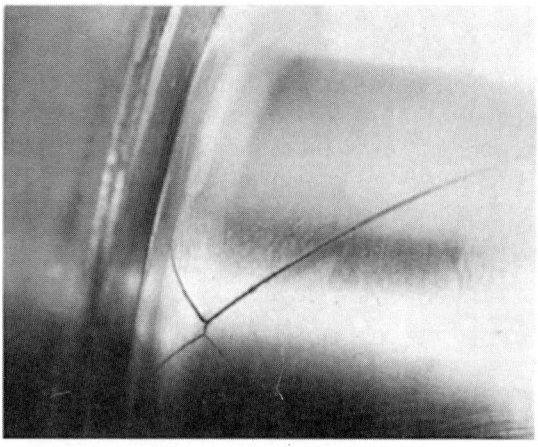

Figure 4.43 Fatigue crack geometry in a shaft fractured by reversed torsion loading. This was a 1045 steel crankshaft. (*After Wulpi.*[6])

Figure 4.44 Fatigue fracture in a steel shaft broken by reversed torsion loading. (*From Peterson.*[10])

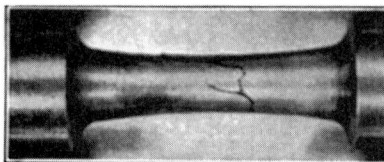

Figure 4.45 Fatigue crack geometry showing axial and radial crack growth due to reversed torsion loading. The material was ductile steel (Dolan). (*From Peterson.*[10])

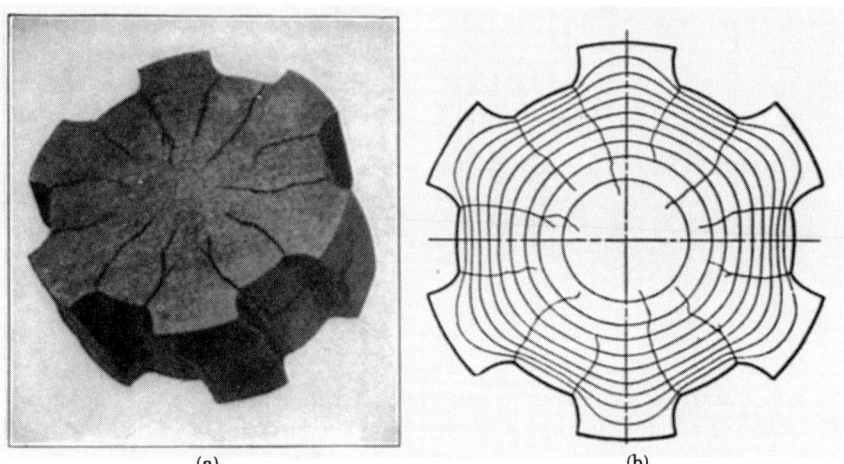

(a) (b)

Figure 4.46 (*a*) Fatigue cracks in a spline due to reversed torsion loading (Oschatz). (*b*) Maximum tensile stress distribution. (*From Peterson.*[10])

(a)

(b)

Figure 4.47 Fracture surface of spline shafts broken due to reversed torsion loading. (a) 6¾-in-diameter spline shaft showing characteristic "starry" pattern of multiple fatigue cracks. Each of the 32 spline teeth has two fatigue cracks, each at 45° to the shaft axis, that form a V-shaped region. In addition there are longitudinal radial fatigue cracks that penetrate nearly to the center of the shaft. (b) Fracture in reversed torsion loading of a 1½-in-diameter spline. The shaft was made of a low-carbon alloy steel with a hardness of 24 Rockwell C in the shaft area. (*From Wulpi.*[6])

show the maximum tensile stress distribution prior to the crack formation and growth. Examples of the fracture surface of splined shafts broken in torsion fatigue are shown in Fig. 4.47.

In cylinders with keyways, the torsion fatigue crack often originates at the root of the keyway, as illustrated in Fig. 4.48, and

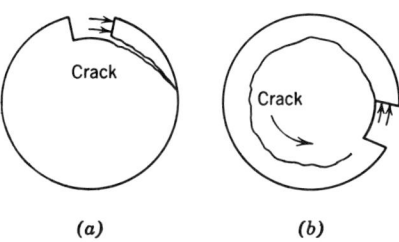

Figure 4.48 Schematic illustration of the formation of fatigue cracks from the root of a keyway of shafts loaded in reversed torsion. (*From Peterson.*[10])

progresses along the circumference from this location. An example is shown in Fig. 4.49.

The features described for cylinders and shafts are also found in components of other cross sections. For the case of rectangular sections, the fatigue fracture appearances are summarized in Fig. 4.50. It is important to realize that the fatigue fracture path depends not only on the geometry of the component and the loading conditions, but also on the structural inhomogeneity of the material. Such variations in structure are common in machine components. They can be a consequence of the processing of that component, such as carburization, induction hardening, and peening of a surface. They also can be due to the anisotropy of mechanical properties caused by microstructural features such as elongated inclusions.

Figure 4.49 Fracture surface due to torsion fatigue loading, as illustrated in Fig. 4.48. The component was a 3⅜-in-diameter keyed tapered shaft of 1030 steel. The fatigue crack originated in corner A on the keyway from pressure of the key aligning a large hub to the shaft. The fatigue crack progressed completely around the shaft, under the keyway, and started around again, all in a clockwise direction, before final separation occurred, and nearly broke off the "wing" at left. The only thing necessary to prevent this type of fracture is to make sure that the large nut which forces the hub to fit tightly against the tapered surface does not come loose. If the nut is loose, the key and keyway, rather than the frictional fit of the conical joint, carry the torsional force. (*From Wulpi.*[6])

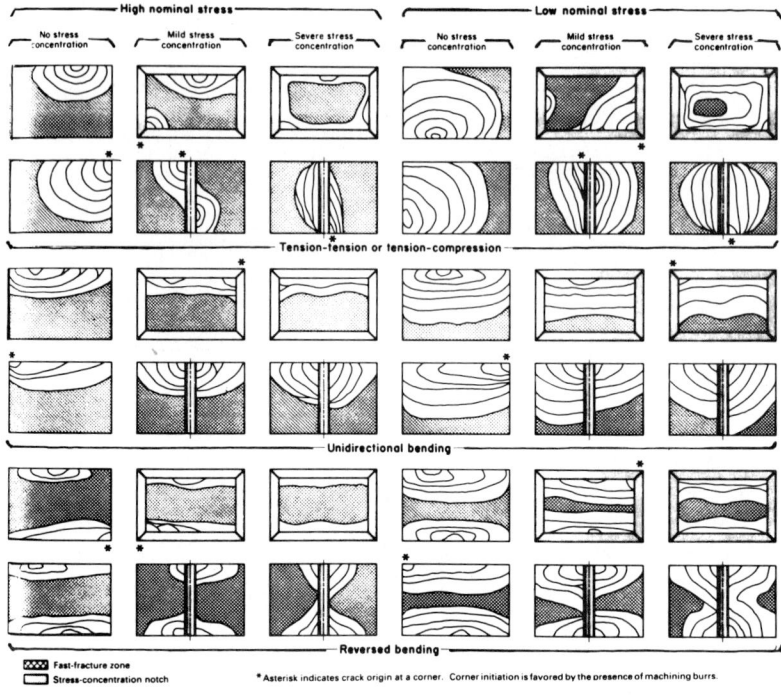

Figure 4.50 Schematic illustration of the effects of the type of loading, stress level, and notch severity on the fatigue fracture surface appearance of components of rectangular cross section. (*After Vander Voort.*[1])

4.6 Correlation of Micro- and Macrofractographic Features

In this section examples of the fracture surfaces of test samples are used to compare the topographic features seen macroscopically and the corresponding microscopic features. These examples also illustrate the sensitivity of the macroscopic fracture features and the fracture mechanism to the microstructure of the material.

Figure 4.51a shows the fracture surface of a tensile sample of a 10B21 steel. The mode of fracture is thus tensile overload, and the high ductility [56 percent reduction in area (RA)] and the large shear-lip zone show that this was a ductile fracture. The features can be compared to those in Fig. 4.4b. In this fracture the radial marks are not prominent. The dimples in the fractographs in Fig. 4.51b and c, including shear dimples in the shear-lip zone (Fig. 4.51d), show that over the entire surface, the mechanism of fracture was void coalescence.

The fracture surface of an impact sample of the same steel is shown

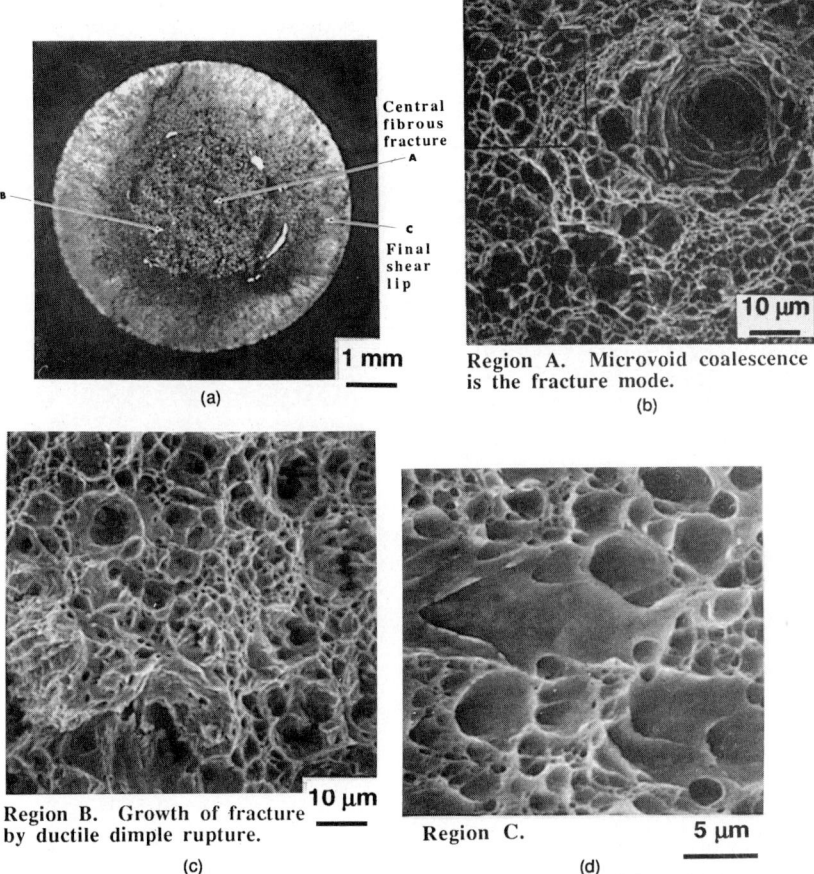

Figure 4.51 Fractographs of a quenched and tempered 10B21 steel broken in tension. The mechanical properties were hardness 43 Rockwell C; yield strength 179,000 lb/in² (1.23 × 10⁹ Pa); tensile strength 218,000 lb/in² (1.50 × 10⁹ Pa); ductility RA (reduction in area at fracture) 56%. (*Adapted from Bhattacharyya et al.*[9])

in Fig. 4.52a. Fine radial marks point back toward the root of the machined notch where fracture initiated. Due to the nature of the test, the mode of fracture is bending overload, and the material fractured in a ductile manner, as evidenced by the large amount of shear lip. The fractographs in Fig. 4.52b and c show that everywhere the mechanism of fracture was by void coalescence.

Figure 4.53a shows the fracture surface of the same 10B21 steel broken in a high-cycle axial fatigue test. In this test the loading frequency was constant and the loading uninterrupted, and thus there are no conchoidal marks. As expected from the loading condition, the

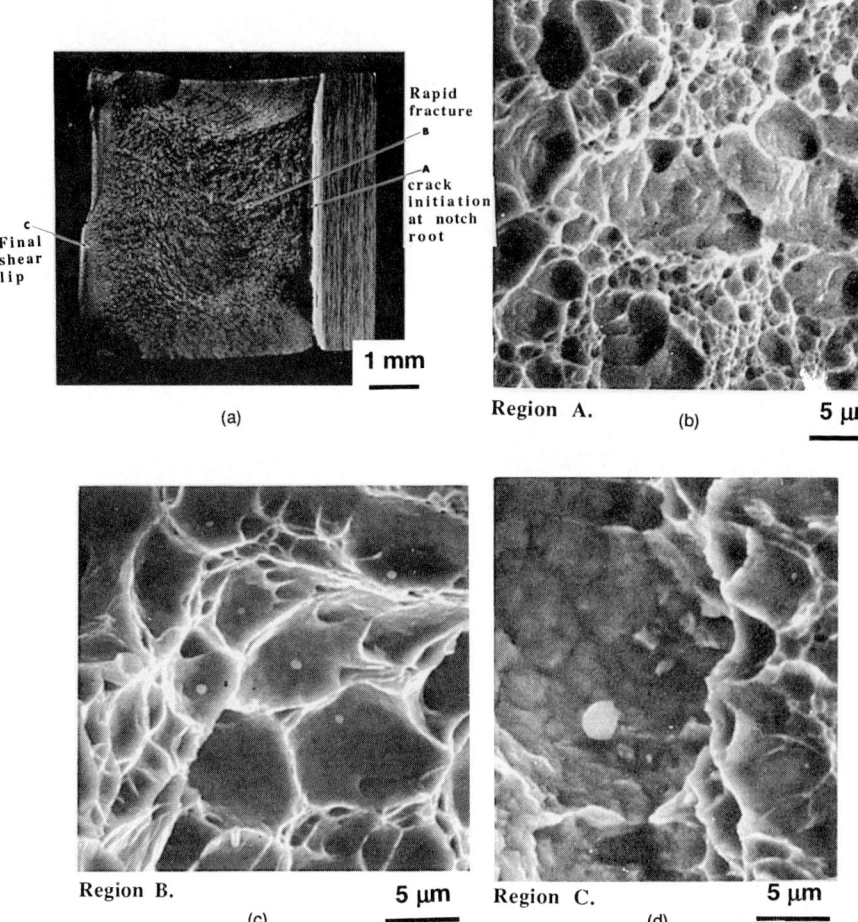

Figure 4.52 Fractographs of a Charpy impact test sample of 10B21 steel. Impact energy 45 ft · lb (61 J). (*Adapted from Bhattacharyya et al.*[9])

fatigue crack initiated at only one location on the surface. Because of the lack of conchoidal marks, the location of the initiation of the fracture must be deduced from the fine radial marks which point clearly to the fatigue crack initiation zone. The fatigue crack propagated in a much coarser fashion in the last few cycles prior to the final overload. The fractograph in Fig. 4.53b purports to show fatigue striations, but they are not prominent, even though this is a fatigue test sample. However, faint striations can be found on the surface (Fig. 4.53c to e). The final overload fracture region shows a dimple structure (Fig. 4.53f), characteristic of void coalescence.

Fracture Modes and Macrofractographic Features 251

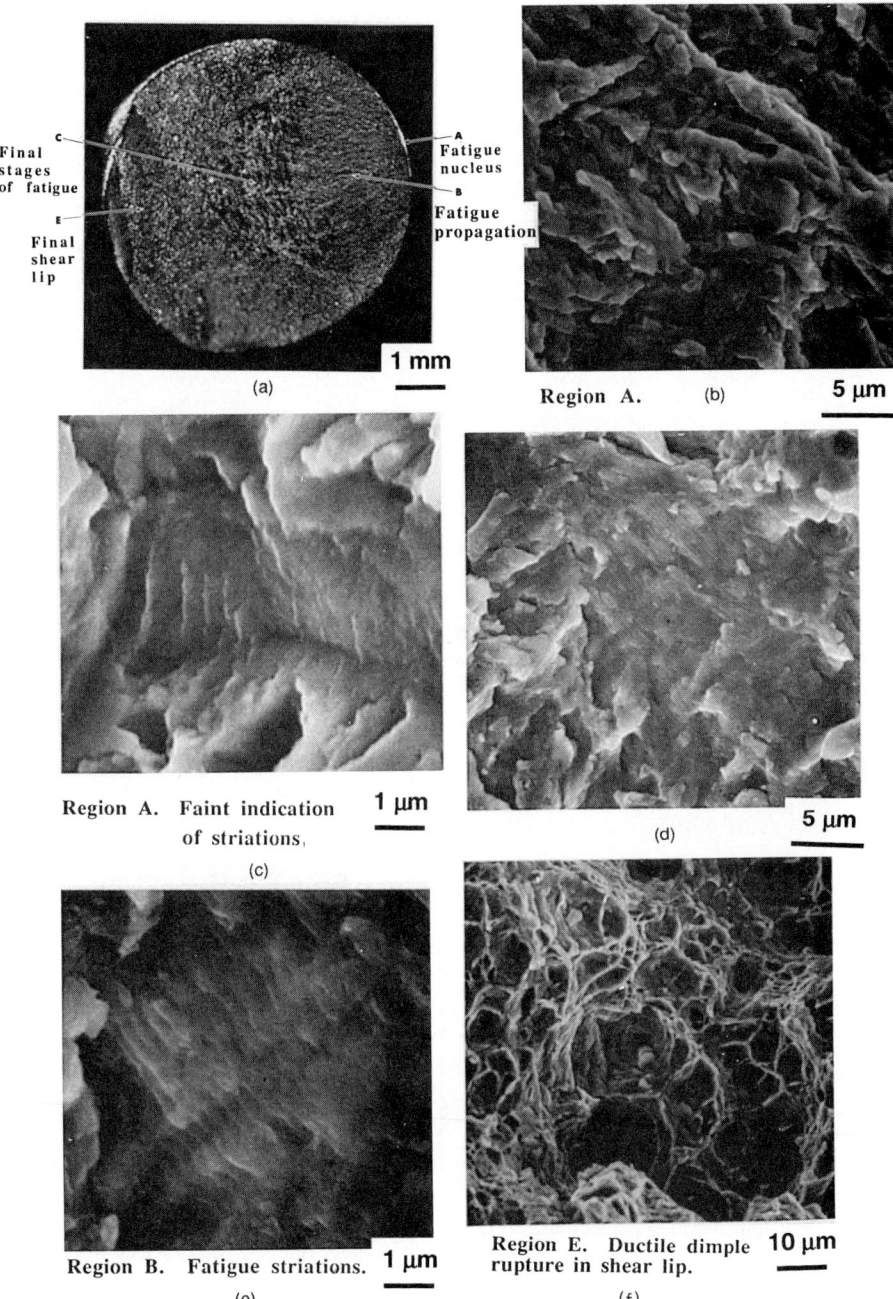

Figure 4.53 Fractographs of a fatigue test sample (high-cycle axial fatigue test). Stress 43,300 to 168,000 lb/in^2 (2.98 × 10^8 to 1.16 × 10^9 Pa); frequency 2600 cycles/min; number of cycles to failure 221,000. (*Adapted from Bhattacharyya et al.*[9])

The next example is a 4140L steel which has been quenched and tempered at 1050°F (566°C). Figure 4.54a shows the fracture surface of a tensile sample. The ductility was high (54 percent RA). A small fibrous zone in the center, coarse radial marks, and a shear-lip zone are present. The radial marks are much coarser than those for the 10B21 steel (Fig. 4.51a). All three regions show fine dimples (Fig. 4.54b to d), and these are much finer (and somewhat less well defined) than those for the tensile sample of the 10B21 steel (Fig. 4.51b). The difference in the morphology of the radial marks and of the dimples is related to the greater amount of carbides in the 0.4 percent carbon 4140 steel and perhaps to a finer carbide distribution.

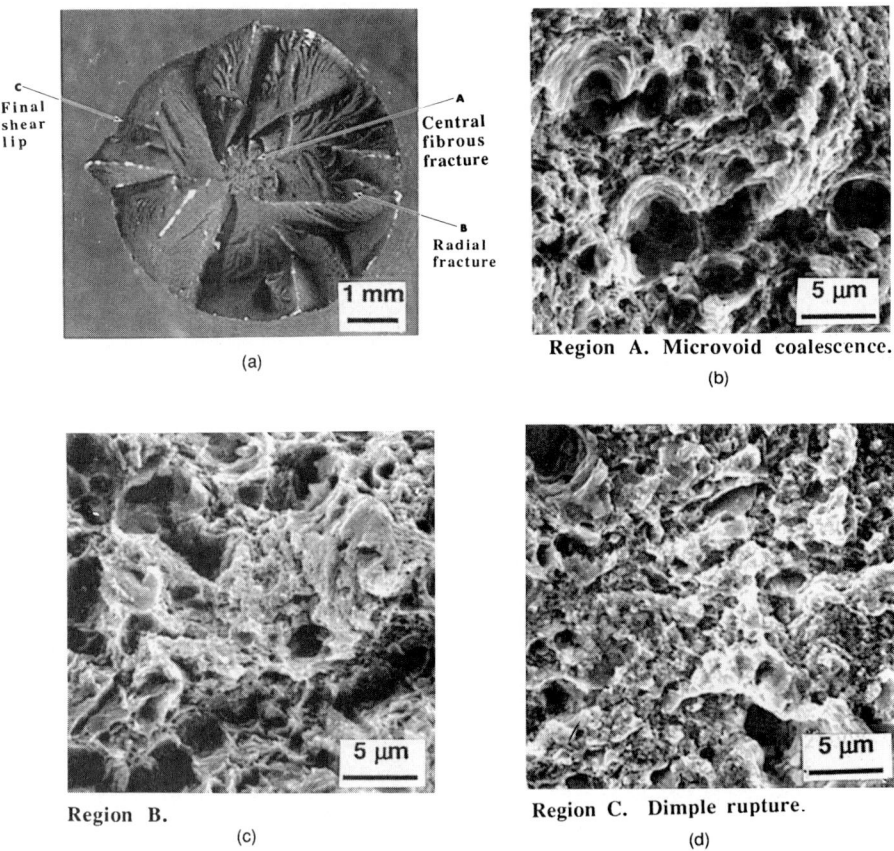

Figure 4.54 Fractographs of a quenched and tempered (1050°F; 566°C for 2 h) 4140L steel tensile sample. The mechanical properties were tensile strength 144,800 lb/in^2 (9.98 × 10^8 Pa); ductility RA 54%. (*Adapted from Bhattacharyya et al.*[9])

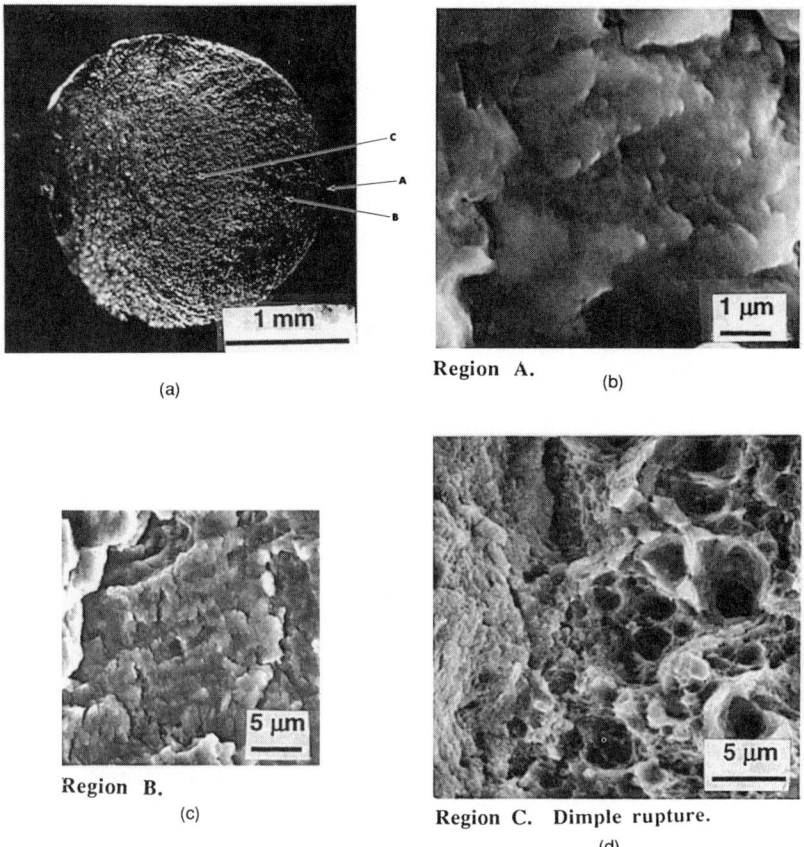

Figure 4.55 Fractographs of a rotating-beam fatigue test sample of 4140L steel. Maximum longitudinal stress 90,000 lb/in^2 (6.20 × 10^8 Pa); frequency 10,000 cycles/min; number of cycles to failure 250,800. (*Adapted from Bhattacharyya et al.*[9])

The appearance of a fatigue fracture surface of this 4140L steel is shown in Fig. 4.55a. Fine radial marks point to the region of the initiation of the fatigue crack. Ill-defined striations are observed (Fig. 4.55b and c), best revealed by the secondary cracking. The final overload region shows fine dimples (Fig. 4.55d).

We now look at several examples of a 4340 steel which has been quenched and tempered at 400°F (204°C). Figure 4.56a shows a tensile sample fracture surface. A small fibrous zone at the center, a region of fine radial marks, and a large shear-lip zone are present. The fine dimple structure of the fracture surface is shown in Fig. 4.56b to d. The morphology of this sample is similar to that of the 10B21 steel (Fig. 4.51).

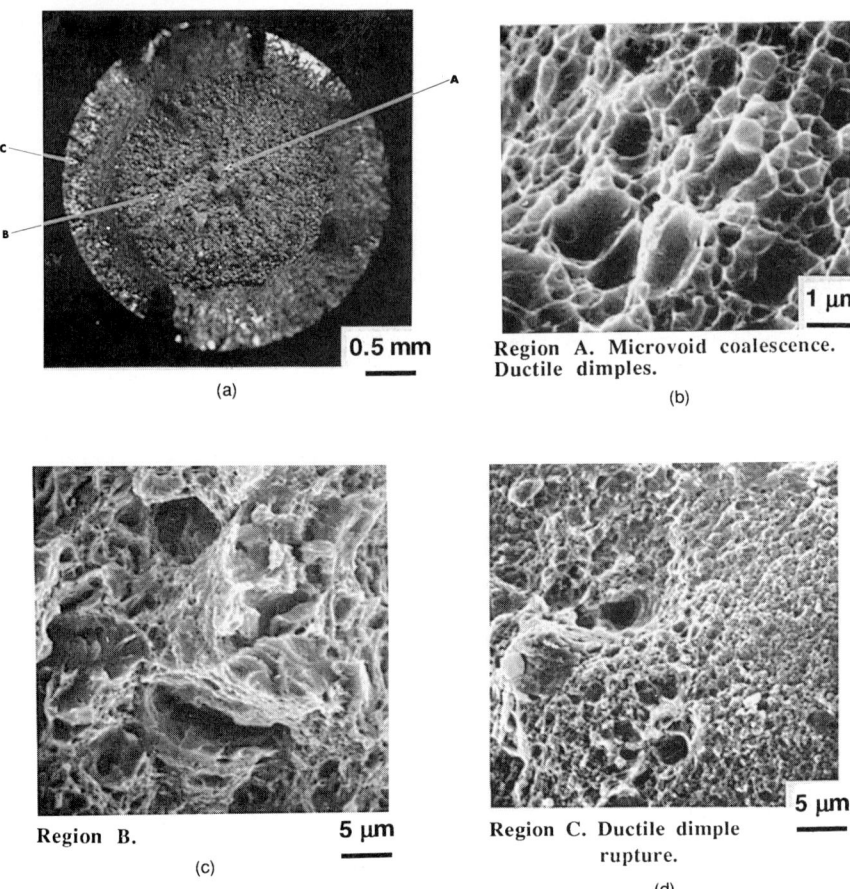

Figure 4.56 Fractographs of a tensile test sample of quenched and tempered (400°F, 204°C, for 1 h) 4340 steel. The mechanical properties were yield strength 198,000 lb/in² (1.36 × 10⁹ Pa); tensile strength 251,000 lb/in² (1.73 × 10⁹ Pa); ductility RA 45%. (*Adapted from Bhattacharyya et al.*[9])

The fracture surface of an impact sample for this steel tested at −40°F (−40°C) is shown in Fig. 4.57a. Radial lines point toward the notch, and there is only a small shear lip. The region of initial crack growth contains dimples (Fig. 4.57b), but the rest of the fracture surface (except for the small shear lip) contains a mixture of dimples and cleavage, or quasicleavage (Fig. 4.57c and d). Figure 4.58a shows the impact fracture surface obtained upon fracture at high temperature, 70°F (21°C). Again, the radial marks point clearly toward the origin of the fracture (the notch), and there is more shear-lip region than for the lower test temperature. For this higher test temperature the entire surface fractured

Fracture Modes and Macrofractographic Features

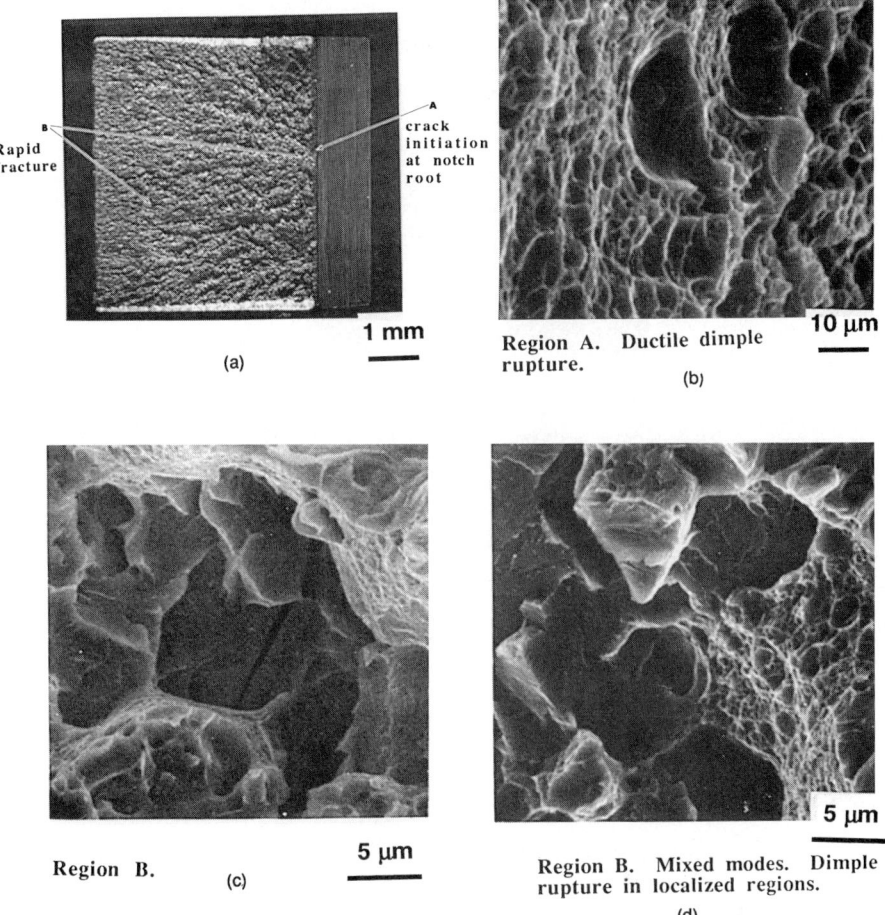

Figure 4.57 Fractographs of a Charpy test sample (−40°F, −40°C, test temperature) of the 4340 steel. Impact energy 9 ft · lb (12 J). (*Adapted from Bhattacharyya et al.*[9])

by void coalescence, as evidenced by the dimple morphology (Fig. 4.58*b* and *c*).

Figure 4.59*a* shows a fatigue test fracture surface for this 4340 steel for a high-cycle axial test. The radial marks point to the origin of the fatigue fracture. Fractography (Fig. 4.59*b* to *e*) shows little clear indication of fatigue striations, although some are revealed by the secondary cracking. The final overload fracture region shows dimples (Fig. 4.59*f*). These characteristics can be compared to the case of low-cycle axial fatigue by examination of Fig. 4.60. The macrofractographic features are similar (Fig. 4.60*a*). However, the

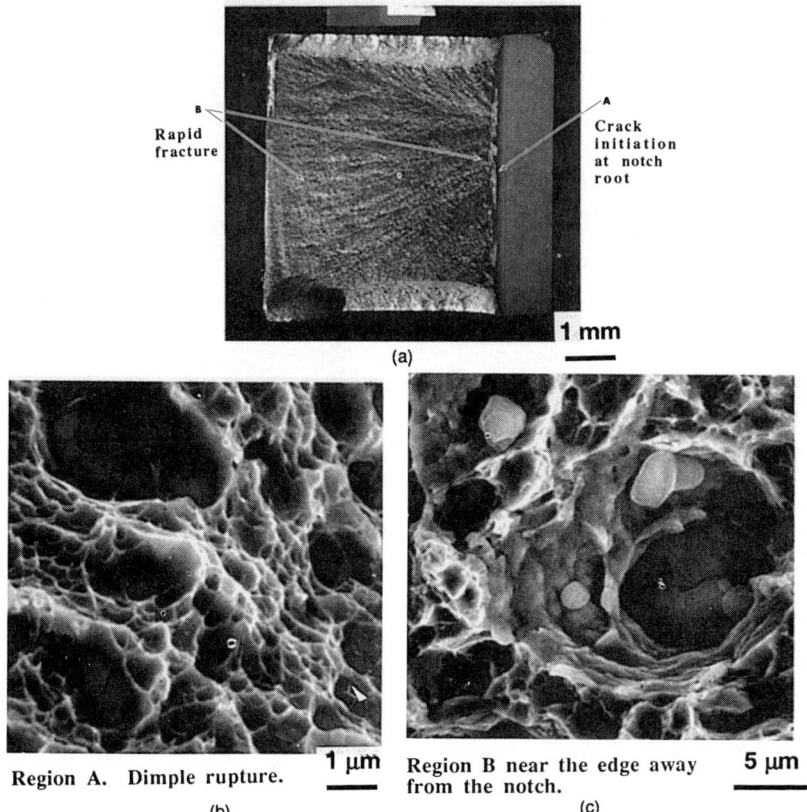

Figure 4.58 Fractographs of a Charpy test sample (70°F, 21°C, test temperature) of the 4340 steel. Impact energy 49 ft · lb (66 J). (*Adapted from Bhattacharyya et al.*[9])

microfractographic features show no evidence of striations (Fig. 4.60b and c).

Figure 4.61a shows the fracture surface of a fatigue sample of this material for tempering at 1050°F (566°C), and hence of lower strength. The fractographs (Fig. 4.61b to d) do not clearly reveal striations, but they appear to be faintly resolved in places (Fig. 4.61b). In Fig. 4.61c there is a striated region, but these may not be fatigue striations, as finer striations are resolved from this region at higher magnification (Fig. 4.61d). The final overload fracture shows dimples (Fig. 4.61e).

The fracture surface of the tensile sample shown in Fig. 4.56 was for the steel quenched and tempered, with a microstructure of fine carbides in ferrite. Figure 4.62 shows the fracture-surface features for

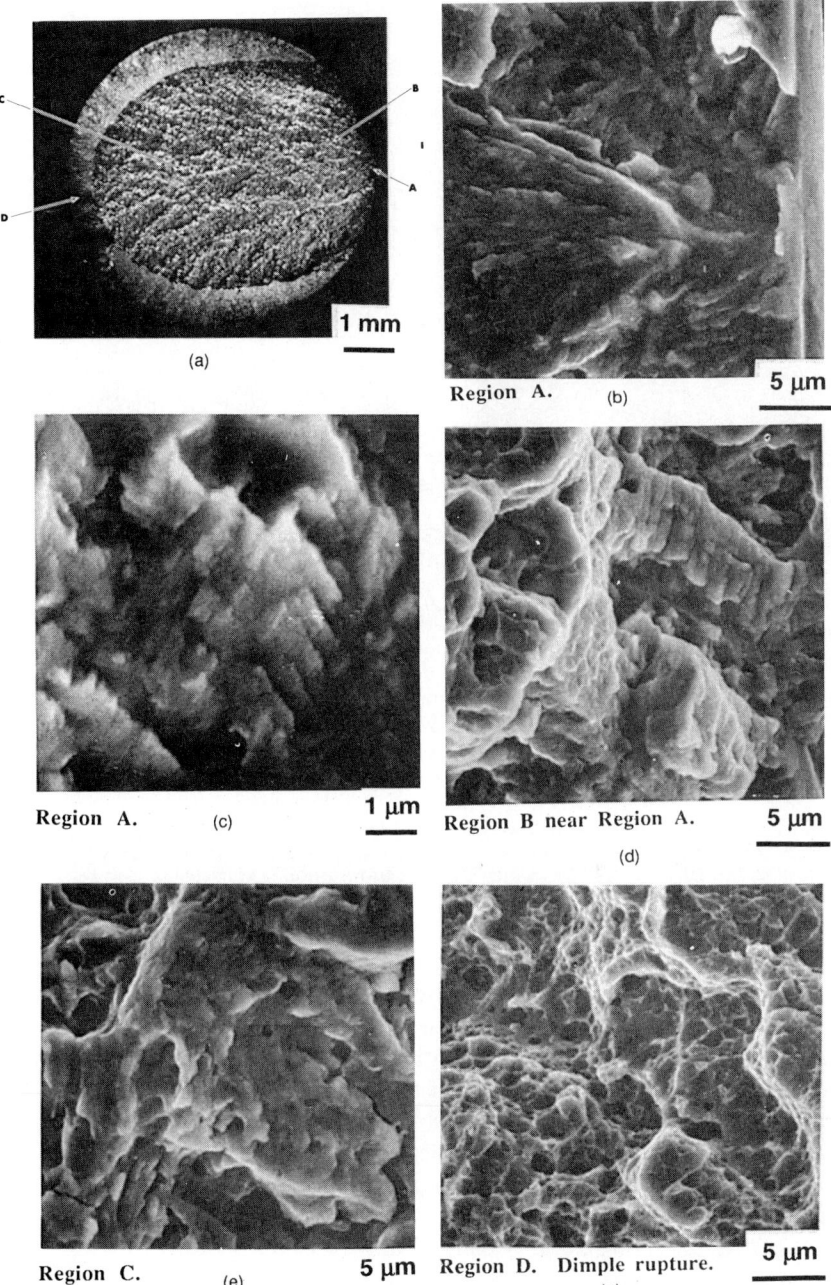

Figure 4.59 Fractographs of a high-cycle axial fatigue test sample of the 4340 steel. Maximum stress 96,000 lb/in^2 (6.61 × 10^8 Pa); minimum stress 0; test frequency 3600 cycles/min; number of cycles to failure 148,000. (*Adapted from Bhattacharyya et al.*[9])

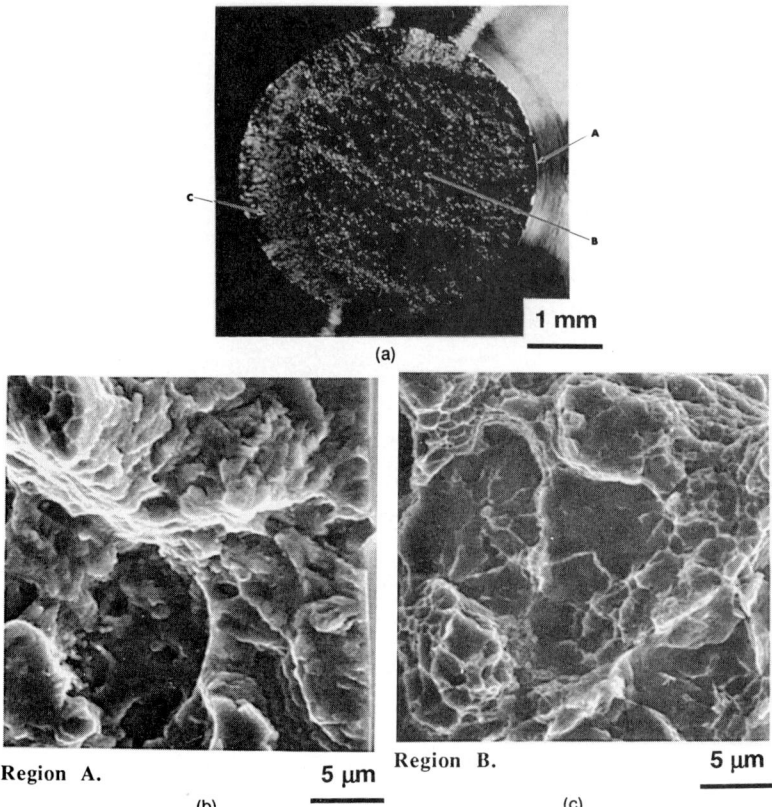

Figure 4.60 Fractographs of a low-cycle axial fatigue test sample of the 4340 steel. Maximum stress 242,400 lb/in^2 (1.67 × 10^9 Pa); minimum stress 7600 lb/in^2 (5.24 × 10^7 Pa); test frequency 6 cycles/min; number of cycles to failure 1317. (*Adapted from Bhattacharyya et al.*[9])

this steel in the normalized condition, which has a heterogeneous microstructure of primary ferrite and pearlite. This is much lower strength than in the quenched and tempered condition. The central fibrous region shows only faint radial marks (Fig. 4.62a). The fractographs in Fig. 4.62b and c show striations, which could be taken as proof of a fatigue fracture. However, this type of fatigue artifact was mentioned in Sec. 3.16 and is due to fracture through the lamellar structure of the pearlite. Figure 4.62d shows dimples, as does the region of the shear-lip zone (Fig. 4.62e).

The fracture surface of an impact test sample of this normalized steel is shown in Fig. 4.63. Faint radial marks point to the origin of fracture (Fig. 4.63a). The crack initiation site at the root of the notch

Fracture Modes and Macrofractographic Features 259

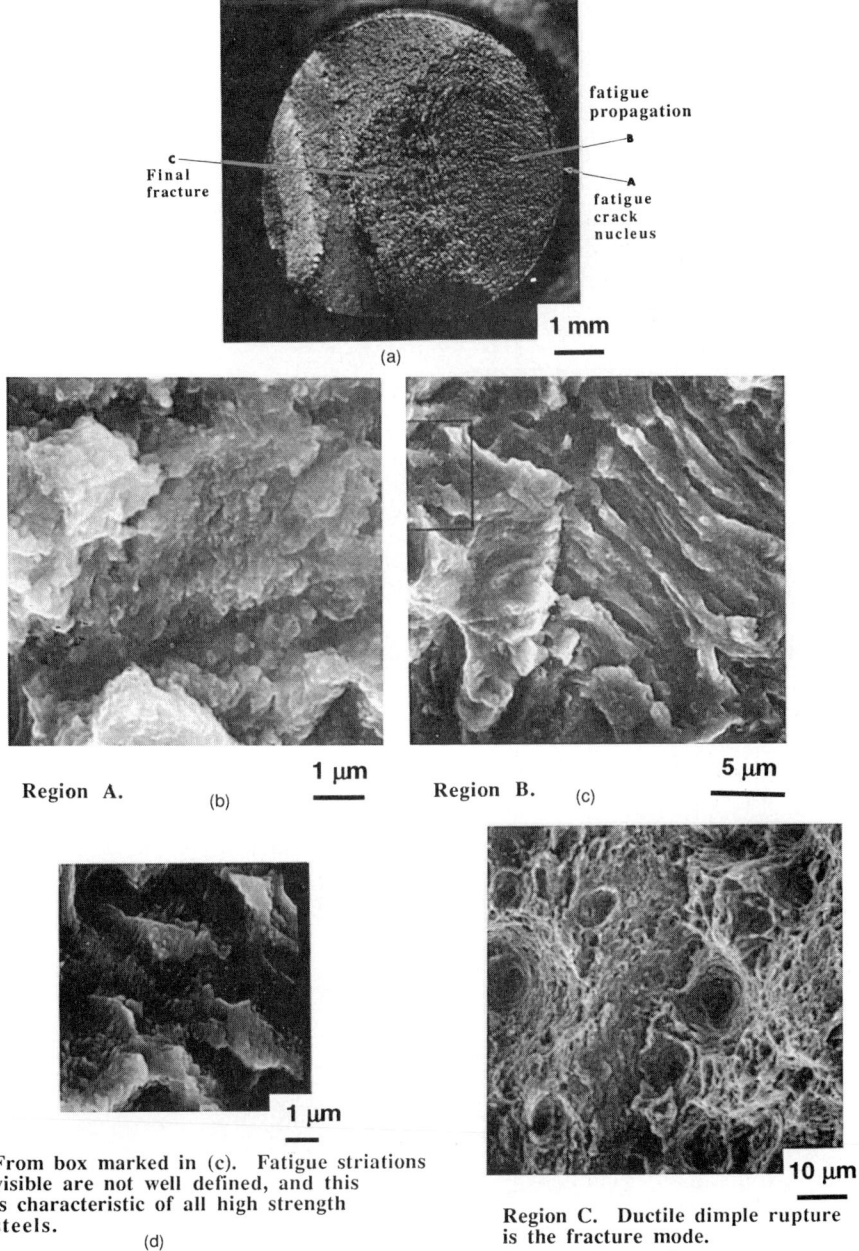

Figure 4.61 Fractographs of a high-cycle axial fatigue test sample of the 4340 steel, tempered at 1050°F (566°C). Maximum stress 89,000 lb/in^2 (6.13 × 10^8 Pa); minimum stress 2500 lb/in^2 (1.72 × 10^7 Pa); frequency 3600 cycles/min; number of cycles to failure 217,000. (*Adapted from Bhattacharyya et al.*[9])

260 Chapter Four

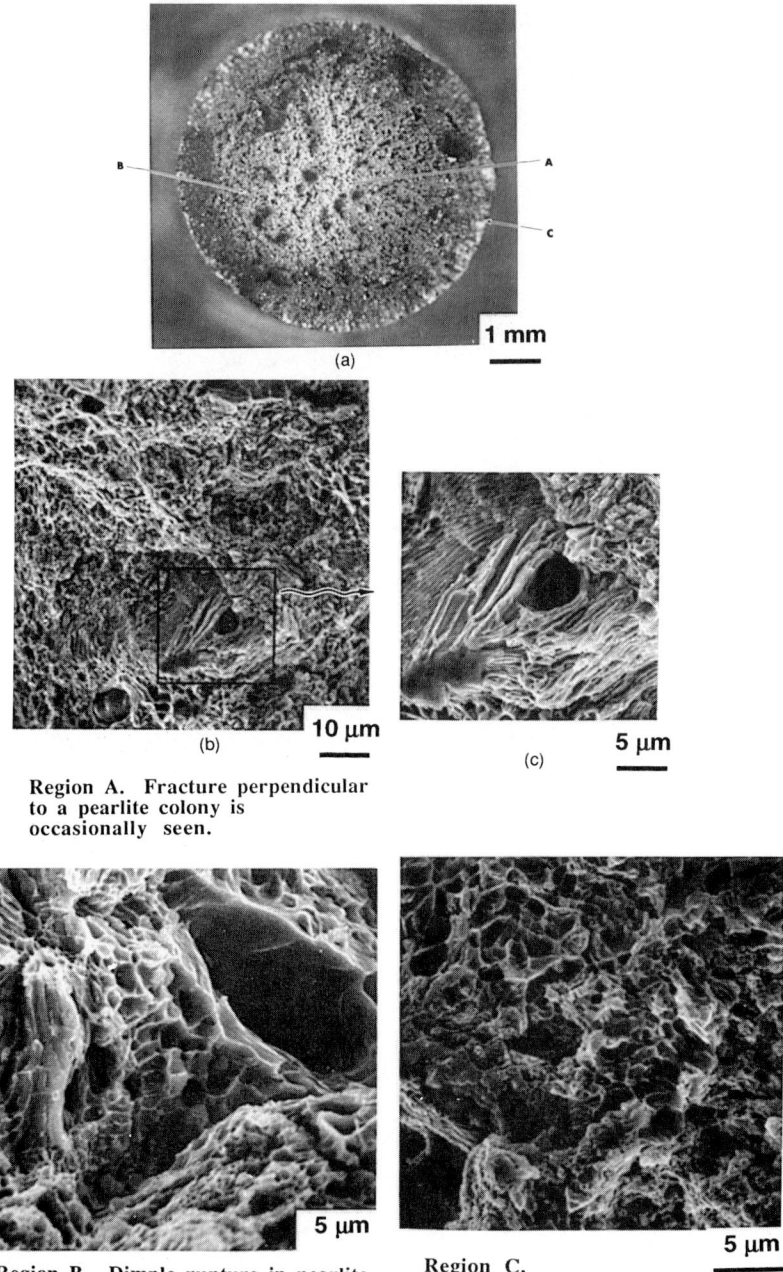

Region A. Fracture perpendicular to a pearlite colony is occasionally seen.

Region B. Dimple rupture in pearlite.   Region C.

Figure 4.62 Fractographs of a 4340 steel tensile test sample in the normalized condition. The mechanical properties were yield strength 57,500 lb/in^2 (3.96 × 10^8 Pa); tensile strength 101,500 lb/in^2 (6.99 × 10^8 Pa); ductility RA 45%. (*Adapted from Bhattacharyya et al.*[9])

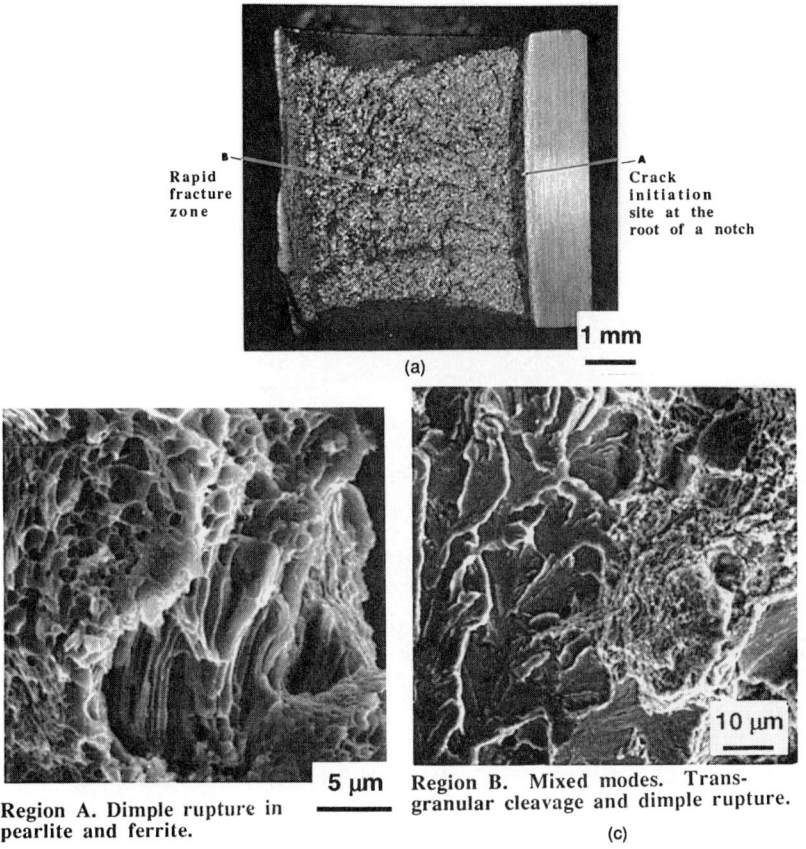

Figure 4.63 Fractographs of a Charpy impact sample (broken at 70°F, 21°C) of the normalized 4340 steel. Impact energy 39 ft · lb (54 J). (*Adapted from Bhattacharyya et al.*[9])

consists of dimples (Fig. 4.63b), and the striated regions are where fracture occurred in the lamellar pearlite. However, most of the fracture propagated by a mixed mechanism of cleavage and void coalescence, as revealed in Fig. 4.63c.

Figure 4.64 shows the features of the quenched and tempered steel fractured in monotonic torsion. The macroscopic swirl pattern is seen clearly (Fig. 4.64a). The final fracture region is that marked D. The microfractography (Fig. 4.64c to e) shows that the surface is covered with shear dimples whose orientation reflects the torsion loading.

For comparison of the microfractographic features of a fatigue fracture in this normalized steel, the fractographs in Fig. 4.65 are presented. Macroscopically, fine radial marks point to the fatigue fracture initiation site (Fig. 4.65a). Faint striations are resolved in places

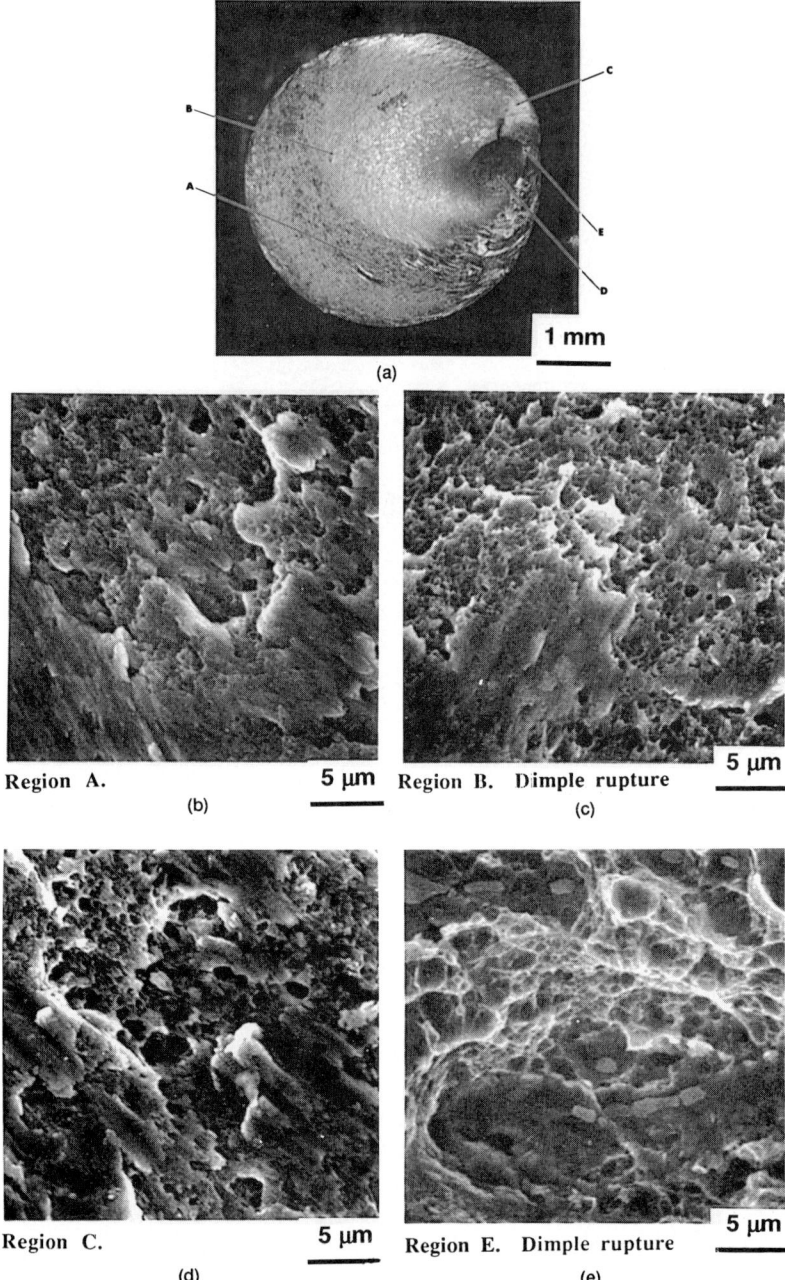

Figure 4.64 Fractographs of a torsion test of a 4340 steel, quenched and tempered (800°F, 426°C, for 1 h). Torque to failure 52 ft · lb (70 J); arc 720°. (*Adapted from Bhattacharyya et al.*[9])

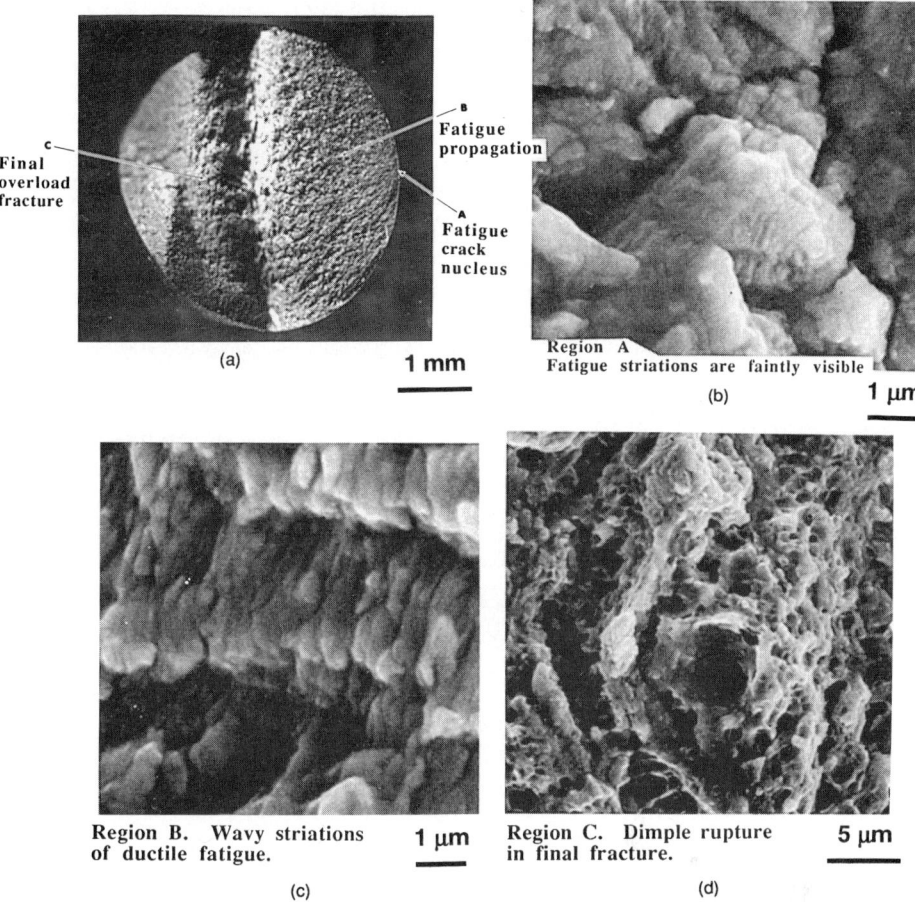

Figure 4.65 Fractographs of a high-cycle axial fatigue test sample of a 4340 normalized steel. Maximum stress 71,000 lb/in^2 (4.89 × 10^8 Pa); minimum stress 0; frequency 2600 cycles/min; number of cycles to failure 853,000. (*Adapted from Bhattacharyya et al.*[9])

on the fatigue crack propagation region (Fig. 4.65b and c). Compare their appearance to that of fracture through the pearlite shown in Fig. 4.62c and d. The final overload fracture occurred by void coalescence (Fig. 4.65d).

We now present the fracture appearance of a higher-strength material, an H-11 steel. The ductility was relatively high (28 percent RA), but about half that of the materials of the previous examples. The tensile test fracture surface is shown in Fig. 4.66a and is similar to that of the sample of Fig. 4.56a. The fracture mechanism was by void coalescence (Fig. 4.66b and c), but the dimples are less de-

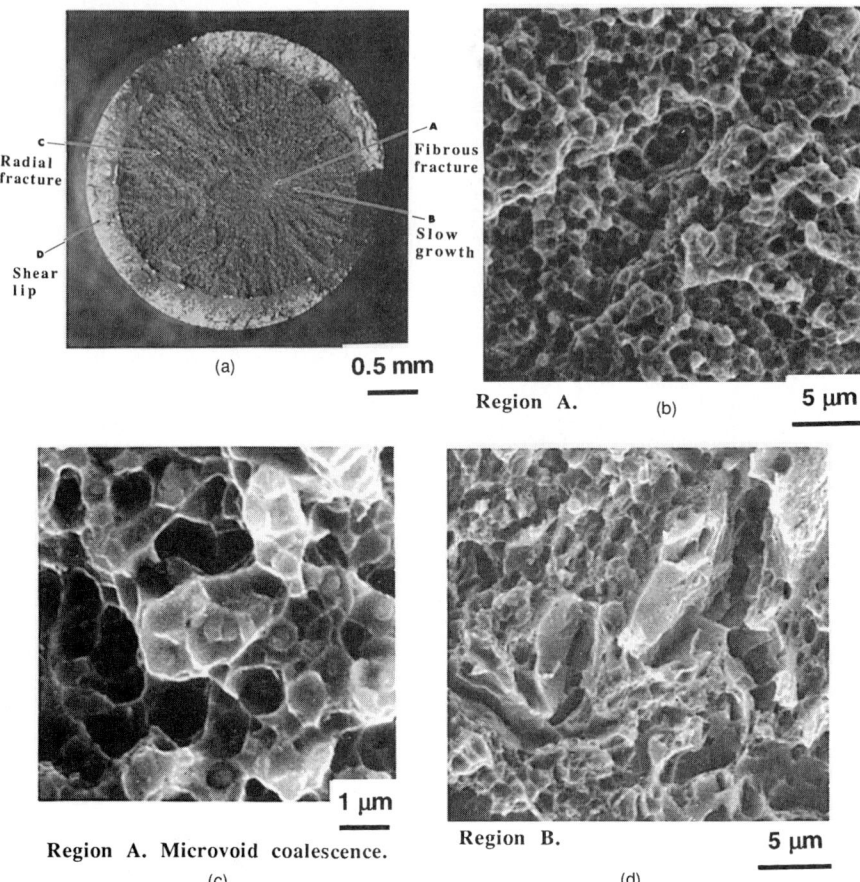

Figure 4.66 Fractographs of a tensile sample of an H-11 steel. The steel was austenitized, air-cooled, then double-tempered at 1000°F (538°C) for 2.5 h. The mechanical properties were yield strength 242,000 lb/in^2 (1.67 × 10^9 Pa); tensile strength 299,000 lb/in^2 (2.06 × 10^9 Pa); ductility RA 28%. (*Adapted from Bhattacharyya et al.*[9])

fined than those shown in Fig. 4.56. The impact fracture surface in Fig. 4.67a reveals well-defined radial marks. The initial stage of fracture propagation occurred by void coalescence (Fig. 4.67b), but the later stages by quasicleavage (Fig. 4.67c). The microstructures of both this high-strength H-11 steel and the 4340L steel consisted of fine carbides in ferrite. Both steels had about the same amount of carbides as the carbon content was about 0.4 percent for each. The different fracture micromechanisms, and the difference in appearance for the same mechanism (such as void coalescence), must be due to the difference in the dispersion of the carbides, which is an

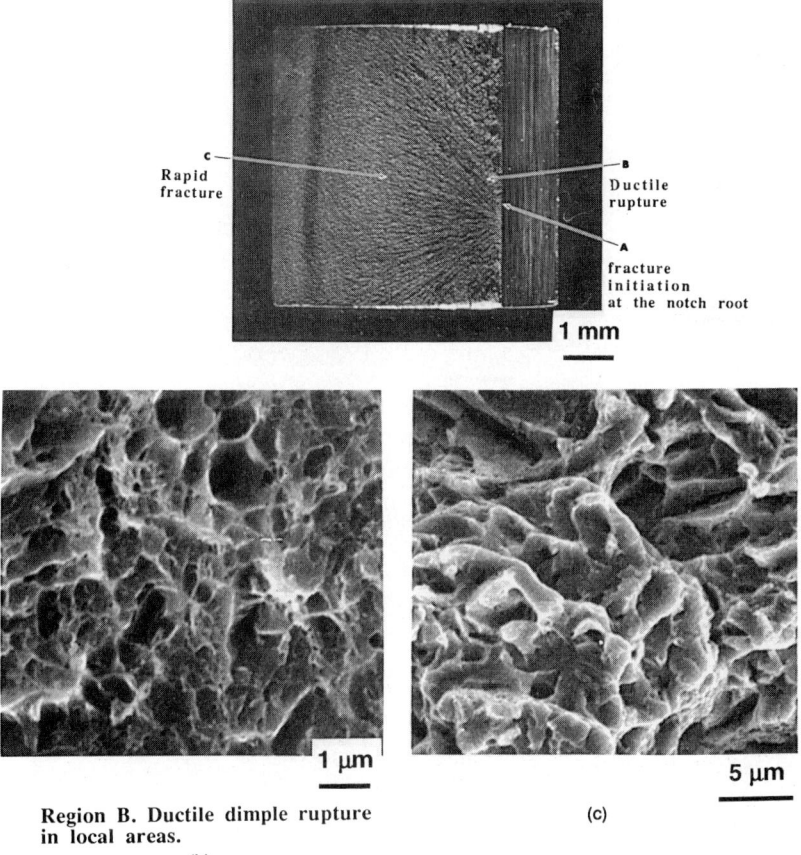

Region B. Ductile dimple rupture in local areas.
(b) (c)

Figure 4.67 Fractographs of a Charpy impact sample of the H-11 steel. Impact energy 5 ft · lb (7 J). (*Adapted from Bhattacharyya et al.*[9])

example of the strong influence of the microstructure on the fracture process and surface morphology.

The previous examples all involved relatively complex microstructures. The next example is a 304 austenitic stainless steel, which is mainly single-phase (except for inclusions). This material was very ductile (80 percent RA). Its fatigue fracture surface is shown in Fig. 4.68a, with radial marks pointing to the origin of fracture. Even though this test was a continuous constant-frequency, constant-maximum-load test, there are faint conchoidal marks (see arrows). Most of the fracture surface shows clear fatigue striations (Fig. 4.68b and c). However, in the region near the final overload fracture, a mixture of coarser fatigue striations and dimples is present (Fig. 4.68c). The final overload fracture occurred by void coalescence (Fig. 4.68d).

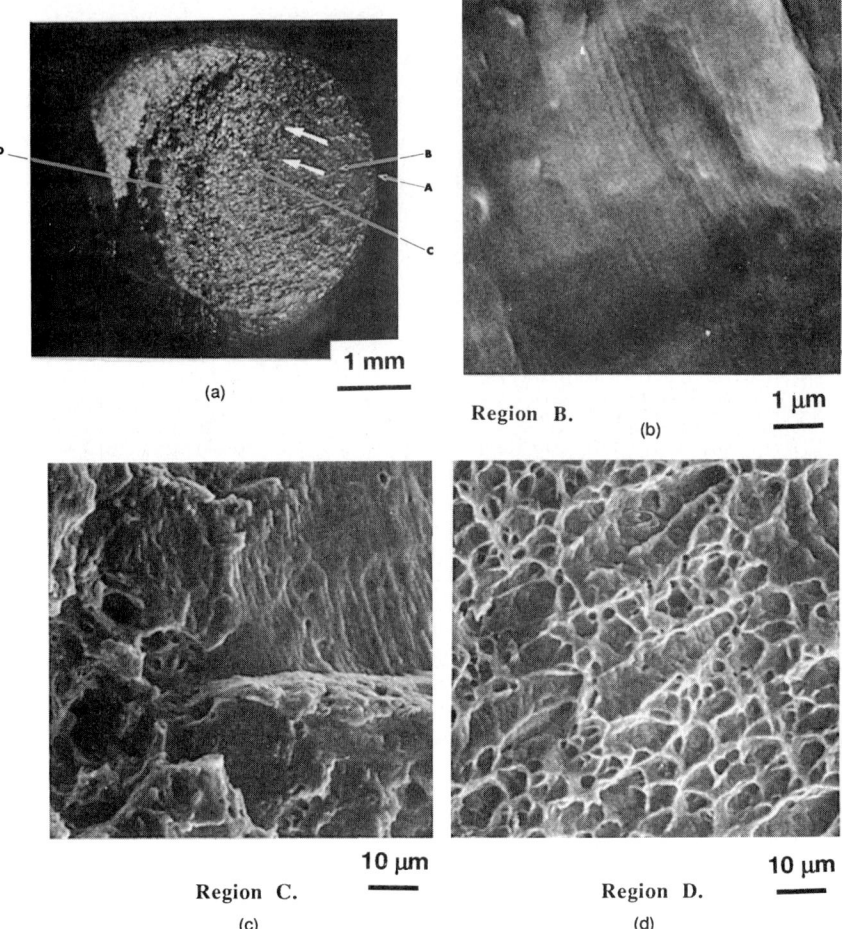

Figure 4.68 Fractographs of a high-cycle axial fatigue test sample of a 304 austenitic stainless steel. Maximum stress 89,000 lb/in^2 (6.13 × 10^8 Pa); minimum stress 4000 lb/in^2 (2.76 × 10^7 Pa); frequency 3600 cycles/min; number of cycles to failure 945,000. (*Adapted from Bhattacharyya et al.*[9])

Figure 4.69a shows the fracture surface of a tensile sample of a brittle material (1008 steel). Note the relatively shiny areas which are present. These are due to reflections from the relatively flat and coarse cleavage planes illustrated in Fig. 4.69b. A similar shiny appearance can be caused by intergranular fracture in coarse-grained material if the grain-boundary facets are smooth. An example is shown in Fig. 4.70.

Fracture Modes and Macrofractographic Features 267

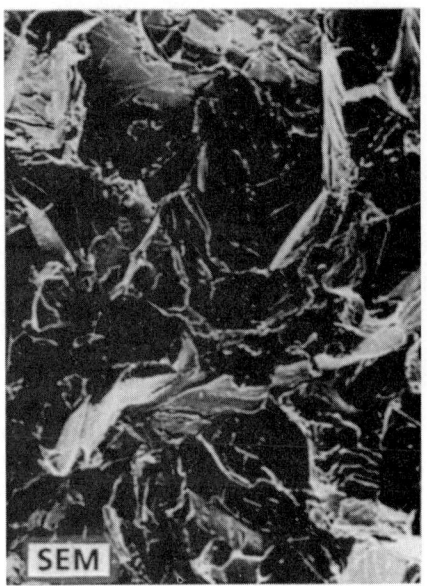
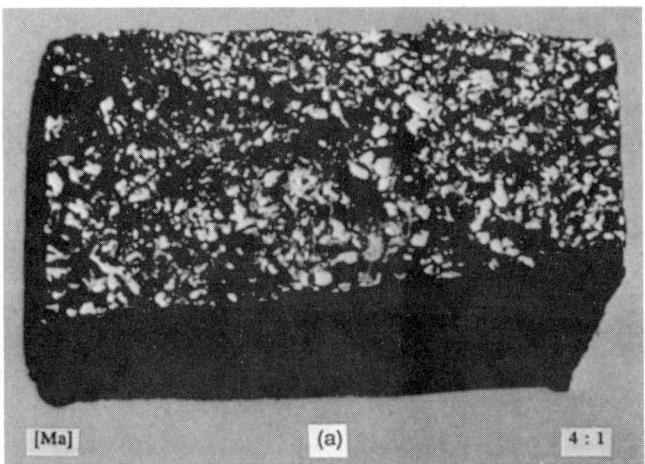

Figure 4.70 Tensile fracture surface of a sample of large-grained iron containing 0.68% P, showing intergranular fracture. (*From Henry and Horstmann.*[8])

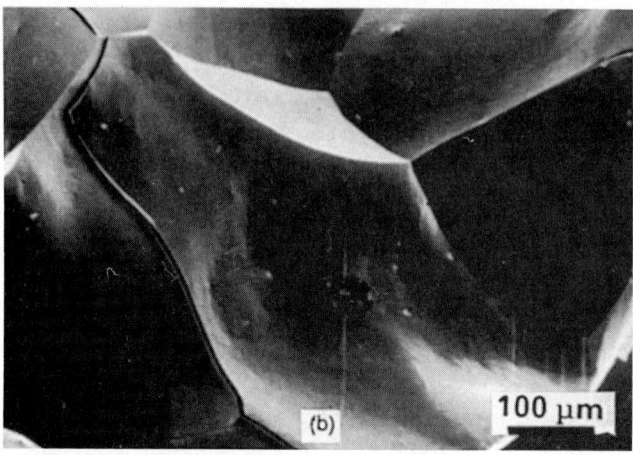

Figure 4.70 (*Continued*) Tensile fracture surface of a sample of large-grained iron containing 0.68% P, showing intergranular fracture. (*From Henry and Horstmann.*[8])

References

1. G. F. Vander Voort, "Visual Examination and Light Microscopy," in *Metals Handbook*, 9th ed., vol. 12: *Fractography*, American Society for Metals, Metals Park, Ohio, 1987, p. 91.
2. *Metals Handbook*, 8th ed., vol. 9: *Fractography and Atlas of Fractographs*, American Society for Metals, Metals Park, Ohio, 1974.
3. L. Engel and H. Klingele, *An Atlas of Metal Damage*, Carl Hanser Verlag, Munich, Germany, 1981.
4. J. Nunes, F. L. Carr, and F. R. Larson, "Macrofractographic Techniques," in R. F. Bunshah (ed.), *Techniques for the Direct Observation of Structure and Imperfections*, pt. 1, Wiley, New York, 1968, p. 379.
5. *Metals Handbook*, 9th ed., vol. 12: *Fractography*, American Society for Metals, Metals Park, Ohio, 1987.
6. D. J. Wulpi, *Understanding How Components Fail*, American Society for Metals, Metals Park, Ohio, 1985.
7. *Metals Handbook*, 8th ed., vol. 10: *Failure Analysis and Prevention*, American Society for Metals, Metals Park, Ohio, 1975.
8. G. Henry and D. Horstmann, *De Ferri Metallographia*, vol. V: *Fractography and Microfractography*, Verlag Stahleisen m.b.H., Düsseldorf, Germany, 1979.
9. S. Bhattacharyya, V. E. Johnson, S. Agarwal, and M. A. H. Howes (eds.), *IITRI Fracture Handbook, Failure Analysis of Metallic Materials by Scanning Electron Microscopy*, IIT Research Institute, Chicago, Ill., 1979.
10. R. E. Peterson, "Interpretation of Service Failures," in M. Hetenyi (ed.), *Handbook of Experimental Stress Analysis*, Wiley, New York, 1950, p. 593.

Bibliography

Cottell, G. A., "Fatigue Failures, with special reference to Fracture Characteristics," in F. R. Hutchings and P. M. Unterweiser, *Failure Analysis: The British Engine Technical Reports*, American Society for Metals, Metals Park, Ohio, 1981.

Lang, G. A. (ed.), *Systematic Analysis of Technical Failures,* DGM Informationsgesellschaft Verlag, Braunschweig, Germany, 1986.
Metals Handbook, 9th ed., vol. 11: *Failure Analysis and Prevention,* American Society for Metals, Metals Park, Ohio, 1986.

Chapter 5

Case Studies

5.1 Introduction

In this chapter we present case studies chosen to illustrate the principles of metallurgical failure analysis. In each study references are made to specific points covered in the previous chapters in order to show how these were used in the analyses.

These studies deal with real failures, in which the scope of the failure analysis was limited by factors such as the time that could be devoted to the problem, the expense of the testing desired, and particularly incomplete information about the failure situation. Thus the reader will find that some of the conclusions are quite tentative, which is a common result of analyses of such real failures.

Microstructural analysis is invariably a part of most metallurgical failure analyses. However, in this book we have made no attempt to review the physical metallurgy of the alloys examined. The reader will find this information in many books and references, such as the two listed in the Bibliography.

5.2 Case A: A Cracked Vacuum Bellows*

5.2.1 Introduction

Tubular metal bellows are useful devices for joining components. One use is to make a vacuum seal for a movable component which must be inserted into a vacuum chamber, such as a shaft that must be moved, either axially or sideways. An example is shown in Fig. 5.1a. This device is used to move the seat, with its rubber gasket, down onto a flat surface to provide a vacuum seal. This is accomplished by turning the

---

*The authors acknowledge the work done by David Dellinger on this case study.

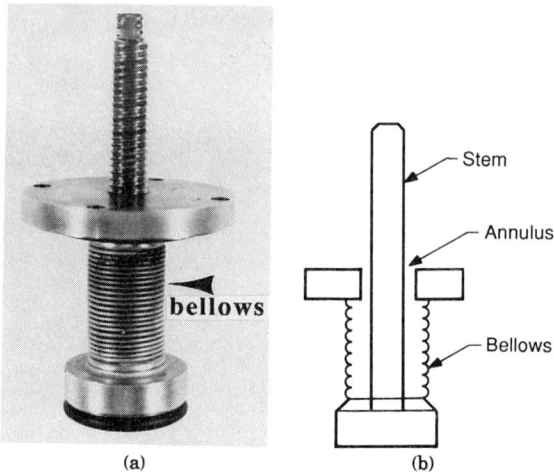

Figure 5.1 (a) Bellows used for a vacuum seal. (b) Cross-sectional view.

shaft, which is connected to a pivot joint, as shown in the cross-sectional drawing in Fig. 5.1b. Thus the atmosphere on the inside of the bellows never enters the chamber except through the open valve. On the shaft there is no sliding O-ring-type seal, which would be subject to wear and require lubrication.

A bellows very similar to that in Fig. 5.1 was on a vacuum system of an arc melter, which was periodically pumped down to about 10^{-4} torr (mechanical pump only). This was done only occasionally (such as a few times a month). The operators noticed over several months that the time required to attain the desired vacuum had increased and eventually reached an unacceptable length, so that the system had to be examined carefully for a source of leaks. Inspection of the bellows revealed a fine crack across one fin (Fig. 5.2). It was barely discernible by eye and had a greenish deposit at each end of the crack.

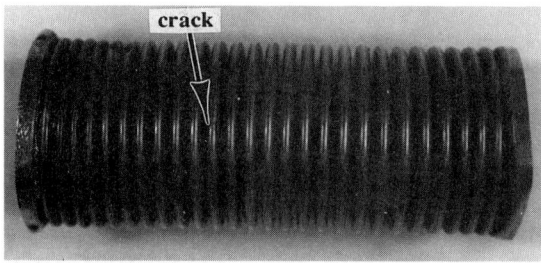

Figure 5.2 Bellows that was analyzed to determine the cause of the fine crack at the location noted by arrow.

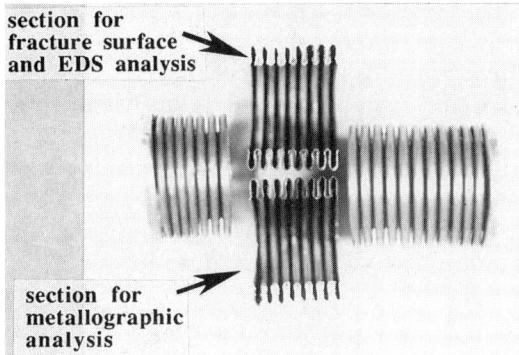

Figure 5.3 Two sections that were removed for analysis.

5.2.2 Experimental procedure

The surface of the tube was examined in a scanning electron microscope to determine the material from which the tube was made and to obtain chemical analyses of the corrosion deposits, using the energy-dispersive spectrometer (EDS) on the scanning electron microscope. Although the EDS analysis was only qualitative (see Chap. 1), it was sufficiently sensitive to distinguish the relative concentration of the elements present.

To examine the inside of the tube, it was sectioned with a thin blade (0.004 mm thick) on a high-speed cutoff wheel using copious water for cooling, followed immediately by rinsing in methanol followed by drying. The sections thus obtained are shown in Fig. 5.3. The section containing the crack was notched on each side of the crack, as shown in Fig. 5.4a,

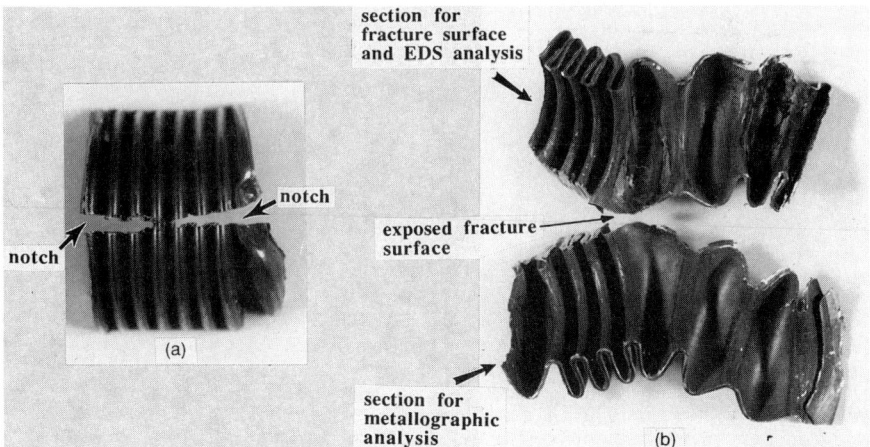

Figure 5.4 (a) Section that was notched on each side of the crack, so that upon breaking, the crack surface was exposed. (b) Sample in (a) after breaking to expose the crack surface.

and then broken to expose the original crack (Fig. 5.4b), which could easily be distinguished from the fresh crack formed while breaking the piece. (The procedure used to break the section stretched the bellows.) One section was examined in the scanning electron microscope. The other section was cut through the crack and one-half embedded in plastic to make a metallographic mount, which allowed examination of the crack itself in cross section. This sample was ground and polished using standard procedures for brass, then etched in an ammonium hydroxide–hydrogen peroxide solution.

5.2.3 Results

Figure 5.5 is a scanning electron micrograph showing the crack on the outside of the tube. The corrosion products on each end of the crack are prominent. On the unaffected part of the surface, the EDS analysis revealed only Cu (strong) and Zn. The corrosion products showed Cu, Zn, S, Cl, Ca, and K. Examination of the inside of the bellows showed that the fin adjacent to the one that had the crack contained similar corrosion products at the same location as the crack. On the surface of other fins many cracks were found which had not pene-

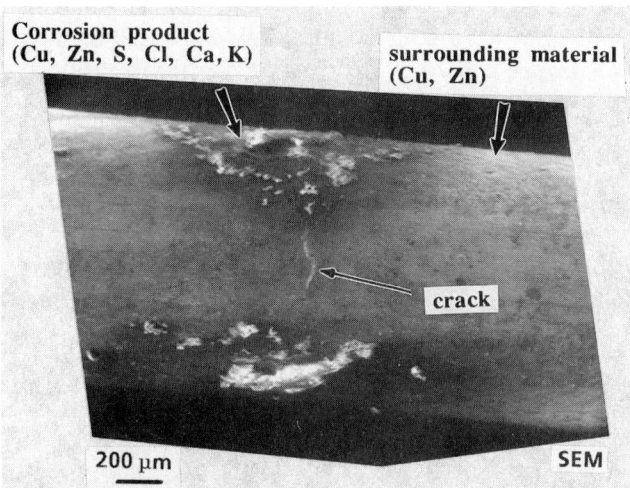

Figure 5.5 Scanning electron micrograph of the crack in the bellows (Fig. 5.2). Corrosion products were on each end, and the EDS analysis of one is given. Also shown is the EDS analysis from the uncorroded surface, revealing only copper and zinc.

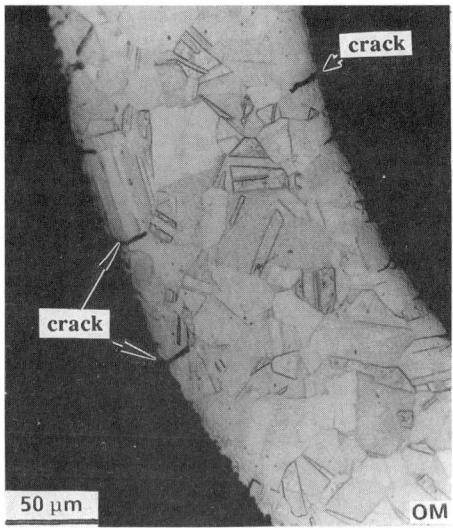

Figure 5.6 Optical micrograph of the cross section of the bellows wall, showing cracks on the surface of the bellows which had not penetrated through the wall thickness.

trated through the wall of the tube. The cross-sectional view of the bellows wall in Fig. 5.6 shows such cracks.

The fracture-surface topography was obscured by corrosion products, which were too tenacious to be removed by three replicating tapes (see Sec. 1.4.1). However, EDS analysis revealed on the fresh fracture (caused by breaking the sample) only Cu (strong) and Zn, whereas on the original fracture surface Cu (strong), Cl, Si, and Al were detected. It is to be noted that no Zn was found on the fracture surface.

The microstructure (Fig. 5.7) revealed numerous deformation bands (see Glossary) and bent annealing twins (see Sec. 3.3), showing that the material was in the cold-worked condition (see Brooks,[1] Chap. 1). This appearance was found on the bend in the fins and on the straight sides. Numerous cracks were found across the wall in the bend in many of the fins, such as that noted by the arrow in Fig. 5.7 and shown in Fig. 5.6. Nearer the location of the crack, several of these regions were found (Fig. 5.8). The cracks seem to be both intergranular and transgranular, perhaps following the deformation bands in the latter case. Figure 5.9b shows an area very near the fracture cross section, and this same area was examined in the scanning electron microscope (Fig. 5.9a). The EDS analysis revealed that some of the regions which are shown in relief in Fig. 5.9a were relatively low in Zn. Upon examination with an optical microscope (Fig. 5.9b) these raised regions had a noticeable copper color to them, compared to the surrounding material.

276 Chapter Five

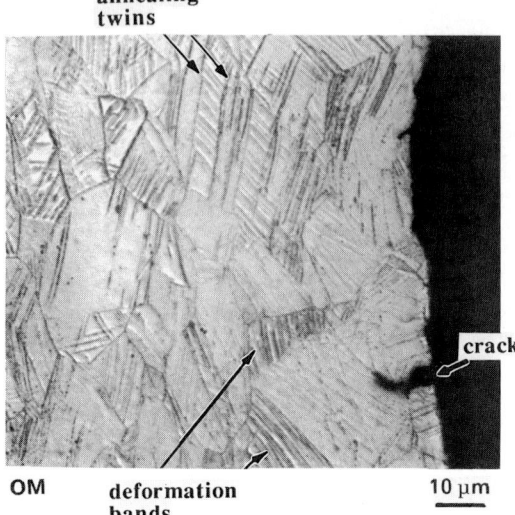

Figure 5.7 Optical micrograph of the cross section of the bellows. Note the deformation bands and bent annealing twins showing that the material was cold-worked.

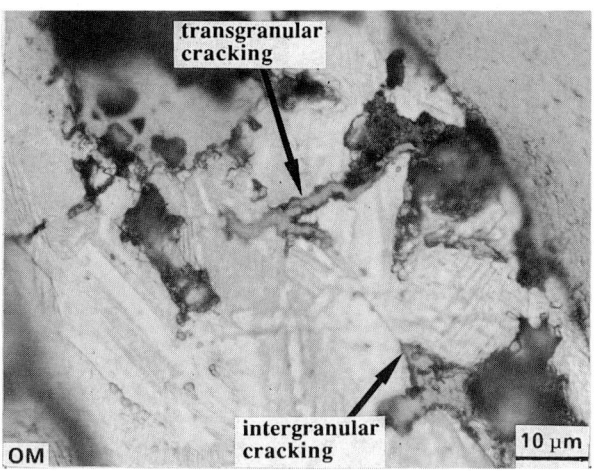

Figure 5.8 Optical micrograph of the cross section near the crack surface, showing numerous cracks at the outside surface.

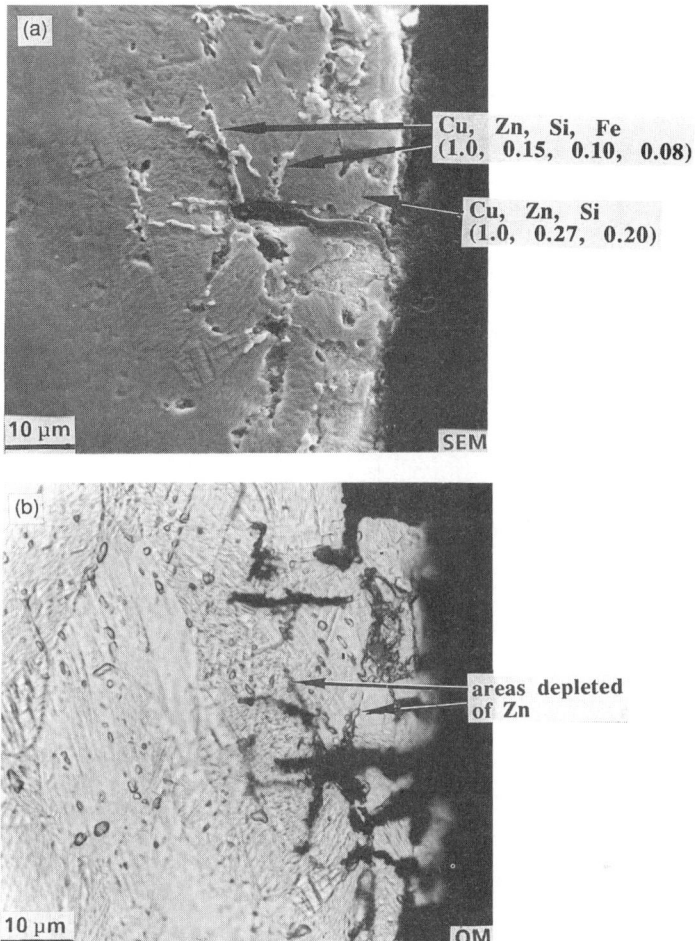

Figure 5.9 (a) Scanning electron micrograph showing a crack penetrating the wall. Note that the EDS analysis of the small regions in relief contained less Zn than that of the adjoining matrix. (The analyses are given in terms of the intensity of the elements to that of Cu.) (b) Optical micrograph of the same area; the areas noted had a copper color.

5.2.4 Discussion

The construction of bellows involves the formation of fins by the simultaneous application of hydraulic pressure to the inner surface and of axial compression to a tube, such as shown in Fig. 5.10. Clearly at the bend there is severe plastic deformation. However, the microstructure of the straight portion of the tube showed extensive plastic

278 Chapter Five

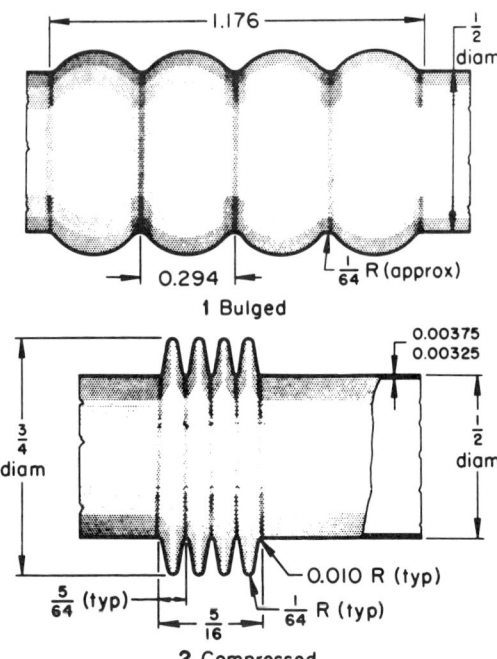

Figure 5.10 Schematic diagram of the process of deforming a tube into a bellows. (*From* Metals Handbook.[2])

deformation (Fig. 5.7), so the starting tube was already cold-worked to give a strong bellows.

The lack of Zn on the crack surface and the copper-depleted grains near the crack surface (Fig. 5.9a) show that the corrosion process involved dezincification. This is selective leaching or dealloying,[3–5] a common corrosion mechanism in brasses containing more than about 15 percent Zn.[3–5] The Zn is dissolved, and a copper-rich region is left behind. This usually leads to porosity,[6] although the microstructure in Fig. 5.9 showed no evidence of it. Note in Fig. 5.11 that the zinc-depleted grains show fine, straight annealing twins, with no sign of deformation bands, which indicates that these grains have formed during dezincification. Such fine, twinned grains were found by Polushkin and Shuldener[6] in brass hot-water pipes which had been corroded in service.

A factor that favors dezincification is surface deposits, which deplete the region under the deposit of oxygen and set up an oxygen concentration cell.[5,7] In the case of this failed bellows, the surface on which the crack started was only in contact with air, but if a deposit formed, then such a mechanism could operate. However, the origin of the deposit is not clear. The deposits contained S, Cl, Si, Ca, Al, and K. The outside of the bellows was untarnished, with no evident corrosion except at the crack itself (Fig. 5.5), but the inside surface was cor-

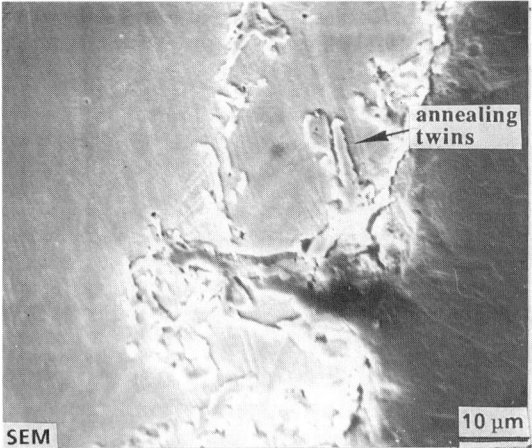
Figure 5.11 Micrograph showing annealing twins in the copper-rich region formed by dezincification.

roded. A heavy corrosion deposit was found on the inside fin adjacent to the one containing the crack and at the same location. It is speculated that these regions contained flux deposits from the brazing process used to join the bellows to the plates. Fluxes used for brazing brass may contain chlorides (see *Metals Handbook*,[8] p. 1035), an element that is known to promote dezincification.[7] This flux could not be easily removed after brazing because of the small clearance in the annulus (Fig. 5.1b). Moisture from air could accumulate in the region because of this and activate the corrosion if it were absorbed by the flux.

Cold-worked brass is susceptible to stress corrosion cracking (season cracking)[4] (see Sec. 3.11), and dezincification can play a key role in such cracking.[5] Transgranular fracture associated with dezincification has been reported in cold-worked brass,[3] which is consistent with the crack appearing to follow the deformation bands in the present study. The many cracks seen on the inside and outside surfaces indicate that the bellows was undergoing generalized stress corrosion cracking, but the process was accelerated at the region of corrosion in which dezincification occurred.

The problem can be avoided by using a brass of no more than 15 percent Zn.[4] If such an alloy will not meet the strength requirements, then another material, such as austenitic stainless steel, will have to be used to make the bellows.

5.2.5 Conclusions

The brass bellows had developed a fine crack through the wall thickness in the cold-worked region of the bend. Other fine cracks were observed which had not penetrated the wall. The crack formation was associated with corrosion, and dezincification occurred in this region. The corrosion was induced by a surface contaminant, probably soldering flux which was not removed during subsequent cleaning of the bellows assembly. The problem may have been aggravated by moisture which condensed in the narrow annulus. The crack probably propagated by stress corrosion cracking (season cracking). Brass containing less than 15 percent Zn usually is not susceptible to this type of attack and cracking.

5.3 Case B: Failure of a Large Air-Conditioning Fan Blade*

5.3.1 Introduction

In some air-cooling systems for large buildings the hot air is passed through a water-cooled chilling tower. The large volume of air requires a relatively large fan, consisting of several blades rotating around a common axis and operating as a turbine fan. A schematic drawing of such a fan is shown in Fig. 5.12. The neck of the blade is

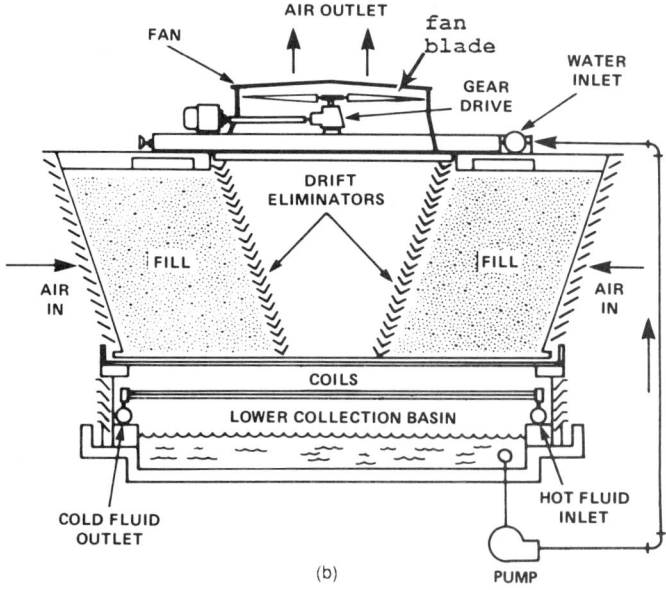

Figure 5.12 Schematic diagram of a centrifugal fan in a cooling tower. (*From* ASHRAE Handbook.[9] *Reprinted by permission of the American Society of Heating, Refrigerating and Air-Conditioning Engineers.*)

---

*The authors acknowledge the work done by Hwa-Perng Kao on this case study.

the location that sustains the greatest centrifugal and bending stress, and the subject of this case study is a fan blade which fractured at that location. No history of the blade nor operating conditions were available, except that the blade rotated at about 550 r/min.

5.3.2 Experimental procedure

A drawing of the broken blade is shown in Fig. 5.13. Samples for analysis were cut from the broken blade. The larger pieces were removed using a hack saw, and then smaller pieces were cut using a water-cooled cutoff wheel. After cutting, each piece was immediately washed in water, then acetone, and then dried.

The metallographic samples were mounted in plastic and then ground and polished using standard methods. The etchant used was

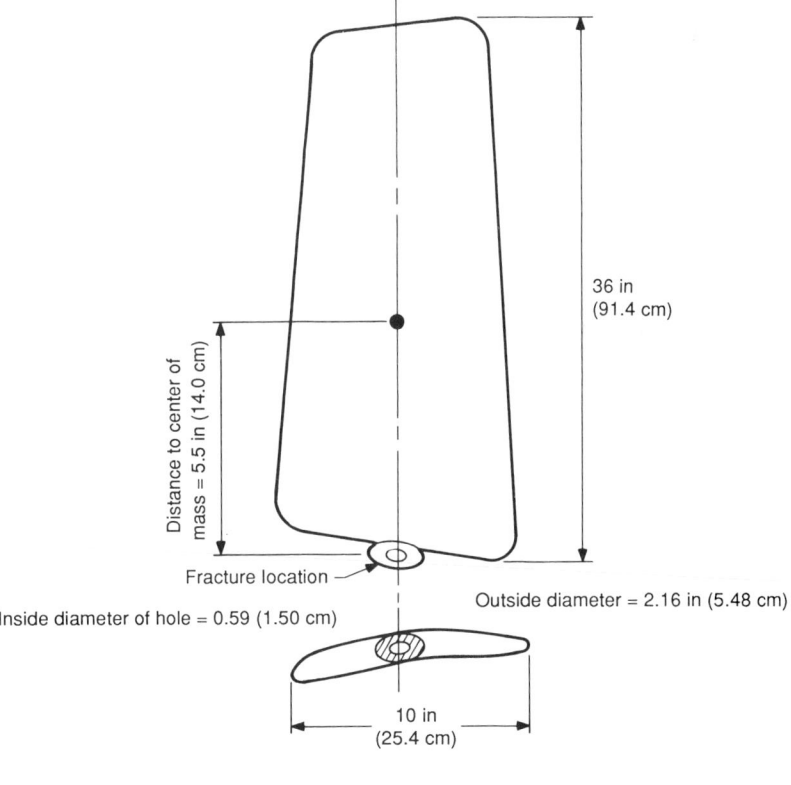

Figure 5.13 Drawing of broken fan blade.

HF in water, although the microstructure was revealed in the unetched condition. The microstructure was examined using optical microscopy and scanning electron microscopy. The EDS on the scanning electron microscope was used to identify the elements in the alloy and in the phases.

Hardness measurements were made using the Rockwell E scale (accurate to about ±2). Subsize tensile samples were made from the blade and tested to fracture in order to obtain tensile mechanical properties. The tensile strength was accurate to about 3.5 MPa (500 lb/in^2) and the elongation at fracture to about 1 percent. Both metallographic and tensile samples were solution heat-treated at 538°C (±5°C) (1000°F) for 12 h, then quenched in water. Some of these were subsequently aged at 155°C (±5°C) (311°F) for 3, 3.5, and 4 h, then air-cooled to 25°C (77°F). These samples allowed a comparison of the effects of heat treatment on the properties and the microstructure.

5.3.3 Results

Fractography and microstructure. Figure 5.14 shows the region of the fracture. Note that there is no obvious plastic deformation. The fracture surface is shown in Fig. 5.15, where the radial marks point back to the origin of the fracture (see Sec. 4.2).

Figure 5.16 shows low-magnification views of the polished section

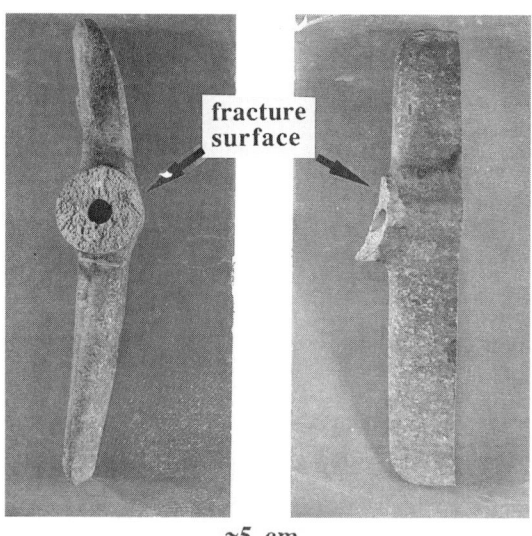

Figure 5.14 Fracture region of broken fan blade.

Case Studies 283

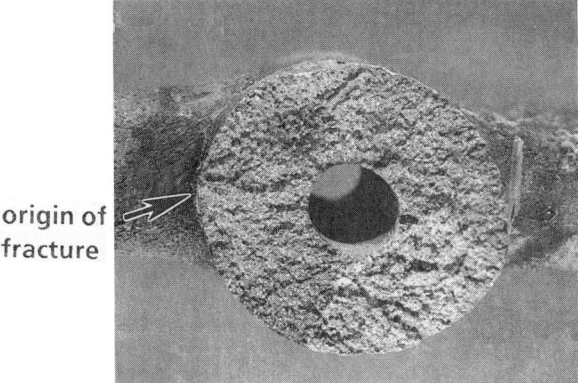

origin of fracture

Figure 5.15 Fracture surface, showing radial marks which point to the origin of fracture.

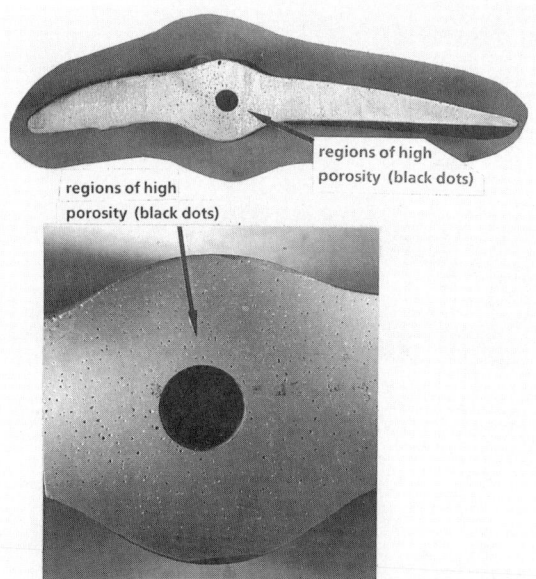

Figure 5.16 Polished section near the fracture surface, showing regions of high porosity in the center.

just below the fracture surface. Note the regions of porosity, which suggest that the blade was made by casting. The spherical areas are due to bubbles of gas that are ejected as the metal freezes and then trapped before they can leave the liquid. These holes are revealed on the fracture surface (Fig. 5.17) and in microstructures, shown in

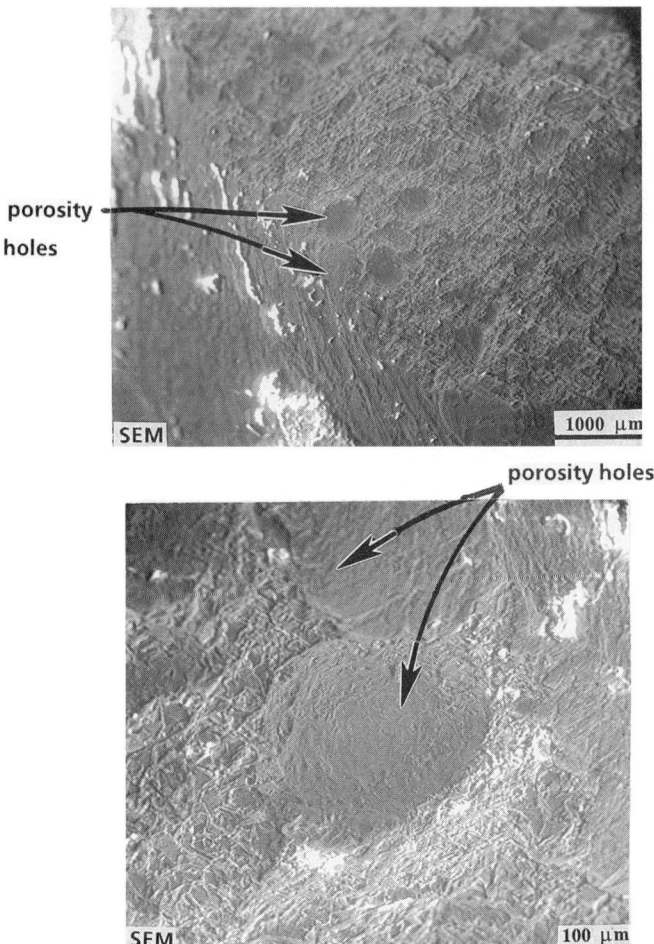

Figure 5.17 Scanning electron micrographs of fracture surface. The spherical holes are where fracture occurred through the holes in Fig. 5.16.

cross section in Fig. 5.18, in a section near the fracture surface. Note that there is also a prominent group of elongated porosity present in this area.

The microstructure in Fig. 5.19 shows interdendritic porosity, which is considerably finer than the spherical holes. Here the dendrites are surrounded by a separate phase. In Fig. 5.20 the phases are identified (based on EDS results). The matrix α is Al, containing a small amount of Si in solid solution. Two types of par-

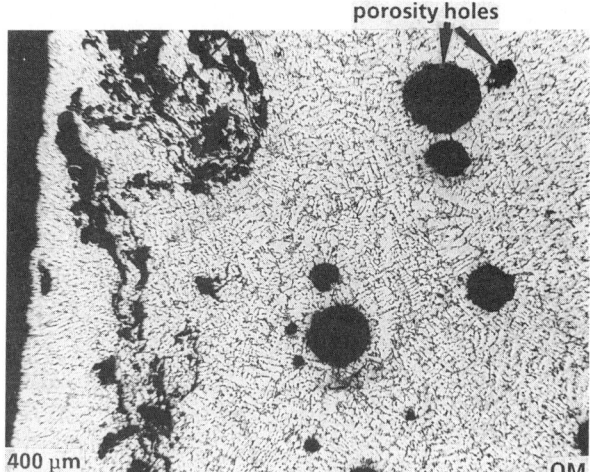

Figure 5.18 Optical micrograph of section near fracture surface, showing porosity.

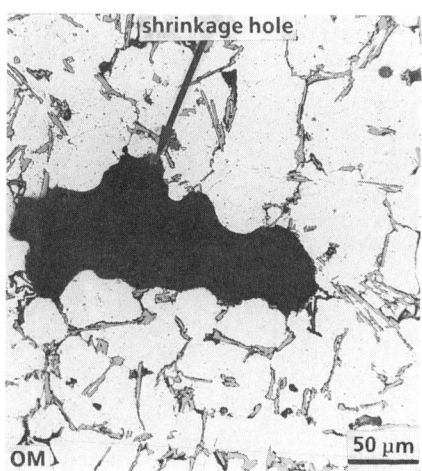

Figure 5.19 Microstructure of region showing interdendritic shrinkage porosity.

ticles are present—the majority is essentially pure Si, the balance an intermetallic (Fe–Al–Mg–Si) compound. These are regions of a eutectic structure, part of which is α.

These results show that the fan blade was made of an Al casting alloy, a likely candidate being alloy 356 (Table 5.1). Based on the binary

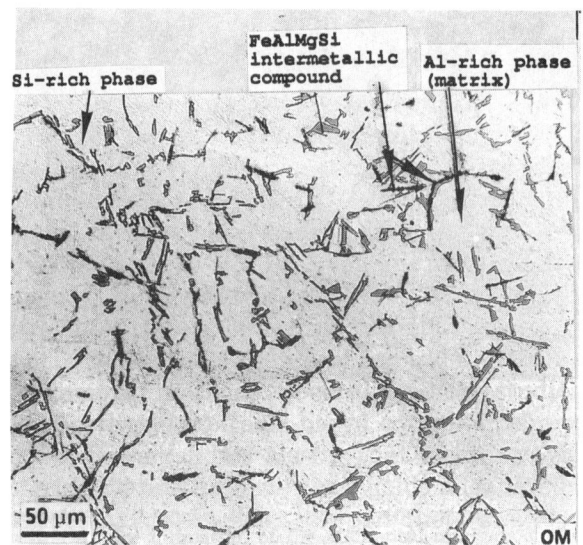

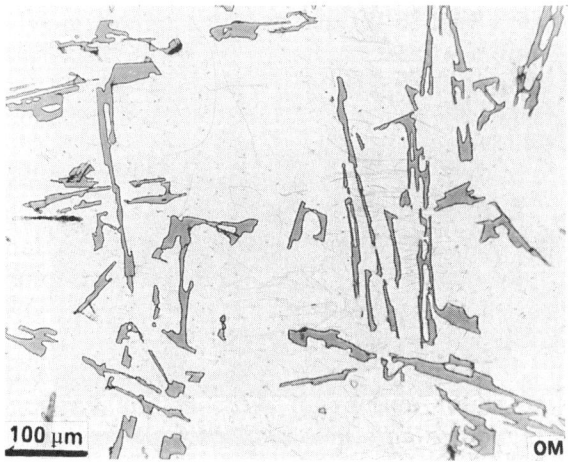

Figure 5.20 Microstructures showing eutectic particles in Al-rich α matrix. Note angular plates with sharp edges.

Al–Si phase diagram (Fig. 5.21), a composition of about 7 percent Si in Al will give the amount of primary α seen in the microstructure (Fig. 5.20a) if the alloy is cooled relatively slowly from the liquid to 25°C (77°F), such as would occur for a sand casting.

Note in Fig. 5.20b that the particles are relatively sharp. Upon

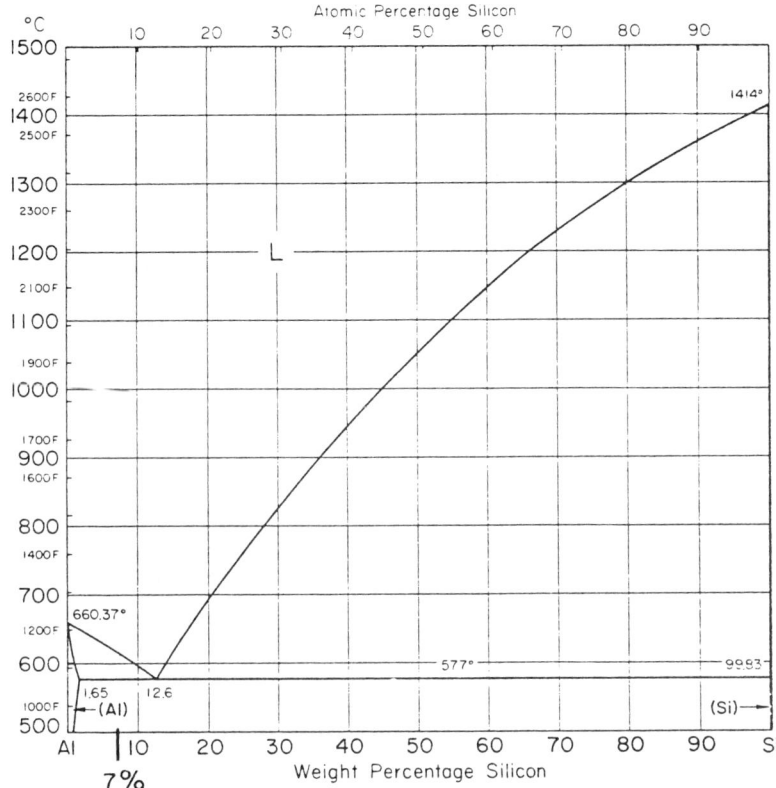

Figure 5.21 Aluminum-silicon phase diagram. The approximate composition of the fan blade alloy is shown at 7% Si. (*From* Metals Handbook.[11])

solution-treating a sample, water quenching, then aging for 3 to 4 h at 155°C (311°F), the particles become rounded (Fig. 5.22). Supposedly this morphology enhances tensile mechanical properties.[12]

Mechanical properties. The hardness of the casting ranged from 45 to 55 Rockwell E, the variation being due to the degree of porosity in the area of the hardness indentation. The tensile strength was about 110 MPa (16,000 lb/in^2), and elongation was about 4 percent (Table 5.2). Samples of the blade were solution heat-treated for 12 h at 538°C (1100°F), which as Fig. 5.21 shows, gives the maximum amount of Si dissolved in the α. The hardness was about the same as in the as-received condition (Table 5.2). Samples were then heated-treated at 155°C (311°F) to induce precipitation hardening,[10] and the results are shown in Fig. 5.23. Note that the hardness increased significantly.

TABLE 5.1 Chemical Composition of Common Aluminum Casting Alloys

Alloys			Composition, %					
AA number	Former AA designation	Former ASTM number	Product(a)	Cu	Mg	Mn	Si	Others
201.0	...	...	S	4.6	0.35	0.35	...	0.7 Ag, 0.25 Ti
206.0	...	...	S or P	4.6	0.25	0.35	0.10(b)	0.22 Ti, 0.15 Fe(b)
A206.0	...	...	S or P	4.6	0.25	0.35	0.05(b)	0.22 Ti, 0.10 Fe(b)
208.0	108	CS43A	S	4.0	...	...	3.0	...
242.0	142	CN42A	S or P	4.0	1.5	...	...	2.0 Ni
295.0	195	C4A	S	4.5	...	...	0.8	...
296.0	B295.0, B195	...	P	4.5	...	...	2.5	...
308.0	A108	SC64A	S or P	4.5	...	...	5.5	...
319.0	319, Allcast	SC64D	S or P	3.5	...	...	6.0	...
336.0	A332.0, A132	SN122A	P	1.0	1.0	...	12.0	2.5 Ni
354.0	354	SC92A	P	1.8	0.50	...	9.0	...
355.0	355	SC51A	S or P	1.2	0.50	0.50(b)	5.0	0.6 Fe(b), 0.35 Zn(b)
C355.0	C355	SC51B	S or P	1.2	0.50	0.10(b)	5.0	0.20 Fe(b), 0.10 Zn(b)
356.0	356	SG70A	S or P	0.25(b)	0.32	0.35(b)	7.0	0.6 Fe(b), 0.35 Zn(b)
A356.0	A356	SG70B	S or P	0.20(b)	0.35	0.10(b)	7.0	0.20 Fe(b), 0.10 Zn(b)
357.0	357	...	S or P	...	0.50	...	7.0	...
A357.0	A357	...	S or P	...	0.6	...	7.0	0.15 Ti, 0.005 Be
359.0	359	SG91A	S or P	...	0.6	...	9.0	...
360.0	360	SG100B	D	...	0.50	...	9.5	2.0 Fe(b)
A360.0	A360	SG100A	D	...	0.50	...	9.5	1.3 Fe(b)
380.0	380	SC84B	D	3.5	...	...	8.5	2.0 Fe(b)
A380.0	A380	SC84A	D	3.5	...	...	8.5	1.3 Fe(b)
383.0	...	SC102A	D	2.5	...	...	10.5	...
384.0	384	SC114A	D	3.8	...	...	11.2	3.0 Zn(b)
A384.0	384	SC114A	D	3.8	...	...	11.2	1.0 Zn(b)

Alloy	Former designation	ANSI	Process	Cu	Mg	Mn/Cr	Si	Others
390.0	390	...	D	4.5	0.6	...	17.0	1.3 Zn(b)
A390.0	A390	...	S or P	4.5	0.6	...	17.0	0.5 Zn(b)
413.0	13	S12B	D	...	...	...	12.0	2.0 Fe(b)
A413.0	A13	S12A	D	...	...	...	12.0	1.3 Fe(b)
4430	43	S5B	S	0.6(b)	...	...	5.2	...
A443.0	43	...	S	0.30(b)	...	...	5.2	...
B443.0	43	S5A	S or P	0.15(b)	...	...	5.2	...
C443.0	A43	S5C	D	0.6(b)	...	...	5.2	2.0 Fe(b)
514.0	214	G4A	S	...	4.0	...	...	...
518.0	218	G8A	D	...	8.0	...	...	...
520.0	220	G10A	S	...	10.0	...	...	...
535.0	Almag 35	GM70B	S	...	6.8	0.18	...	0.18 Ti
A535.0	A218	...	S	...	7.0	...	...	...
B535.0	B218	...	S	...	7.0	0.18	...	0.18 Ti
712.0	D712.0, D612, 40E	...	S or P	...	0.6	...	...	5.8 Zn, 0.5 Cr, 0.20 Ti
713.0	613, Tenzaloy	ZC81A,B	S or P	0.7	0.35	...	...	7.5 Zn, 0.7 Cu
771.0	Precedent 71A	ZG71B	S	...	0.9	...	...	7.0 Zn, 0.13 Cr, 0.15 Ti
850.0	750	...	S or P	1.0	...	...	...	6.2 Sn, 1.0 Ni

(a) S = sand casting, P = permanent mold casting, D = die casting. (b) Maximum.

SOURCE: From *Metals Handbook*.[10]

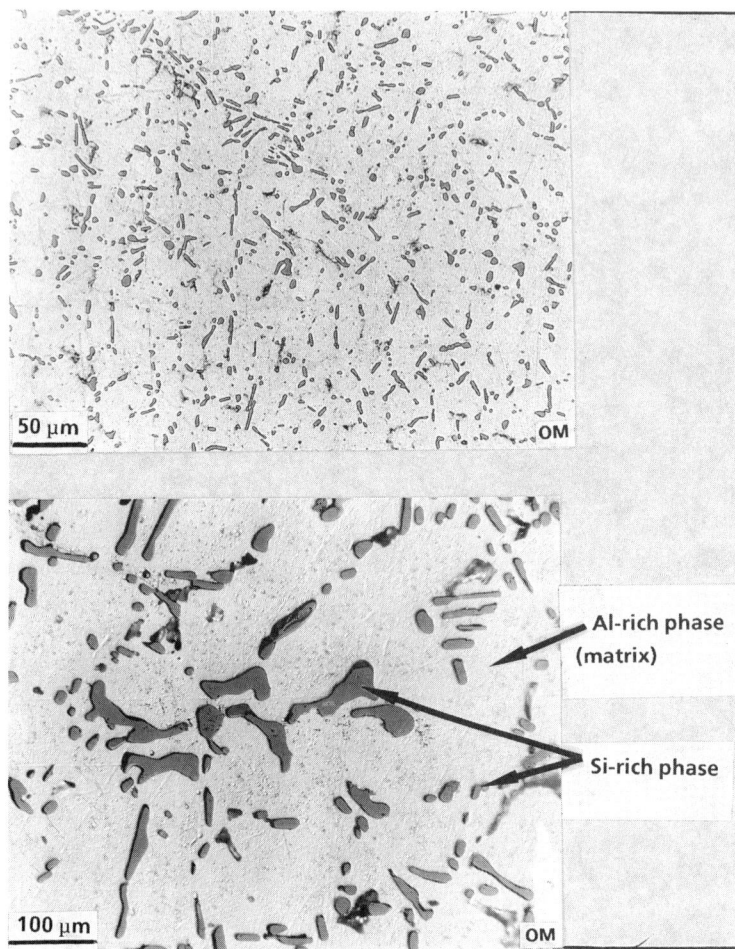

Figure 5.22 Microstructure of material after aging heat treatment. Note that the particles are rounded.

The tensile strength is similar to that listed for the heat treatment T6 for this alloy (Table 5.3).

Stress analysis. There are four forces to consider in calculating the stress on the blade—gravitational, lift, drag, and centrifugal. The gravitational force is obtained from the mass of the fan, which was 24 kg (30 lb). The lift force L is given by[14]

$$L = C_L \rho bc \left(\frac{u^2}{2g}\right)$$

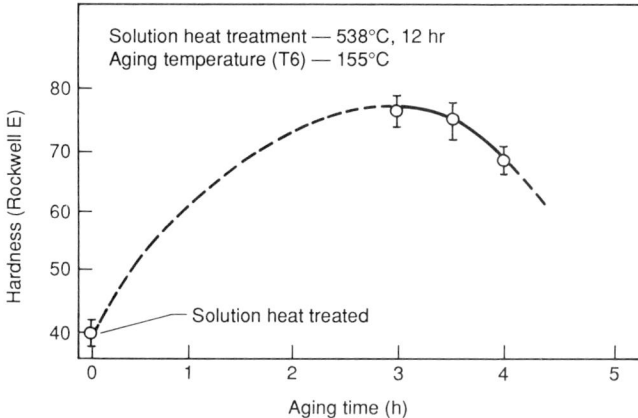

Figure 5.23 Effect of aging at 155°C (311°F) on the hardness of the fan blade material.

TABLE 5.2 Effect of Heat Treatment on Hardness, Tensile Strength, and Elongation at Fracture for Fan-Blade Material

	(a) Results of tensile tests			
		Heat-treatment condition		
		As received		Solution treatment + T6
		T1	T2	T3	T4
Sample number		T1	T2	T3	T4
Hardness, Rockwell E		49–55	45–55	77–78	77–78
Tensile strength, lb/in^2		16,200	12,000, low strength, lots of defects	25,900	28,000
Elongation, %		4.1	—	—	2.5
Reduction of area, %		3.6	—	(Broke out of gage)	2.2
	Typical values			
Tensile strength, lb/in^2		16,000		27,000	
Elongation, %		4		3	
	Nominal mechanical properties of 356-T6			
Tensile strength, lb/in^2	38,000				
Elongation, %	5				

(b) Effect of heat treatment of hardness	
Heat-treatment condition	Hardness, Rockwell E
Solution treatment (ST): 538°C, 12 h, water-cooled	42, 32, 38, 42, 37
S.T. + T6 (155°C, 3 h)	77, 77, 79, 74
S.T. + T6 (155°C, 3.5 h)	73, 76, 78, 75
S.T. + T6 (155°C, 4 h)	67, 66, 72
As-received, broken blade	45–55

TABLE 5.3 Typical Mechanical Properties of Common Aluminum Sand Casting Alloys

AA No.	Temper	Tensile strength MPa	ksi	Tensile yield strength(b) MPa	ksi	Elongation in 50 mm or 2 in., %	Hardness(c), HB	Compressive yield strength(b) MPa	ksi	Shear strength MPa	ksi	Endurance limit(d) MPa	ksi	Modulus of elasticity(e) kPa × 10⁶	psi × 10⁶
201.0	T43	414	60	255	37	17.0	...	...	...	...	...	...	...	...	...
	T6	448	65	379	55	8.0	130	386	56	290	42	...	...	...	...
	T7	469	68	414	60	5.0	...	...	...	...	...	97	14	...	...
A206.0	T4	431	62	264	38	17.0	...	285	41	292	42	...	...	...	...
	T7	436	63	347	50	11.7	...	372	54	257	37	...	...	...	...
213.0	F	207	30	165	24	1.5	85	172	25	165	24	66	9.5	...	...
222.0	T52	241	35	214	31	1.0	100	214	31	172	25	...	...	...	...
	T551	255	37	241	35	<0.5	115	276	40	207	30	59	8.5	74	10.7
	T65	331	48	248	36	<0.5	140	248	36	207	30	62	9	74	10.7
238.0	F	207	30	165	24	1.5	100	207	30	165	24	...	...	...	...
242.0	T571	276	40	234	34	1.0	105	234	34	207	30	72	10.5	71	10.3
	T61	324	47	290	42	0.5	110	303	44	241	35	66	9.5	71	10.3
249.0	T63	476	69	414	60	6.0	...	...	...	...	...	...	...	...	...
296.0	T7	278	62	359	52	9.0	...	414	60	276	40	55	8.0	72	10.5
	T4(f)	255	37	131	19	9.0	75	138	20	207	30	66	9.5	70	10.1
	T6	276	40	179	26	5.0	90	179	26	221	32	69	10	70	10.1
	T7	270	39	138	20	4.5	80	138	20	207	30	63	9	70	10.1
308.0	F	193	28	110	16	2.0	70	117	17	152	22	90	13	...	...
319.0	F	234	34	131	19	2.5	85	131	19	165	24	...	...	...	...
	T6	276	40	186	27	3.0	95	186	27	...	...	...	...	...	...
324.0	F	207	30	110	16	4.0	70	...	...	...	...	...	...	...	...
	T5	248	36	179	26	3.0	90	...	...	...	...	...	...	...	...
	T62	310	45	269	39	3.0	105	...	...	...	...	...	...	...	...
332.0	T5	248	36	193	28	1.0	105	...	...	...	...	...	...	...	...
333.0	F	234	34	131	19	2.0	90	131	19	186	27	100	14.5	77	11.2
	T5	234	34	172	25	1.0	100	172	25	186	27	83	12	...	...
	T6	290	42	207	30	1.5	105	207	30	228	33	103	15	...	...
	T7	255	37	193	28	2.0	90	193	28	193	28	83	12	...	...
336.0	T551	248	36	193	28	0.5	105	193	28	193	28	93	13.5	...	...
	T65	324	47	296	43	0.5	125	296	43	248	36	...	...	...	...

Alloy	Temper													
355.0	T51	205	30	165	24	2.0	75	165	24	165	24		69	10	...
	T6	290	42	185	27	4.0	90	185	27	235	34		69	10	...
	T61	310	45	275	40	1.5	105	275	40	250	36		69	10	...
	T7	275	40	205	30	2.0	85	205	30	205	30		69	10	...
	T71	250	36	215	31	3.0	85	215	31	185	27		69	10	...
C355.0	T61	303	44	234	34	3.0	90	248	36	221	32		97	14	...
356.0	F	179	26	124	18	5.0									10.5
	T51	186	27	138	20	2.0									10.5
	T6	262	38	186	27	5.0	80	186	27	207	30		90	13	72	10.5
	T7	221	32	165	24	6.0	70	165	24	172	25		76	11	72	
	T71	283	41	207	30	10.0	90	221	32	193	28		90	13	72	
A356.0	T61	193	28	103	15	6.0									
357.0	F	200	29	145	21	4.0									
	T51	359	52	296	43	5.0	100	303	44	241	35		90	13		
	T6	359	52	290	42	5.0	100	296	43	241	35		103	15		
A357.0	T6	345	50	290	42	5.5								110	16	82	11.9
359.0	T62	200	29	200	29	<1.0	110									
A390.0	F	200	29	200	29	<1.0	110									
	T5	310	45	310	45	<1.0	145	414	60				117	17		
	T6	262	38	262	38	<1.0	120	359	52				100	14.5		
	T7	159	23	62	9	10.0	45	62	9	110	16		55	8	71	10.3
443.0	F	165	24	76	11	13.0	44		11							
444.0	F	159	23	69	10	21.0	45	76	11	110	16		55	8		
	T4															
513.0	F	186	27	110	16	7.0	60	117	17	152	22		69	10		
711.0	F	241	35(g)	124	18(g)	8.0(g)	70(g)						76	11	76	11.0
850.0	T5	159	23	76	11	12.0	45	76	11	103	15		62	9	71	10.3
851.0	T5	138	20	76	11	5.0	45	76	11	97	14		62	9	71	10.3
852.0	T5	221	32	159	23	5.0	70	159	23	148	21		76	11	71	10.3

(a) Tension and hardness values determined by tests on standard 13-mm (1/2-in.) diam test specimens, without surface machining, each cast in permanent mold. (b) At 0.2% offset. (c) 500-kg (1102-lb) load on 10-mm (0.4-in.) ball. (d) Endurance limits based on 500 million cycles of completely reversed stresses using rotating beam–type machine and specimen. (e) Average of tension and compression moduli; compression modulus is about 2% greater than tension modulus. (f) Properties of T4 approach those of T6 after standing for several weeks at room temperature. (g) Tests made approximately 30 days after casting.

SOURCE: From *Metals Handbook*.[13]

where C_L = lift coefficient, = 1
ρ = density of air, = 1.2 kg/m³ (0.043 lh/in³)
b = span length, = 0.91 m (36 in)
c = chord length, = 0.25 m (10 in)
u = peripheral velocity of blade,

$$u = \left(\frac{r/\min}{60}\right)\pi D = \left(\frac{550}{60}\right)(\pi)(72 \text{ in}) = 53 \text{ m/s } (2073 \text{ in/s})$$

Then

$$L = (1)(0.042 \times 10^{-3})(36)(10)(2073^2)\left[\frac{1}{(2)(9.8)(3)(12)}\right]$$

$$= 375 \text{ N } (84 \text{ lb})$$

The drag force is negligible since $D \ll L$ (see Wallis[14]). The centrifugal force is given by

$$C = \left(\frac{w}{g}\right)\left(\frac{v^2}{r}\right)$$

where r = distance from center of mass, = 0.36 m (14 in)
v = velocity of center of mass, = $(550/60)(\pi)(28 \text{ in})$ = 20.5 m/s (806 in/s)
w = mass, = 13.6 kg (30 lb)

Then

$$C = 16{,}000 \text{ N } (3600 \text{ lb})$$

From these values the stress at the neck was calculated. The centrifugal forces cause a tensile stress σ_c of

$$\sigma_c = \frac{3600}{(\pi)(1.08^2) - (\pi)(0.30^2)} = 7.3 \text{ MPa } (1060 \text{ lb/in}^2)$$

The bending stress σ_b is caused by the gravitational force and the lift force. The maximum bending occurs at the location at which the fracture originated (see Fig. 5.13). This is given by

$$\sigma_b = \frac{My}{I}$$

where M = moment, = $(30 + 84)(14)$ = 180 N·m (1596 in·lb)
I = moment of inertia, = $(\pi/4)(R_o^4 - R_i^4)$ = $(\pi/4)(1.08^4 - 0.30^4)$
= 4.40×10^{-7} m⁴ (1.07 in⁴)
$y = R_o$ = 0.027 m (1.08 in)

So σ_b = 11.2 MPa (1620 lb/in²)

Thus the tensile stress at point a is 11.2 + 7.3 = 18.5 MPa (2700 lb/in²).

5.3.4 Discussion

The tensile strength of the broken blade was about 110 MPa (16,000 lb/in²) (Table 5.2). Since this material is relatively "brittle," as evidenced by the low tensile ductility (Table 5.2), a simple design criterion is to require the tensile strength to be equal to the maximum imposed tensile stress. (Note that in this brittle material, the yield strength is not reported.) However, the loading is both static and variable, the latter involving both variations in the centrifugal stress and machine vibrations. According to Table 5.2, in the aged condition (such as T6) alloy 356 has a fatigue strength (endurance limit) (see Sec. 2.13) of about 60 MPa (8500 lb/in²), which is about one-third of the tensile strength listed. Thus the fatigue strength of the blade would be about 110/3 = 37 MPa (5300 lb/in²). The hardness measurements varied by about 50 percent (Table 5.2) due to porosity, and thus the fatigue strength may be as low as 37 × 0.5 = 19 MPa (2600 lb/in²). Due to the roughness of the surface of the casting, this value will be reduced further by about a factor of 2. Thus the expected fatigue strength is about 10 MPa (1300 lb/in²). This is less than the calculated maximum tensile stress of 19 MPa (2700 lb/in²), which is too low.

The 356 alloy is precipitation-hardenable,[1,15] but the blade had received no such heat treatment. The measured tensile strength, after aging, of 186 MPa (27,000 lb/in²) is similar to that reported for this alloy in this condition (Table 5.3). Even if the alloy had been precipitation-hardened, the strength would only have been similar to the applied stress, which is too low a "safety factor" for a dynamic machine part.

5.3.5 Conclusions

The failure was due to the use of material which did not have requisite fatigue strength. Even if the blade had been heat-treated to strengthen it, the safety factor would have been marginal. Thus, consideration should have been given to the use of another material.

296 Chapter Five

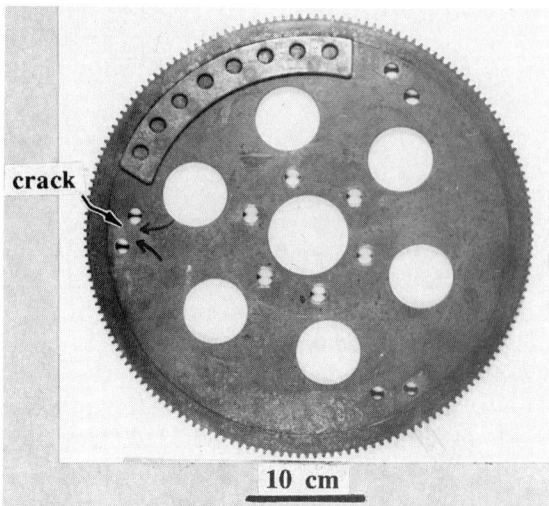

Figure 5.24 Cracked flywheel. The crack is noted by arrow.

5.4 Case C: A Cracked Automobile Flywheel Flex Plate*

5.4.1 Introduction

Automobile parts are subjected to complex vibration loading, and thus fatigue is a commonly observed failure mode in flex plates. This case study involved a cracked automobile flywheel from a 1978 Oldsmobile Custom Cruiser Station Wagon. Figure 5.24 shows the flywheel; the location of the crack is indicated by the arrow. There was no obvious plastic deformation, such as bending. Figure 5.25 shows the relation of such a flywheel to the drive train of an automobile. It serves as a mechanical link between the engine and the transmission, the torque developed by the engine being transferred to the transmission through the flywheel. The owner stated that there had been no major repairs of the automobile prior to the flywheel failure. The automobile had been driven in excess of 200,000 mi, a reasonable life for such a part.

5.4.2 Experimental procedure

The section containing the crack (Fig. 5.26), and another uncracked section for comparison, were removed by a band saw. From these,

---

*The authors acknowledge the work done by David Dellinger on this case study.

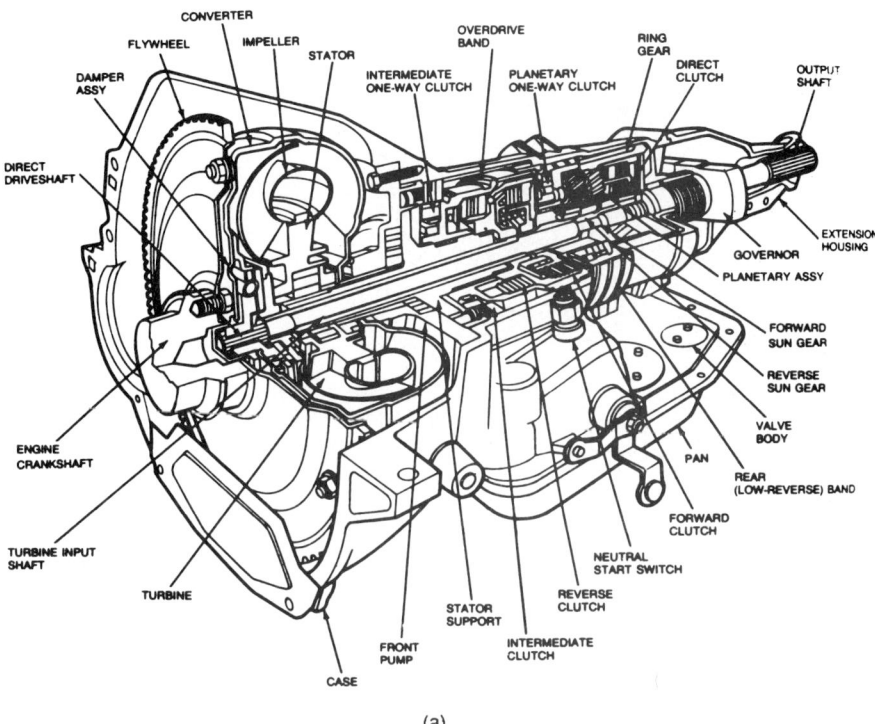

Figure 5.25 Diagram showing the function of a flywheel in an automobile. Note bolts which connect flywheel to torque converter. (*From* Motor Auto Repair.[16])

samples were cut for metallography. These were prepared by mounting them in plastic and then grinding and polishing. The microstructure was revealed by etching in 2% nitric acid in methanol.

It is seen in Fig. 5.26 that the crack extended from the edge of one hole to just below the adjacent hole. To examine the fracture surface, the sample was cut between the holes, which exposed part of the fracture surface. However, this surface showed extensive rubbing (see Sec. 3.16), so the remaining portion was notched opposite the end of the crack (Fig. 5.27) and broken to expose it. This allowed examination of the entire crack surface in the scanning electron microscope and EDS analysis of it.

5.4.3 Results

Fractography. Low-magnification fractographs are shown in Fig. 5.28. There was no indication of gross plastic deformation. The fracture re-

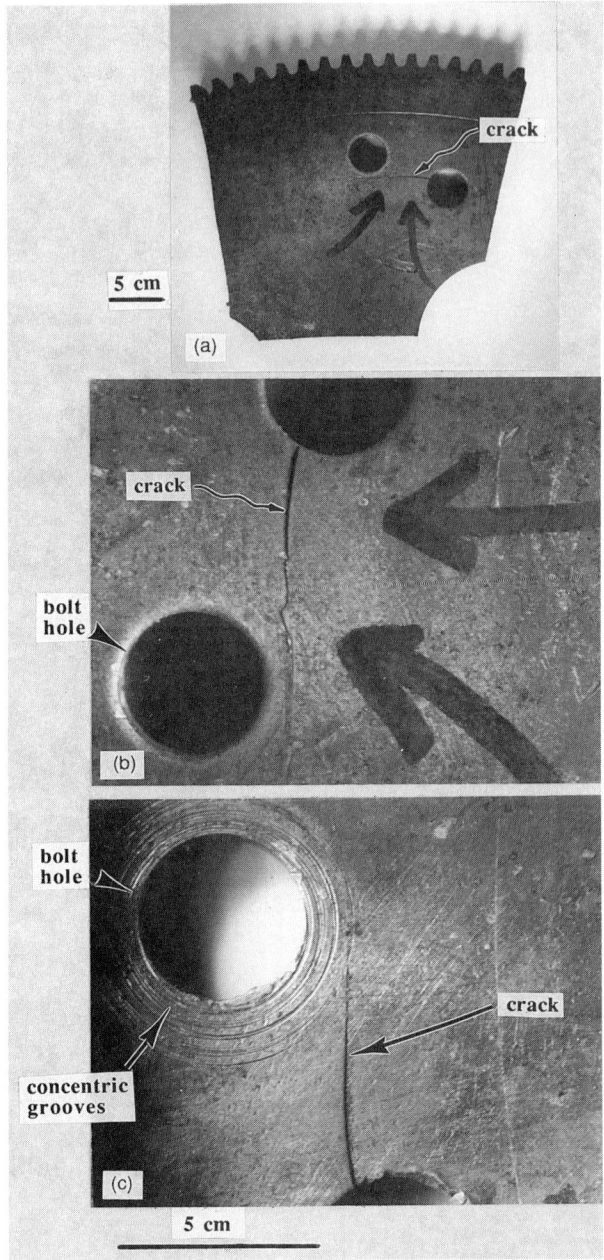

Figure 5.26 Section containing crack.

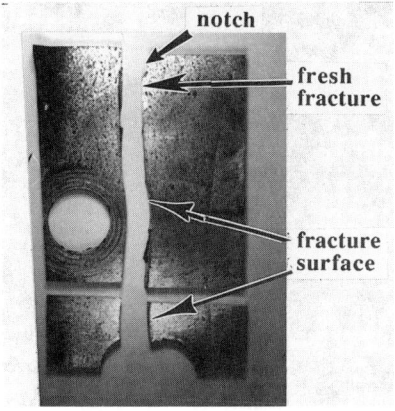

Figure 5.27 Sections cut to reveal fracture surface after breaking the notched part.

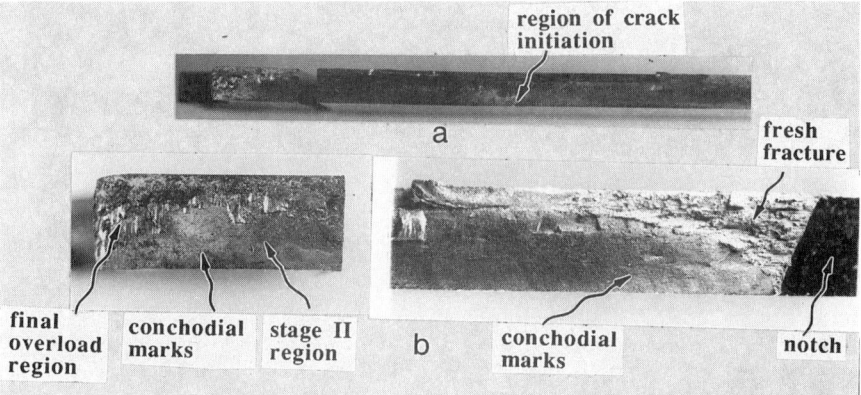

Figure 5.28 Low-magnification scanning electron micrographs of fracture surface.

gion extending inward from the left edge through about one-quarter of the thickness is coarse and irregular, and the remainder of the fracture surface is smooth. On the smooth part, conchoidal marks are visible; thus the fracture mode was fatigue (see Sec. 4.5). The coarse fracture surface appears to be the final overload separation area, and this extends to the edge of the hole. Thus the fatigue crack did not originate at the edge of the hole, but somewhere along the right edge. This was confirmed by examination of the fracture surface, which was revealed by breaking the sample. The curvature of the conchoidal marks shows that the origin of the fatigue crack was directly below the other bolt hole (Fig. 5.29). This coincides with one of the concentric grooves surrounding the bolt hole (Fig. 5.30).

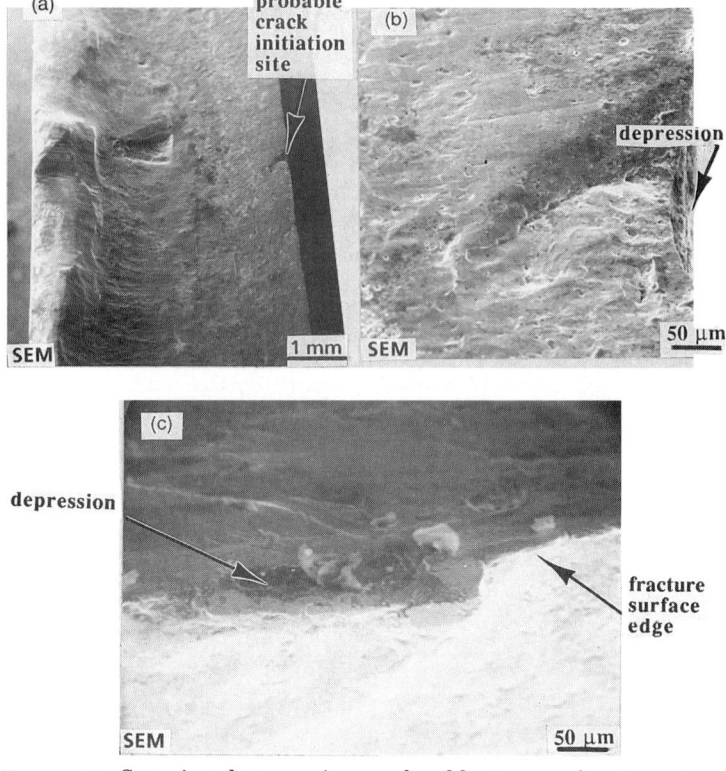

Figure 5.29 Scanning electron micrographs of fracture surface in region of crack origin.

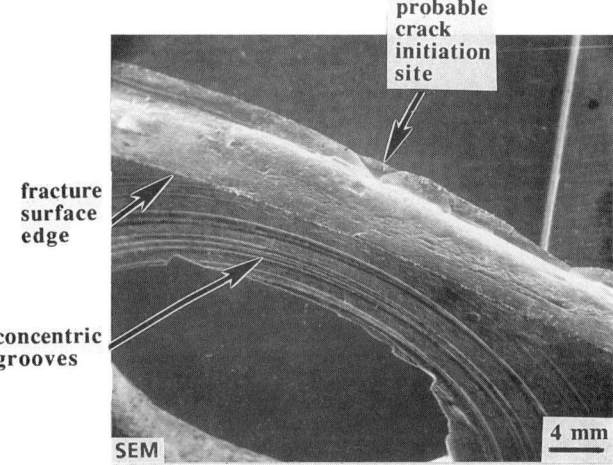

Figure 5.30 Scanning electron micrograph showing damage to the surface from the bolt near the fracture origin.

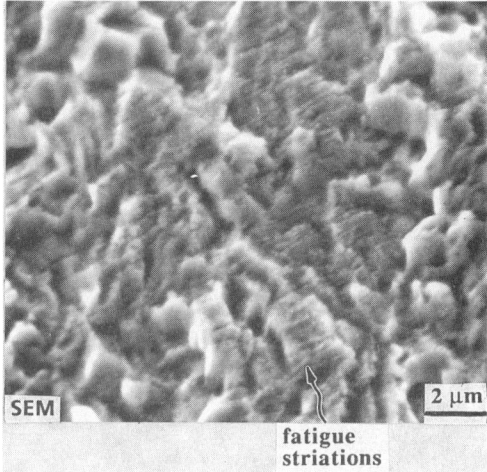

Figure 5.31 Scanning electron micrograph of the fatigue fracture surface showing fatigue striations.

The fracture surface near the end which exited at one of the holes was extensively rubbed (but still showed conchoidal marks) (Fig. 5.28b), but the other end had minimal rubbing damage, and here fatigue striations were found (Fig. 5.31). (As pointed out in Sec. 3.9, fatigue striations are not often so apparent in steels of complex microstructures. However, in this case the structure consisted mainly of single-phase ferrite of uniform grain size.) The fracture surface created by fracture after notching, shown in Fig. 5.32, has a dimpled morphology, showing that this region separated by void coalescence (see Sec. 3.5).

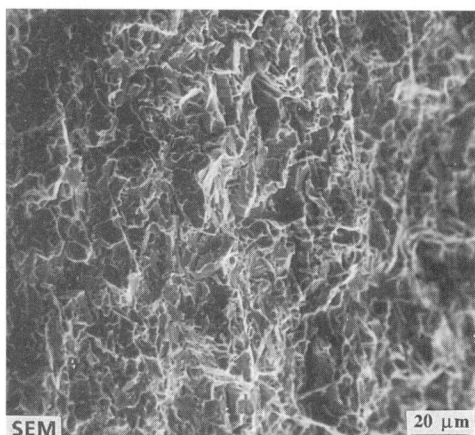

Figure 5.32 Scanning electron micrograph of the fracture surface created by breaking the notched sample.

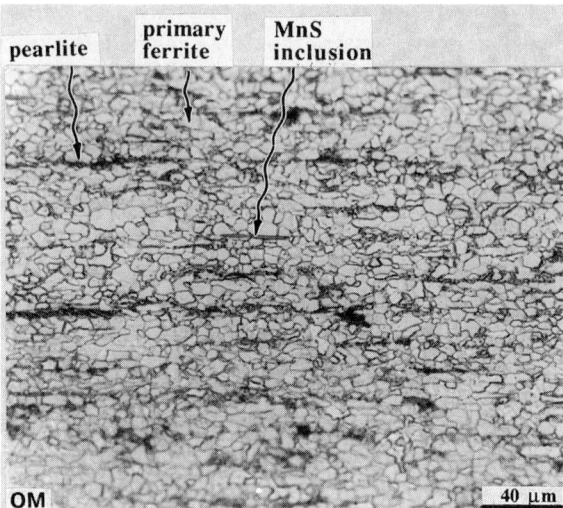

Figure 5.33 Microstructure of the flywheel, showing fine-grain ferrite.

Microstructure. The microstructure is shown in Fig. 5.33. EDS analysis revealed only Fe with traces of Mn and Si. The microstructure consisted of a relatively fine ferrite grain size, with a very small amount of pearlite. Thus the plate was probably made of a low-carbon, plain-carbon steel. There was a rather high density of elongated inclusions (Fig. 5.34), and these were also visible on the fracture surface (Fig. 5.35). EDS analysis showed that these were MnS.

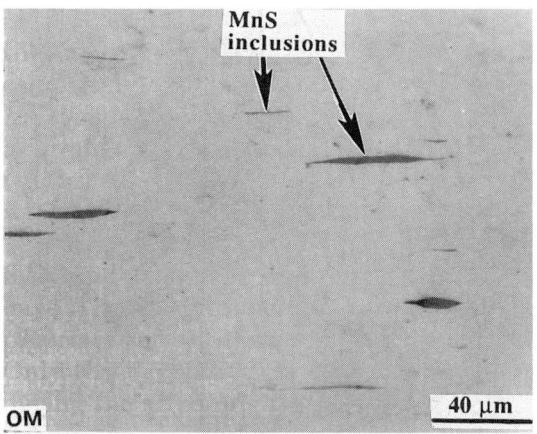

Figure 5.34 Unetched microstructure showing elongated inclusions.

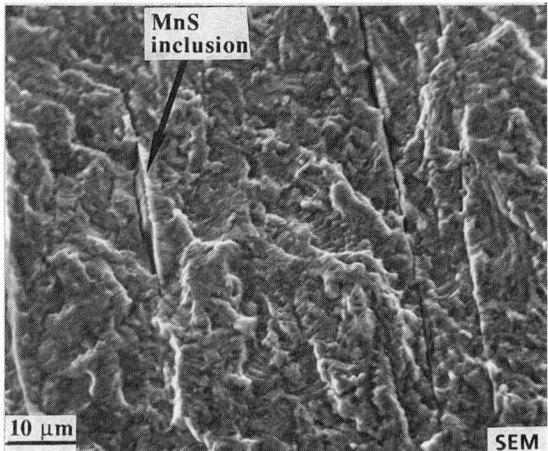

Figure 5.35 Scanning electron micrograph of the fatigue fracture region showing elongated inclusions, identified by EDS as MnS.

5.4.4 Discussion

The flex plate was made of plain-carbon steel having a fine-grain ferrite structure. The service life in excess of 200,000 mi attests to the basic design and choice of material of the plate being sound. In spite of the rather high density of elongated inclusions (Fig. 5.34), they may not have any effect on the fatigue life of such plate material,[17] and cracking does not appear to have originated at the inclusions revealed on the fracture surface (Fig. 5.35). Instead, the origin of the fatigue crack appears to be associated with surface damage due to the bolt or its washer, perhaps when the bolt was tightened during installation.

Although 200,000-mi service is reasonable, a clearer picture of the situation is obtained by estimating the number of cycles of fatigue crack growth. This can be made by dividing the crack length by the fatigue striation spacing (Sec. 3.9). From Fig. 5.28 the distance from the origin to the end of the crack is about 0.5 cm, and from Fig. 5.31, the striation separation is about 0.5 μm. Thus the number of cycles that the fatigue crack grew is about 10^5. On the average, approximately every mile of travel the flex plate received a loading sufficiently large to advance the crack one cycle. In fatigue design, the desired fatigue life is usually about 10^6 cycles or greater.

5.4.5 Conclusions

The flex plate cracked in a fatigue mode of fracture. The fatigue crack initiated in a region of the surface damaged by a bolt or its washer. It should also be remembered that the formation of the fatigue crack in the flex plate may have been symptomatic of excessive vibrations in the automobile from other causes.

5.5 Case D: Failed Welded Railroad Rails*

5.5.1 Introduction

A common method of making railroad rails into very long sections is to weld pieces together by the thermite process. This is illustrated in Fig. 5.36, and the procedure is shown in Fig. 5.37. A mold is placed around the joint between two rails, and a mixture of aluminum and iron oxide powder in a crucible above the mold is ignited. The iron oxide is reduced by the aluminum, forming aluminum oxide. The highly exothermic reaction creates sufficient heat to melt the iron, which then is allowed to flow into the mold. The sensible heat in the molten steel is sufficient to melt part of each rail, and upon cooling, the molten material solidifies to a weld which connects the two rails. The mold is then removed, and the top and sides of the rail are ground to fit the rails on either side. This case study examines fractures which occurred in several thermite welds in railroad rails after only a short service life (for example, 6 months). Figure 5.38 shows a typical broken rail.

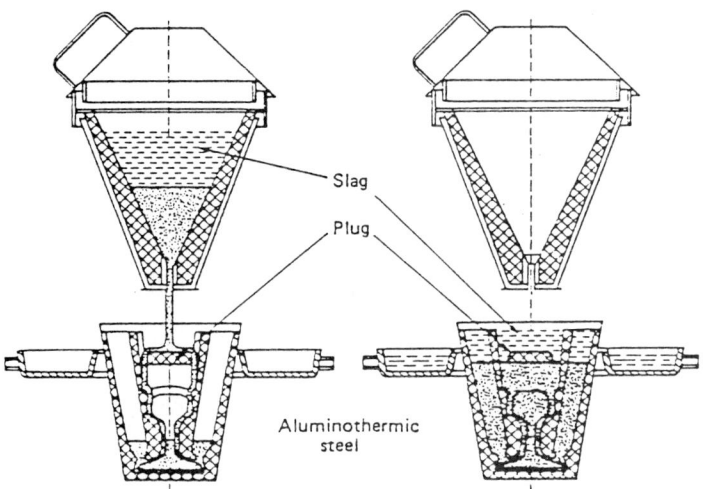

Figure 5.36 Schematic diagram of the process for joining railroad rails by the thermite welding process. (*From H. D. Fricke, in* Metals Handbook,[8] *p. 695.*)

---

*The authors acknowledge the work done by Clayton Crouse on this case study.

Figure 5.37 Thermite welding process for joining railroad rails. (*From H. D. Fricke, in* Metals Handbook,[8] *p. 695.*)

Figure 5.38 Pair of broken rails. Note the radial marks which point to the origin of the fracture.

5.5.2 Experimental procedure

To obtain samples for fractography and metallography, a broken rail was sectioned by a band saw, as shown in Fig. 5.39. The fracture surfaces were cleaned in warm water containing a detergent in an ultrasonic cleaner. A soft brush was also used on the surface after soaking in detergent. After washing in water, the surfaces were cleaned in acetone agitated by ultrasound, then dried (see Sec. 1.4.1). The metallographic sample was mounted in epoxy, then ground and polished by standard methods. The surface was etched in a solution of 2 percent nitric acid in methanol.

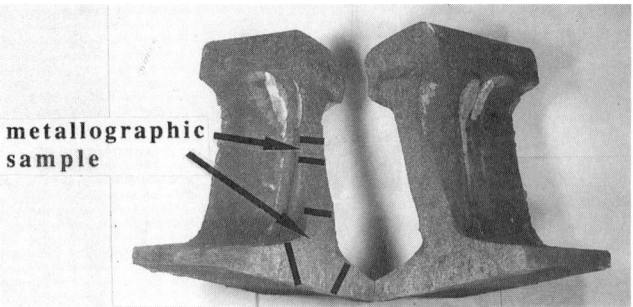

Figure 5.39 Location of samples removed for examination.

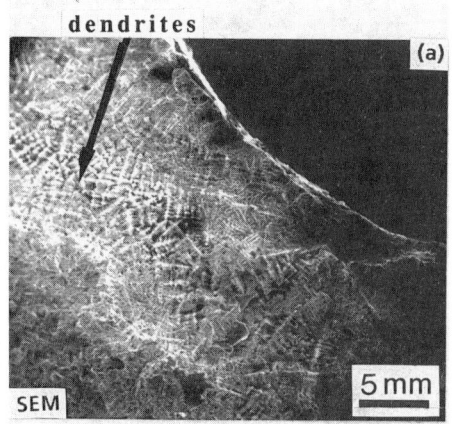

Figure 5.40 Scanning electron micrographs of one of the large cavities at the root of the web, showing dendrites.

Microhardness measurements using a 500-g load were made on the unetched metallographic sample. The diamond pyramid hardness (DPH) values were accurate to about 10 DPH.

5.5.3 Results

Fractography. Fracture occurred across the weld and not in the adjacent heat-affected zone. The fracture surface in Fig. 5.38 shows radial marks pointing to a fracture origin (Sec. 4.2) on each side of the neck of the bottom web. The crack propagated from these two locations through the cross section of the rail, then to the final failure near the top of the rail. The crack origins contained large cavities, the surface of which consisted of dendrites (Fig. 5.40) (see Sec. 3.10). In these regions cracks could be seen propagating away from the fracture surface. There were similar smaller cavities at various locations on the fracture surface. The fracture surface outside of the cavity areas showed that fracture occurred by cleavage (see Sec. 3.4), as illustrated in Fig. 5.41.

Microstructure. The metallographic sample removed from the web (see Fig. 5.39) allowed examination of the microstructure in cross section, from the fracture surface (area that was molten) through the heat-affected zone and into the unaffected base metal. The locations of microstructures from various areas are shown in Fig. 5.42. The base

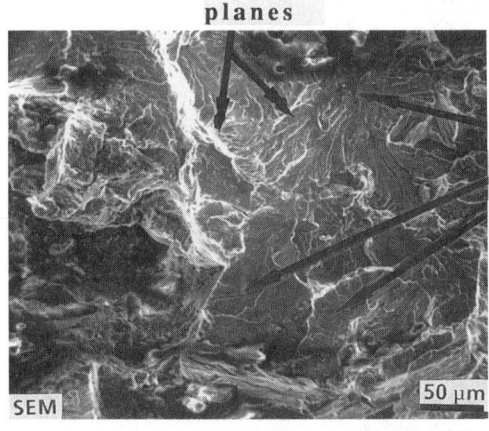

Figure 5.41 Scanning electron micrograph of the fracture surface in the central part of the rail web near the bottom flange. Fracture occurred by cleavage.

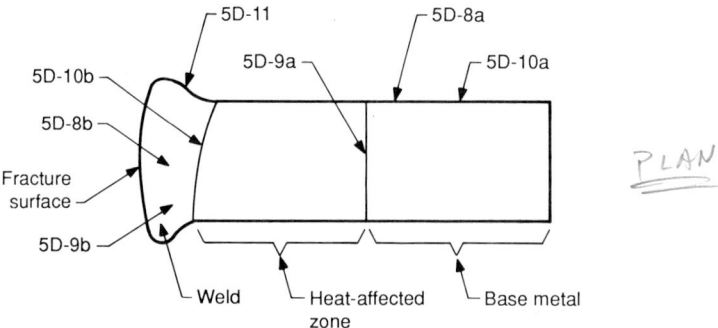

Figure 5.42 Sketch of metallographic sample showing weld and heat-affected zone, and location and figure numbers of higher-magnification micrographs.

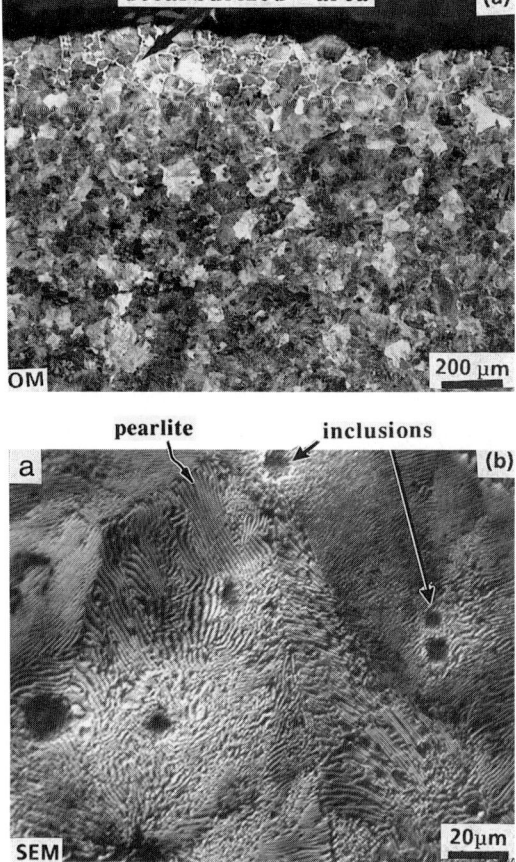

Figure 5.43 Micrographs of base metal.

metal consisted of fine pearlite (Fig. 5.43) with a small amount of primary ferrite. Thus the carbon content is about 0.7 to 0.8 percent. At the surface, decarburization was present, a consequence of fabricating and heat-treating the rail and not associated with the welding process. In the heat-affected zone there was a refinement of the prior austenite grains (Fig. 5.44a), but in the weld metal these grains grew larger, giving a coarser final structure (Fig. 5.44b).

In the base metal there was a relatively high density of inclusions

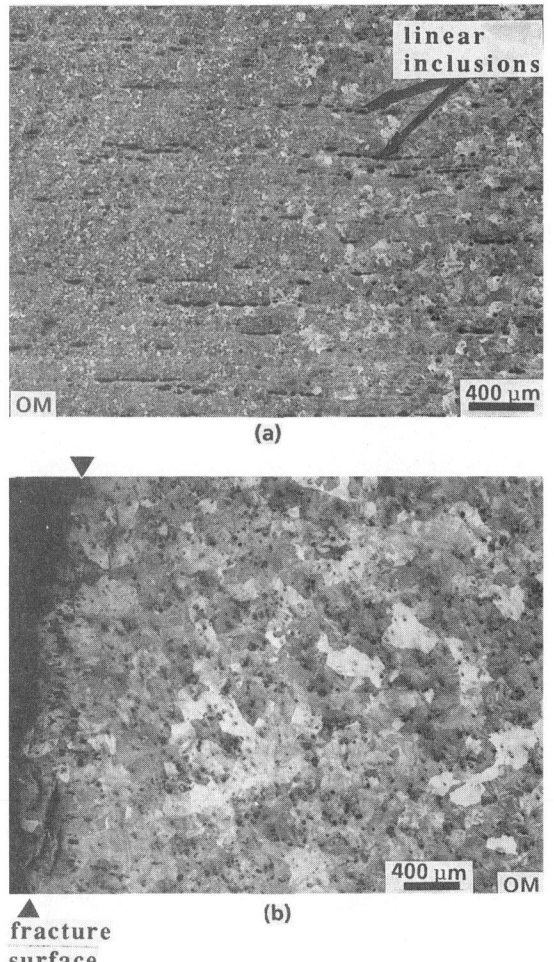

Figure 5.44 Optical micrographs of (a) the heat-affected zone and (b) the weld metal. The difference in the fineness of the structures is due to the prior austenite grain size, which is larger in the weld metal region.

310 Chapter Five

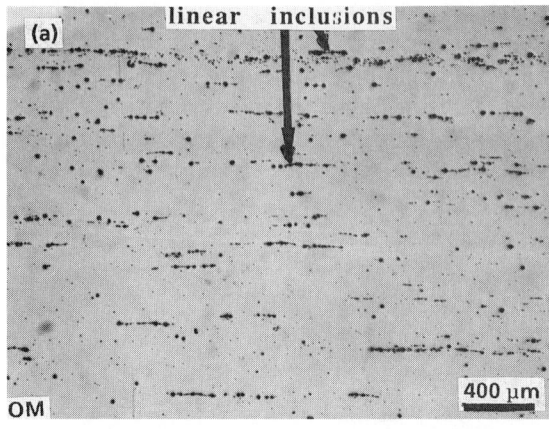

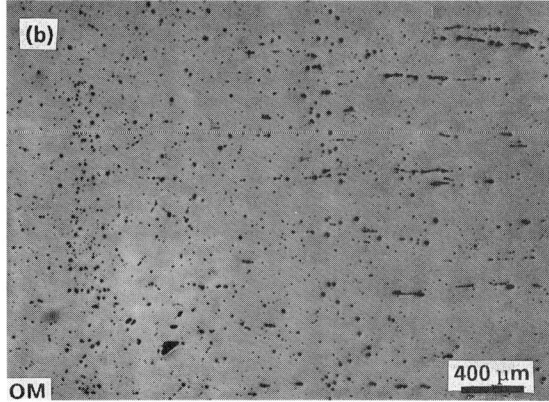

Figure 5.45 Unetched microstructure of (a) the base metal, showing elongated inclusion stringers, and (b) the weld metal showing rounded inclusions.

elongated parallel to the rail axis (a consequence of hot rolling the rail shape). These were probably MnS inclusions (Fig. 5.45a). In the weld zone the inclusions were spherical (Fig. 5.45b). Near the edge of the fracture surface, numerous cracks were found (Fig. 5.46).

Hardness measurements. Figure 5.47 shows the hardness distribution across the weld. The DPH values were converted to Rockwell C.[18]

5.5.4 Discussion

The microstructure of the welds is generally that expected of welding a steel of this carbon content. The hardness distribution across the weld and into the base metal (Fig. 5.47) gives reasonable values for a

Case Studies   311

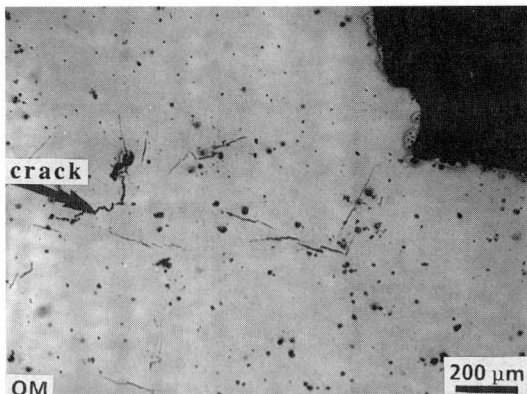

Figure 5.46 Unetched microstructure near the fracture surface showing the presence of numerous cracks.

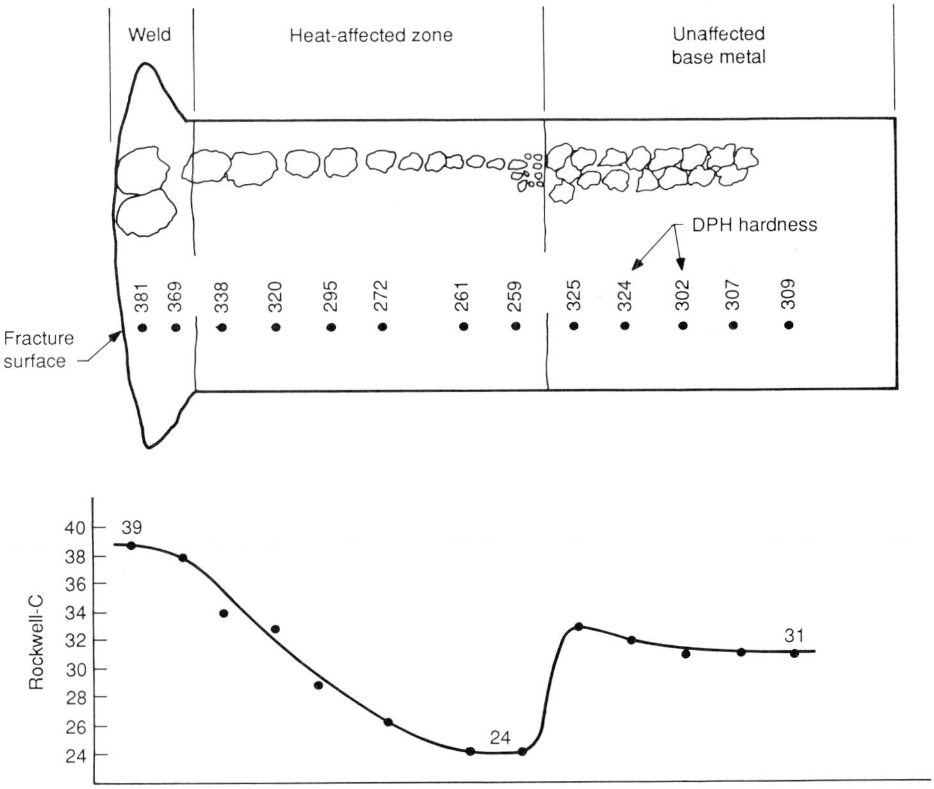

Figure 5.47 Hardness traverse across weld. The locations of these microhardness measurements are approximated.

steel processed in this manner. These observations do not indicate a basic problem with the choice of steel.

The basic cause of the failure of the rails is the presence of the large cavities that formed during solidification, and which acted as crack initiation sites. These cavities formed due to inadequate feeding of the liquid steel during solidification, caused by an improperly designed mold. The railroad company has modified its mold design to alleviate this problem, but field service examination to assess the success of the new design has not been completed.

Another aspect of concern here is the presence of cracks found in the cross section near the fracture surface (Fig. 5.46). These were in the region which had been molten, and most likely are caused by thermal and transformation stresses generated during cooling after solidification. If these cracks existed throughout the weld, then they provided an easy path for the crack to propagate. Establishing the cause of the failure is complicated by the fact that the railroad company had in the previous 2 years changed from a non-heat-treated rail, which had been used for many years, to one which was heat-treated to increase the surface hardness and hence the rail life.

5.5.5 Conclusions

The cracks which led to failure of the rails began at large cavities at the bottom of the web. These cavities were due to inadequate feeding of liquid metal. To prevent these required a change in the design of the mold. However, even with proper feeding, the numerous cracks found in the microstructure of the weld are of concern, indicating that limited life may still be a problem. If field testing of the welds using the redesigned molds reveals a continuing fracture problem, then the relation between welding, solidification, and the chemistry of the new rails should be examined carefully.

5.6 Case E: Broken Stainless-Steel Wires from an Electrostatic Precipitator

5.6.1 Introduction

Gas effluents from paper plants contain fine particles which are sometimes removed by an electrostatic precipitator. In these systems thin wires are charged to several thousand volts so that dust particles from flowing gases are attracted to the wires and the surrounding dust collector plates. Periodically, the wires and plates are vibrated to remove these particles into receptacles at the bottom (Fig. 5.48). Several of these wires failed in one such precipitator in an unexpectedly short time. The wires are manufactured in coils which are then straight-

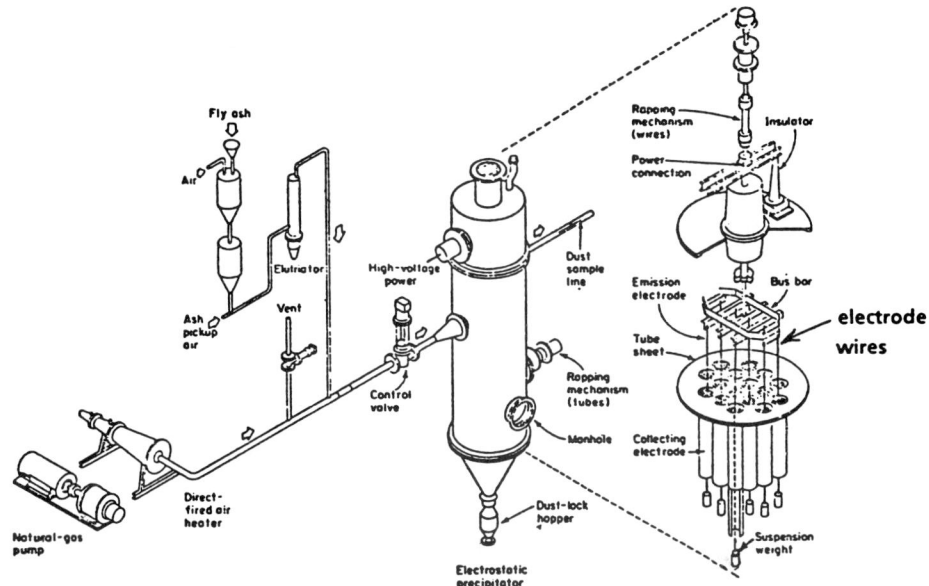

Figure 5.48 Schematic diagram of an electrostatic precipitator system, showing location of the wires. (*From Shale and Faschig.*[19])

ened by attachment to a lower support by weights (Fig. 5.49). It was known that the wires were made in Sweden, corresponded to a 316L stainless steel, and had a specification of "3/4 hard."

5.6.2 Experimental procedure

Several broken wires were received. These were cut to a size appropriate for scanning electron microscopy and optical microscopy evaluation. Samples were cleaned in acetone using ultrasonic agitation. Visual examination, macrofractography, and scanning electron fractography were carried out. In addition, microhardness measurements were made on polished metallographic specimens.

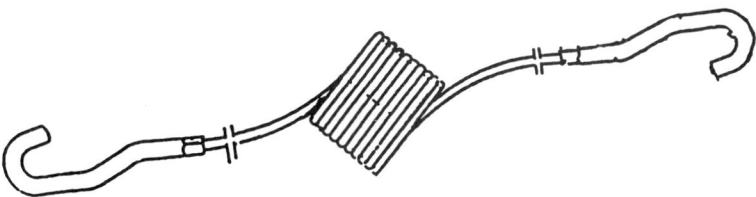

Figure 5.49 Drawing of electrostatic precipitator wire configuration coil before installation.

5.6.3 Results

Visual examination. Visual examination showed the wire surfaces to be relatively clean, with only superficial corrosion products present. The general plane of the fracture was normal to the wire axis, and there was no obvious indication of plastic deformation.

Hardness measurements. Several microhardness measurements were made and converted to the Rockwell C scale.[20] The average hardness was determined to be 37 Rockwell C (±2).

Fractography. At low magnification the fracture surfaces revealed conchoidal marks (Fig. 5.50) (see Sec. 4.5). From this fractograph the

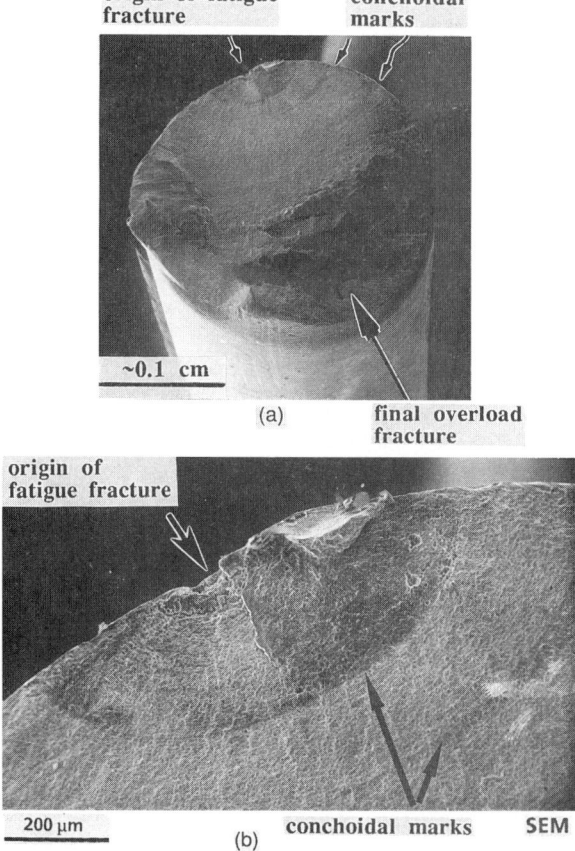

Figure 5.50 Scanning electron micrograph of the fracture surface of one fractured wire, showing that fracture occurred by fatigue.

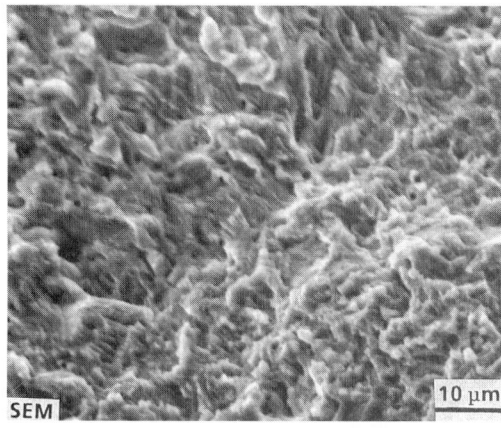

Figure 5.51 Scanning electron micrograph at high magnification of the fatigue fracture propagation region. Note the absence of fatigue striations.

origin of the fatigue fracture is immediately evident. At high magnification (Fig. 5.51), fatigue striations were not seen, but they may have been masked by corrosion products formed on the fracture surface. The final overload region of the fracture surface showed dimples (Fig. 5.52).

A detailed examination of the fracture origin (Fig. 5.50b) showed a pit (Fig. 5.53). Such pits were seen not only at the origin of the fracture on several wires, but also along the surface of the wires (Fig. 5.54). These pits had an appearance similar to that seen on stainless steel surfaces in which corrosion pitting occurred.[21]

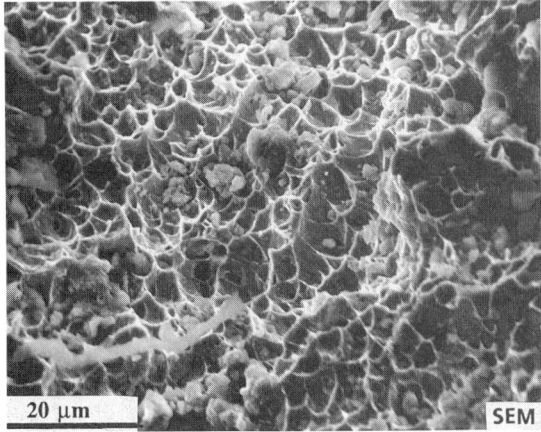

Figure 5.52 Scanning electron micrograph at high magnification of the overload fracture region. The surface shows dimples and some corrosion debris.

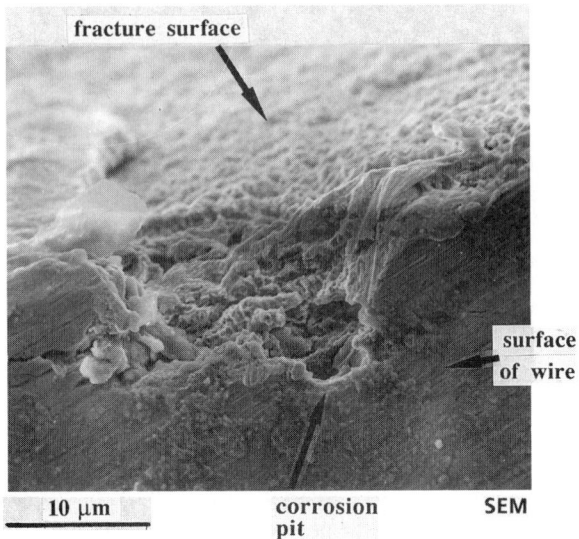

Figure 5.53 Scanning electron micrograph of a corrosion pit at the origin of the fatigue fracture.

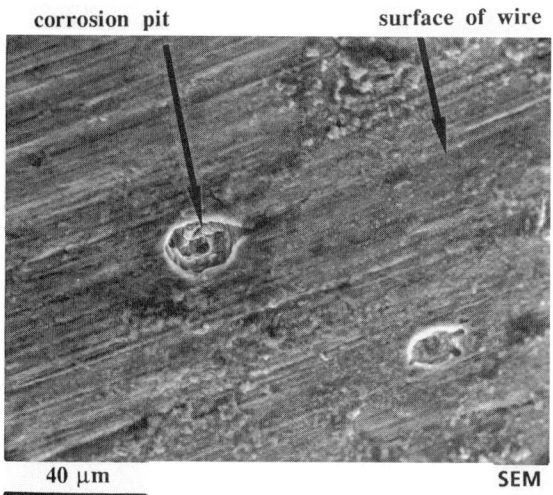

Figure 5.54 Scanning electron micrograph of the surface of a wire showing corrosion pits.

5.6.4 Discussion

The fractography clearly established that the fracture mode was fatigue. It was also clear that the origin of the fatigue cracks was corrosion pits. Hence the inability of the wires to inhibit pitting corrosion was of concern.

TABLE 5.4 Comparison of Chemical Composition of Swedish and U.S. Steels

Swedish steel SIS 2343, wt. %[†]	U.S. steel 316 L, wt. %[*]
0.05C max	0.030C max
1.00Si max	1.00Si max
2.00Mn max	2.00Mn max
0.045P max	0.045P max
0.030S max	0.030S max
16.00–18.00Cr	16.00–18.00Cr
10.50–14.00Ni	10.00–14.00Ni
2.50–3.00Mo	2.00–3.00Mo

SOURCE: *From *Metals Handbook*.[23]
†From D. Peckner and I. M. Bernstein, *Handbook of Stainless Steels*, McGraw-Hill, New York, 1977, p. A1-53.

The chemical composition of the steel used to make the wires is given in Table 5.4, which shows the steel to be basically a 316L stainless steel. The properties of cold-drawn 316 stainless-steel wire are given in Table 5.5. This shows that the hardness in the "3/4 hard" condition should be 29–32 Rockwell C. The measured value of the failed wires (37 Rockwell C) is higher than this and corresponds to the "full hard" condition.

The purpose for specifying that wire in the cold-worked condition be used was not known. The wires are under static load as well as under additional loading from the vibration of the wires, and hence resistance to fatigue fracture is required. Since a common correlation is that the higher the tensile strength, the higher the fatigue strength, then the use of cold-worked stainless steel appears reasonable. However, data (Fig. 5.55)[24] for smooth and notched cold-worked 304 stainless-steel specimens indicate that this steel may be very notch-sensitive in the cold-worked condition. Similar results have been reported for 301 stainless steel (see *Metals Handbook*,[25] p. 765). The fatigue curves in Fig. 5.56[26] for annealed 302 stainless steel show that it is not very notch-sensitive in the annealed condition. Also, data in *Metals Handbook*[25] show that with a stress concentration factor K_t of 3, "extra hard" 301 and annealed 304L stainless steel have about the same fatigue strength. Thus the cold-worked 316L steel may have been notch-sensitive.

It thus appears that while the fatigue strength was improved by cold working, the material may have been rendered very notch-sensitive. In the chemically aggressive environment of paper-plant effluents the wires were subjected to pitting corrosion. These pits provided the notches for lowering the fatigue strength of the wires.

It is also important to recognize that the fracture of the electrode wires may indicate improper operation of the system, causing excessive vibration and static loading. This aspect was an unknown in this investigation.

TABLE 5.5 Nominal Mechanical Properties of 316 Austenitic Stainless Steel

Condition	Diameter, in	Rockwell C hardness	Yield strength (0.2% offset), 1000 lb/in^2	Tensile strength, 1000 lb/in^2	Elongation (in 2 in), %
¼ hard	0.002–0.020	23–28	100–140	130–170	15–20
	0.021–0.125		100–130	125–165	15–25
	0.126–0.375		80–100	110–130	20–30
½ hard	0.002–0.020	29–32	135–170	165–205	5–17
	0.021–0.125		130–165	160–200	11–18
	0.126–0.375		105–125	135–155	12–20
¾ hard	0.002–0.020	33–37	185–210	205–230	3–10
	0.021–0.125		175–205	200–220	6–12
	0.126–0.375		135–165	165–195	8–12
Full hard	0.002–0.020	38–43	230–260	235–300	1–2
	0.021–0.125		215–245	230–275	2–5
	0.126–0.375		155–185	185–215	3–6

SOURCE: From Ref. 22.

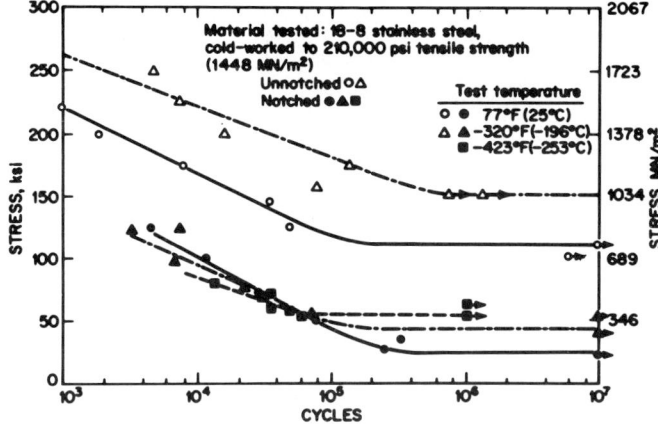

Figure 5.55 Fatigue curves for cold-drawn 304 austenitic stainless steel, showing the notch sensitivity in the cold-worked condition. (*From Spretnak et al.*[24])

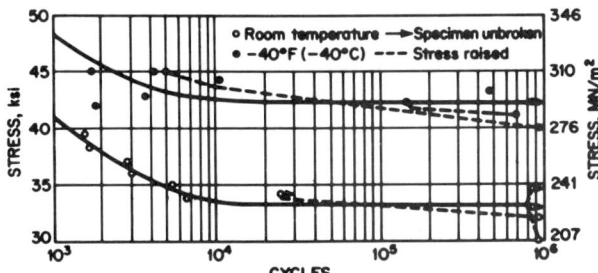

Figure 5.56 Fatigue curves for annealed 302 austenitic stainless steel, showing the lack of notch sensitivity in this condition. (*From Henke.*[26])

5.6.4 Conclusions

The basic cause of the fracture was pitting corrosion, which provided sites for the initiation of fatigue cracks. While the use of material in the cold-worked condition does provide for a better resistance to static overload, the notch sensitivity may be higher for such material. Material with a higher Mo content (such as 317L) is recommended for better resistance to pitting corrosion.

5.7 Case F: Broken Wire Cutters*

5.7.1 Introduction

Diagonal wire cutters are used frequently in a variety of operations, mainly involving electrical wiring. A broken wire cutter was received

---

*The authors acknowledge the work done by Brian Kruse on this case study.

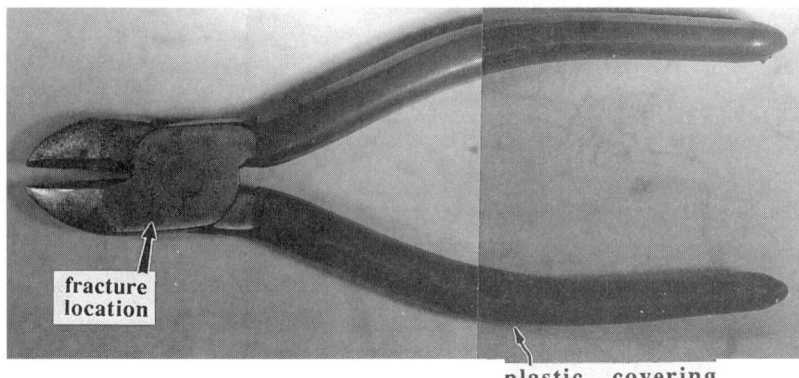

Figure 5.57 Failed wire cutters.

with the failure location at the point where the handle meets the jaw. Figure 5.57 shows an overall view of the broken cutter (with the failure location indicated). The only visible markings on the broken cutter were: "2214" and "DROP FORGED." The manufacturer's trademark was not found, nor was the country of manufacture.

It was learned that the cutters failed while they were being used to cut copper wire of the size normally used in residential house wiring. Wire this size would usually be cut with a larger tool with greater mechanical advantage to reduce the force required by the operator to effect the cut. When received for failure analysis, the cutting edges were in the fully closed position and a screwdriver was used to pry the jaws open. This suggests that the hinge was not functioning properly and this contributed to the excess force required to make the cut when the tool failed.

A survey of manufacturing standards revealed that there were no U.S. standards regarding the manufacture of hand tools. Military standards were not checked. The British Standards Institute and the Japanese Industrial Standards do have standards governing the manufacture of such hand tools.

5.7.2 Experimental procedure

Figure 5.58 shows a low-magnification photograph of the two fracture surfaces and Fig. 5.59 schematically shows the location and orientation of the two fractographs in Fig. 5.58. In the latter, two distinct regions were identified. The outer surface, which is the region of maximum fiber stress during loading, exhibited very little evidence of

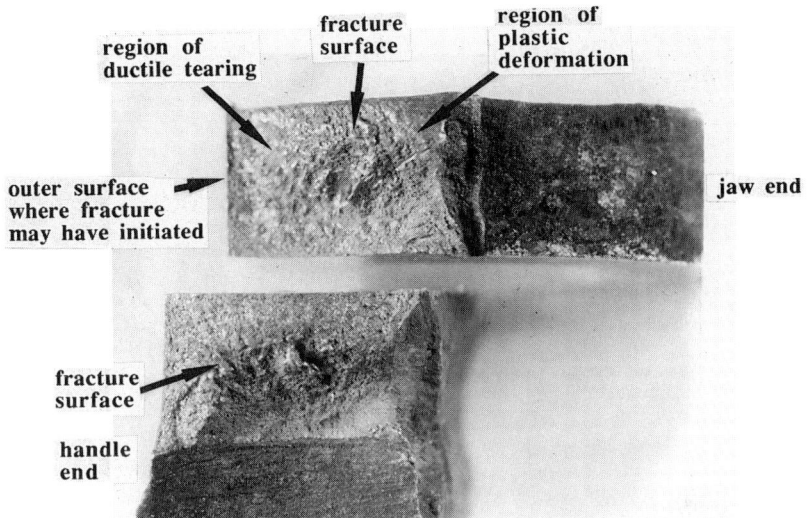

Figure 5.58 Fracture surfaces of the broken pair of cutters. The lower photograph is from the handle and the upper from a location near the hinge pin.

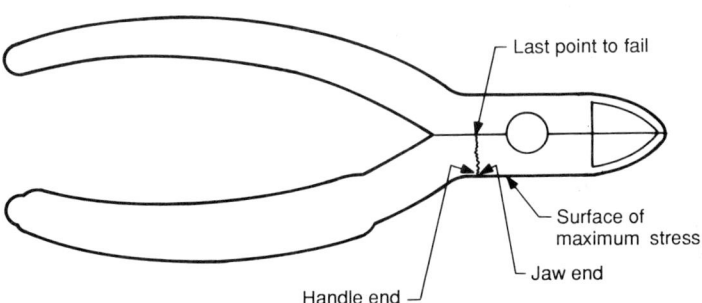

Figure 5.59 Schematic diagram of the wire cutters showing the location of fracture and the location of the features marked in Fig. 5.58.

dimensional change, while some was observed in the central region. In addition, some plastic deformation was evident on the surface that is presumed to be the last point to fail (see Sec. 4.4).

The plastic coating on the handles was cut away to expose the handle area. The chromium plating on the tip of the tool ended at the plastic, and the handles had a sooty substance on them. The tool must have been plated after the plastic was installed, and either the plastic left the residue, or it was left from a heat-treating process.

Following the visual examination, analysis for determining the cause of failure was carried out in three steps. The first step consisted in an evaluation of the stress levels involved in the wire cutting oper-

322   Chapter Five

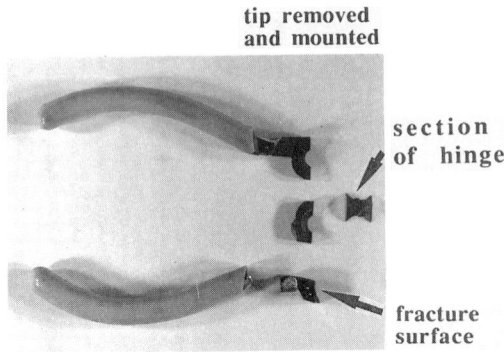

Figure 5.60 Sectioning scheme used for obtaining specimens for microstructural evaluation.

ations for which the tool was designed. This was followed by a metallographic evaluation of the material used and its heat treatment. Lastly, a detailed fractographic examination was conducted using the scanning electron microscope.

Figure 5.60 shows the failed part after it was sectioned for microstructural evaluation. The tip portion of the cutters was mounted for metallography to expose both a plane parallel to the fracture surface and a cross section of the hinge at its centerline. Rockwell hardness measurements were made using the C scale at the location indicated in Fig. 5.61 on both the failed and an unbroken pair of cutters.

5.7.3 Results

Evaluation of stresses. A pair of unbroken cutters of dimensions similar to those of the broken one was used to measure the force required in a typical cutting operation. The cutters were placed between the compression dies of a Tinius Olsen testing machine and the load required to cut a 10 TW [0.102-in (0.259-cm)-diameter] single-strand insulated copper wire was measured. A force of 436 N (98 lb) was re-

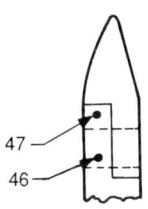
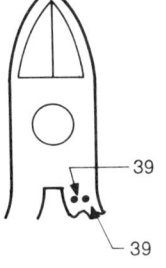

Figure 5.61 Values of hardness measurements (Rockwell C) made on both the unbroken (left) and the failed (right) pair of cutters.

quired to cut the wire, and the cutters appeared to flex elastically during the test (see Sec. 2.2). There was no evident damage to the cutters at the completion of the test. The maximum outer fiber stress imposed on the tool may be approximated by using a simple beam formula,

$$\sigma = \frac{Mc}{I}$$

where M = moment, = 6.4 cm × 436 N (2.5 in × 98 lb)
c = 0.34 cm (0.269/2 in)
I = moment of inertia, = 0.0112 cm^4 [(0.166 × 0.269^3)/12 in^4]

Hence σ = 8.4 × 10^4 Pa (122,000 lb/in^2)

This value now has to be compared to the mechanical properties of the material used.

The results of the hardness measurements are shown in Fig. 5.61. The failed cutters had a somewhat lower hardness. The average value was found to be 40 Rockwell C. This corresponds to an ultimate tensile strength of about 1.24 × 10^9 Pa (180,000 lb/in^2),[13] and a yield strength of about 1.03 × 10^9 Pa (150,000 lb/in^2).[27] It would appear then that the strength was well above the maximum fiber stress calculated. Gross overload may thus be overruled as a possible cause for failure.

Microstructural evaluation. Figure 5.62 is a low-magnification micrograph showing the cross section of the fracture surface from the jaw end. From the graduation in the extent of etching between the outer

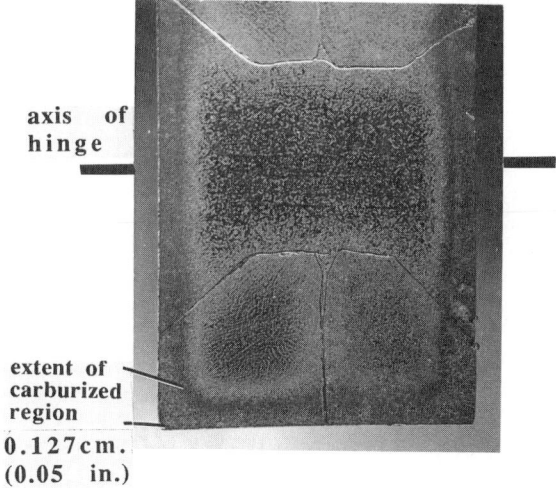

Figure 5.62 Low-magnification micrograph from the fracture region showing the extent of the carburized region.

edge and the center of this micrograph, it appears that the tool was carburized. At higher magnification the microstructure is revealed to be tempered martensite, with undissolved carbides outlining the prior austenite boundaries, as shown in Fig. 5.63. This structure persists to a depth corresponding to about 3 grains (of prior austenite) from the surface, below which the grain-boundary carbides are no longer observed. The prior austenite grain size is estimated to be ASTM 3. There is no evidence of decarburization at the surface. All these points indicate that the material was carburized. In order to confirm this, a section of the handle was removed and austenitized at 980°C (1800°F)

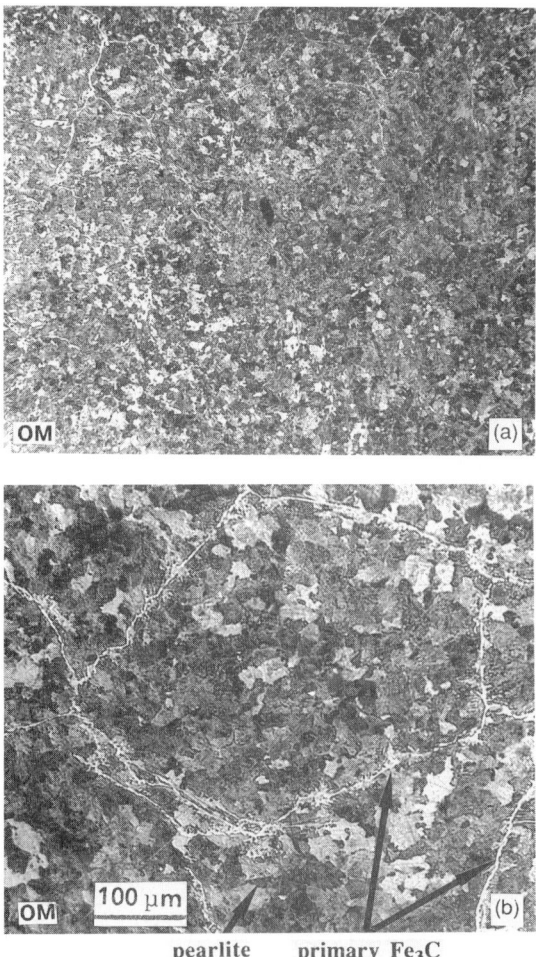

pearlite   primary Fe_3C

Figure 5.63 (a) Low- and (b) high-magnification micrographs of the carburized region showing a carbide network along the prior austenite grain boundaries.

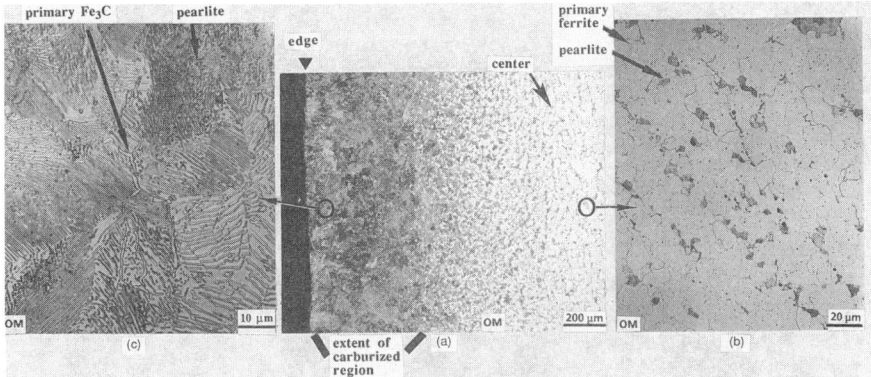

Figure 5.64 (a) Microstructure of the heat-treated specimens. (b) Microstructure from the center showing pearlite and ferrite. (c) The edge shows pearlite and a carbide network along the prior austenite grain boundaries, indicating a carburized structure.

for 2 h, then furnace-cooled. The edge-to-center microstructure of this piece is shown at low magnification in Fig. 5.64a and clearly reveals the carburized zone. Figure 5.64b shows that the center of the specimen was not carburized. This micrograph shows about 10 percent pearlite in a ferrite matrix, indicating a carbon content of approximately 0.1 percent for the steel from which the tool was made. In contrast, the microstructure near the edge (Fig. 5.64c) shows pearlite with carbides along the prior austenite boundaries, from which it is deduced that the carbon content of this region is greater than 0.8 percent.

SEM fractography. The fracture surface of the jaw end (Fig. 5.58) was examined in a scanning electron microscope. As shown in the low-magnification fractograph of Fig. 5.65, the outer region is brittle in appearance while the central area shows ductile tearing (see Sec. 3.4) with some secondary cracking. The higher-magnification fractography of Fig. 5.66 is from the outer carburized region where a mixed mechanism composed of cleavage (see Sec. 3.4) and microvoid coalescence (see Sec. 3.5) is evident. The central region shows only dimples, as seen in the high-magnification fractograph of Fig. 5.67, taken adjacent to one of the secondary cracks. The features at the extreme outer edge (region of maximum fiber stress) were masked by the presence of some corrosion products which could not be removed by standard surface-cleaning procedures (see Sec. 1.4.1).

5.7.4 Discussion

The observed hardness of the tool surface near the fracture area correlated to a yield strength that was significantly higher than the im-

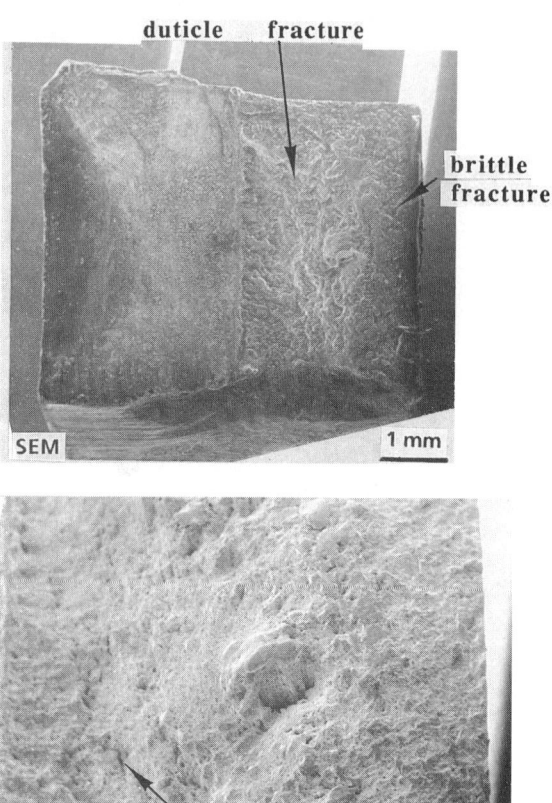

Figure 5.65 Low-magnification fractograph of the fracture surface.

posed stresses. This fact can be misleading, since a microstructure of carbides at prior austenite grain boundaries lowers the fracture toughness of the material.[28]

From the presence of a sooty material on the surface it is suspected that pack carburizing was used. Also, either the part was quenched directly from the carburizing treatment and the carbon content was such that it was beyond the A_{cm} line, thus leading to carbides along the austenite boundaries, or the quench rate was inadequate to suppress the formation of grain-boundary carbides. Since the prior austenite grain size is large (approximately ASTM 3), it is unlikely that the part was reaustenitized. It was in all probability tempered after being quenched directly following carburization.

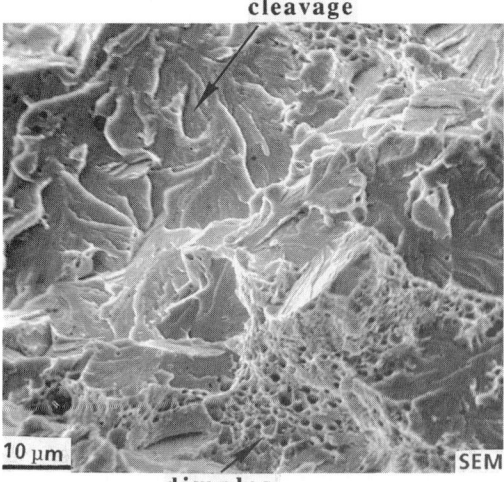

Figure 5.66 SEM fractograph near edge (see Fig. 5.65) showing mixed-mechanism fracture.

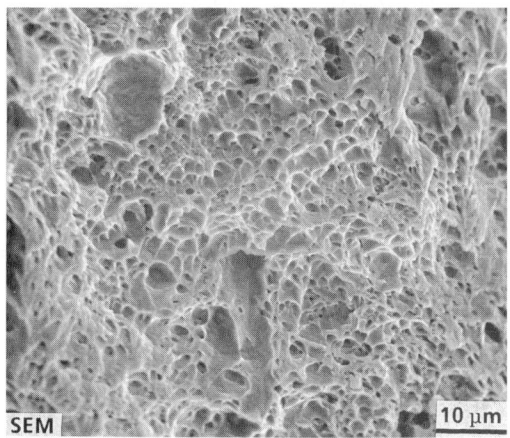

Figure 5.67 SEM fractograph near the secondary crack shown in Fig. 5.65 showing dimples.

While the possibility of hydrogen embrittlement (see Sec. 3.11) due to improper plating techniques cannot be completely ruled out, the cleavage fracture features observed at the outer edges are probably the result of the grain-boundary carbide network in this region. If the tool had been carburized prior to its assembly with the hinge pin, there would be a carburized area evident at the hinge pin surface in Fig. 5.62. Thus it is deduced that the tool was carburized after assembly. Uneven thermal expansion of the assembled part during heat treatment could also account for the binding of the hinge pin mentioned as part of the initial inspection.

5.7.5 Conclusions

It appears that the tool failed because of a defective microstructure of a network of primary carbide along prior austenite grain boundaries which lowered the allowable stresses on it. Also the prior austenite grain size was too large. Both of these problems were due to an improper heat treatment.

5.8 Case G: Broken Steel Punch*

5.8.1 Introduction

Pin punches are used to start holes in concrete blocks. A broken 3/32-in (0.238-cm) punch (Fig. 5.68) was received with the fracture being quite close to its working point. The punch was known to be made of steel, and it was clearly plated, possibly with nickel. The purpose of the investigation was to determine the cause of failure.

In the course of the investigation, the material was determined to be a high-carbon, plain-carbon steel with a tempered martensite structure. There are two well-known causes that can lead to the embrittlement of these steels. Temper embrittlement occurs if the steel is held at or slowly cooled through the 250 to 540°C (482 to 1004°F) temperature range.[29] Tempered martensite embrittlement (TME) occurs after tempering a martensitic structure between 260 and 370°C (500 to 698°F).[29]

TME is aggravated by exposure of the material to hydrogen, as might occur during an electrolytic plating process.[30,31] Higher-strength steels are more susceptible to hydrogen embrittlement.[31] One way to alleviate the problem is to bake the plated components at around 200°C (392°F) for 3 to 4 h. This normally allows the hydrogen to effuse except in cases where the plating itself acts as a diffusion barrier. The steel in this investigation had a relatively high carbon content, and upon quenching from austenite to 25°C (77°F) to harden the material, austenite may have been retained due to the martensite-finish temperature being below 25°C (77°F). During tempering this retained austenite may have decomposed to microconstituents, which are susceptible to hydrogen embrittlement.

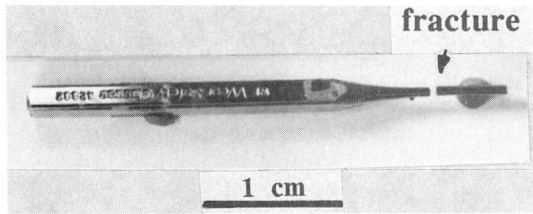

Figure 5.68 Broken punch.

---

*The authors acknowledge the work done by Bruce D. Cutler on this case study.

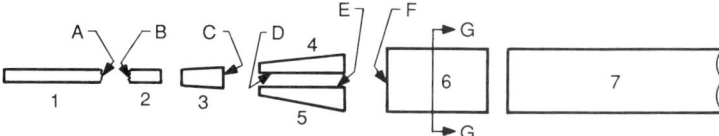

Figure 5.69 Schematic representation of sectioning scheme used to obtain specimens from the broken punch.

5.8.2 Experimental procedure

The first step was a visual examination of the part and photographic documentation of the entire tool (Fig. 5.68). Figure 5.69 shows a schematic representation of how the part was sectioned for the investigation. The fracture surface on section 1 (Fig. 5.69) was examined in a scanning electron microscope to show not only the overall fracture-surface features (low magnification) but also the variation in fine features across the cross section of the fracture surface. In addition, x-ray fluorescent spectroscopic (EDS) analyses were obtained from the bulk material, from an inclusion, and from the plating.

Rockwell C hardness measurements were made on surface F of section 6 (Fig. 5.69). Microhardness measurements were used to obtain a hardness profile across the metallographically polished surface of section 3 (surface C). Microhardness values were converted to Rockwell C values.[18] Microstructural information was obtained by polishing and etching surfaces D and E of sections 4 and 5, respectively, using standard metallographic procedures.

The determination of the carbon content of the steel was accomplished by austenitizing section 6 at 900°C (1652°F) for 1 h followed by furnace cooling over 24 h. Prior to austenitization most of the plating was ground off. Following austenitization and annealing, section 6 was further sectioned, as shown in Fig. 5.69, to yield surface G for metallographic evaluation. From this examination the pearlite and proeutectoid (ferrite or carbide) content of the structure was measured to estimate the carbon content of the steel.

To determine the existence of retained austenite in the material, section 3 was subcooled by immersion in a mixed methanol and dry-ice bath −79°C (−96°F) for approximately 25 min, followed by warming to room temperature in a vacuum desiccator. The microhardness profile across surface C of this specimen was then measured and compared to the profile from the as-received specimen (mentioned earlier).

5.8.3 Results

Preliminary information. Very little preliminary information was available as to the exact conditions under which the failure occurred.

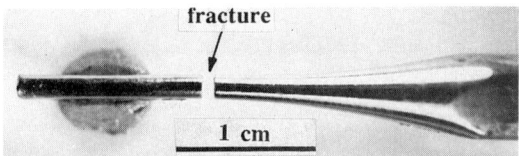

Figure 5.70 Closeup of fracture region of the punch showing lack of gross plastic deformation.

It was assumed that the tool broke under normal working conditions, which would be impact by a hammer. The package of a similar punch mentioned that the tip was double tempered and the punch itself was nickel-chrome plated. Figure 5.70 shows a closeup of the failed tool at the failure location. There is no evidence of gross plastic deformation or of a pronounced shear lip. Hence the fracture occurred in a brittle manner. The lip visible on the fracture surface (mentioned later) is the plating, which incidentally did not fracture at exactly the same location as the tool itself. It is assumed that the fracture of the plating had no effect on the failure of the tool.

SEM fractography and EDS analysis. Figure 5.71 shows low-magnification (overall) fractographs of the two fracture surfaces. At a slightly higher magnification (Fig. 5.72) these regions are more clearly defined, and so is the plating. The central dull fibrous region is surrounded by an outer rim of the bright, brittle appearing fracture. Figure 5.73 shows in the center region classic dimple fracture features. In contrast, the stereo-pair fractographs in Fig. 5.74 of the edge region show predominantly a "rock candy" appearance of intergranular fracture, mixed with areas showing dimples.

EDS analyses from the bulk material of the fracture surface indicated that the material is plain-carbon steel. The analysis of the plating showed an inner nickel layer and an outer chromium layer.

Microstructural information. Figure 5.75 shows the microstructure from the central region of the punch; it is composed entirely of

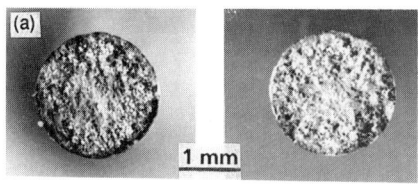

Figure 5.71 Low-magnification photographs of two fracture surfaces of broken punch. (a) Fracture surface of the smaller piece. (b) That of the larger piece.

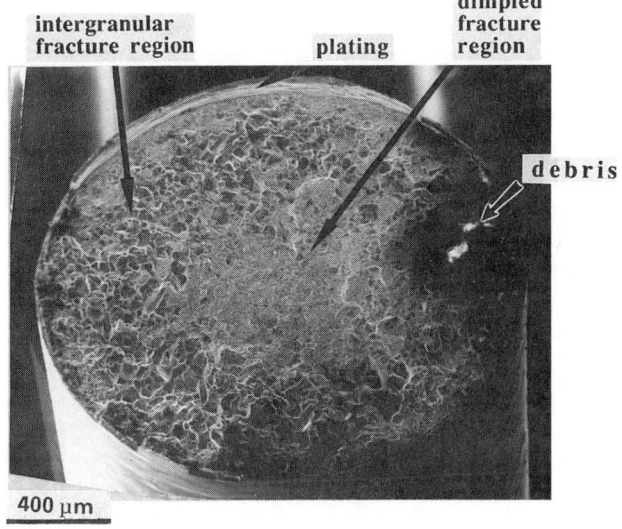

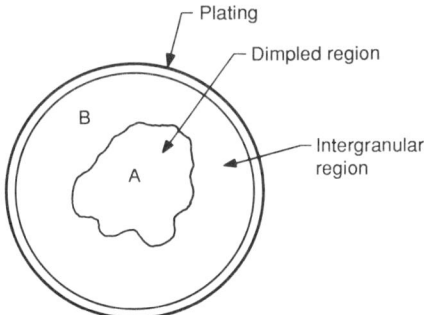

Figure 5.72 Locations of the plating and the dimpled and intergranular areas on the fracture surface of section 1.

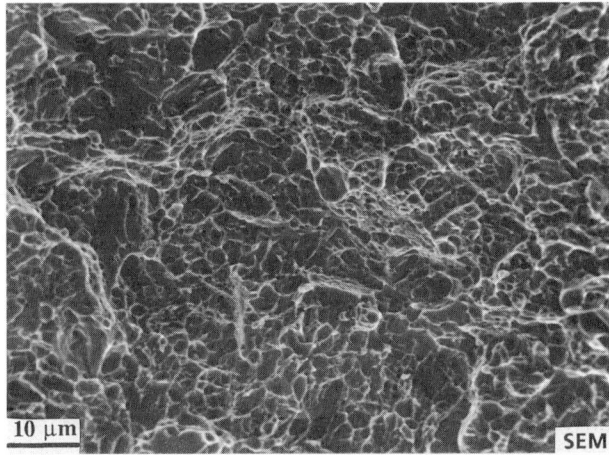

Figure 5.73 Fractograph from center region in Fig. 5.72 showing dimpled fracture features.

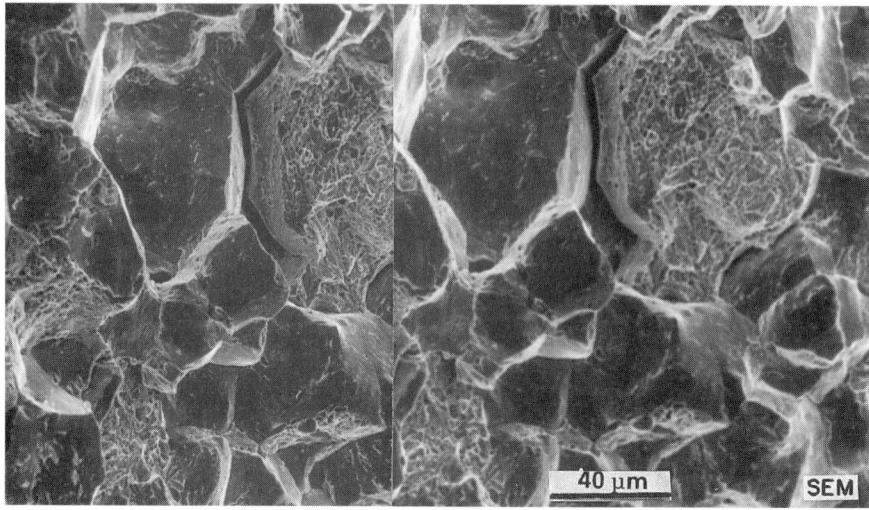

Figure 5.74 Stereo-pair fractographs from region B in Fig. 5.72 showing intergranular fracture.

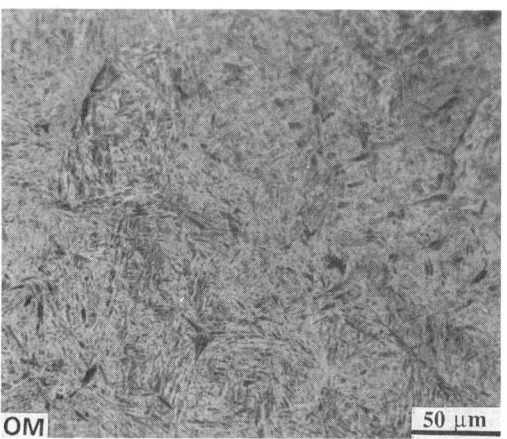

Figure 5.75 Microstructure from central region of punch.

martensite. The lower-magnification micrograph of Fig. 5.76a shows a white constituent at and near the surface. The higher-magnification micrograph of this region (Fig. 5.76b) shows that this constituent is probably retained austenite in martensite. If it is, then the carbon content near the surface must be higher than for the bulk (no retained austenite).

The carbon content of the material (bulk) can be estimated from the microstructure in Fig. 5.77, which is from the central region of the heat-treated specimen. In this micrograph the large dark and grey

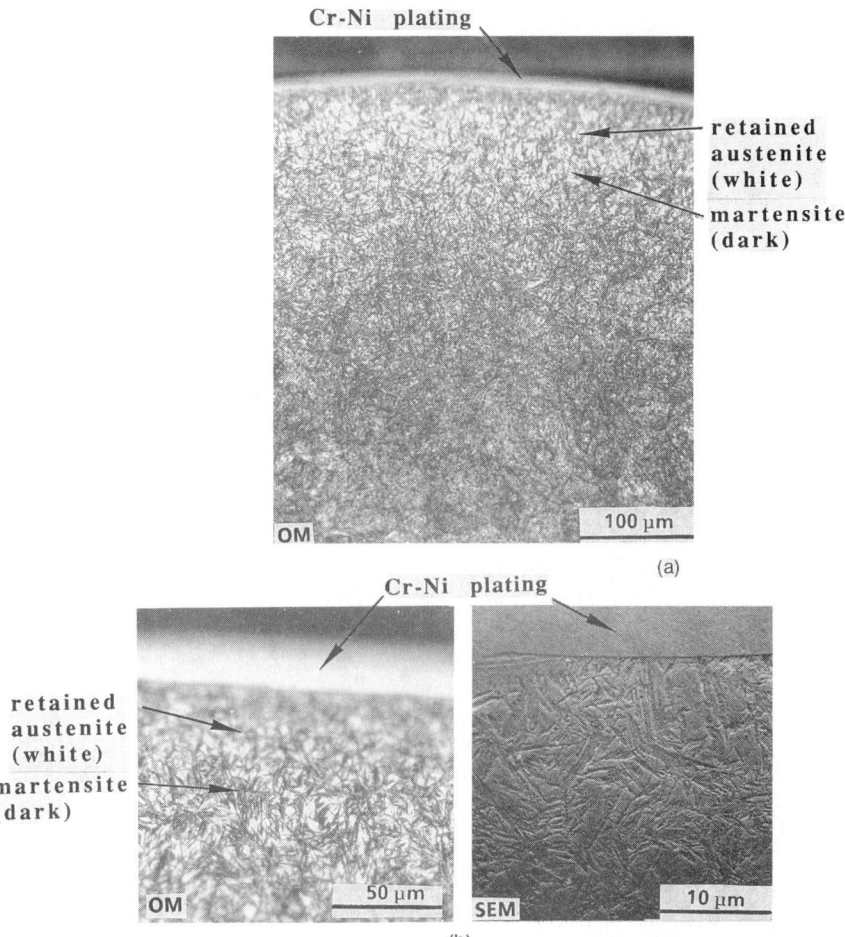

Figure 5.76 Microstructure near the surface of the punch (a) at low and (b) at high magnifications. The high-magnification micrograph clearly shows retained austenite.

areas are pearlite (about 90 percent), whereas the small bright areas are proeutectoid ferrite (about 10 percent) decorating the prior austenite grain boundaries. From this the carbon content of the material is estimated to be about 0.7 percent.

Hardness measurements. Microhardness readings of the as-received tool obtained from the center of the cross section of the tool averaged about 57 Rockwell C, which is consistent with the values obtained from macrohardness measurements. Readings taken close to the edge, however, varied from 50 to 60 Rockwell C, which could imply varia-

Figure 5.77 Microstructure from the central region of the heat-treated specimen showing pearlite and proeutectoid ferrite.

tions in the local microstructure encountered by the indenter. After the section had undergone the subcooling treatment mentioned earlier, the readings at both the surface and the center were consistently near 60 Rockwell C, indicating a transformation of retained austenite to martensite at the outer edges of the tool.

5.8.4 Discussion

This tool clearly failed by a mechanism of intergranular fracture along prior austenite boundaries. It appears that the tool was made from high-carbon (about 0.7 percent carbon), plain-carbon steel. The surface was carburized to a higher carbon content, which led to the presence of retained austenite during subsequent quenching. This in itself may not lead to embrittlement of the material. However, the prior austenite grains were large (ASTM 3), which is known to be a factor in susceptibility to intergranular embrittlement.[29] This could have been aggravated by the introduction of some hydrogen in the surface region during the plating process.

5.8.5 Conclusions

It appears that the tool failed because of a defective microstructure, which was also very sensitive to hydrogen embrittlement.

5.9 Case H: Broken Stainless-Steel Hinge for a Check Valve

5.9.1 Introduction

A broken steel hinge for a check valve in a saltwater line was received from a petroleum manufacturer. The exact configuration of the broken

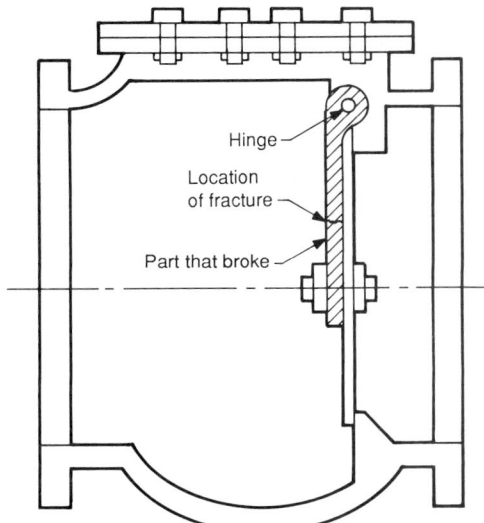

Figure 5.78 Schematic representation of the possible assembled configuration of the hinge and check valve.

part in the check-valve assembly was not known, but a schematic representation of a possible assembled configuration is shown in Fig. 5.78. A cover letter accompanying the two broken pieces stated: "These check valves are in service in produce water (saltwater) pipelines. If the pump goes down, these valves close to prevent the high pressure from the discharge line from getting into the low-pressure suction line. When this valve failed, the suction line was pressurized causing considerable damage. Approximate service life 1 year; material is stainless steel; the impact on these valves, when they close, is considerable."

It is to be noted that the following pieces of useful information could not be obtained:

1. The operating environment of the hinge itself, although it was probably the same as for the valve.
2. The configuration of the hinge in the assembled condition. This information would be invaluable in determining the loading conditions for the hinge.
3. The magnitude of the loading.
4. The grade or class of the stainless steel used.

The broken hinge was received in two parts. Figure 5.79 shows the two parts in their relative operating positions, Figs. 5.80 and 5.81 give the dimensions of the two parts.

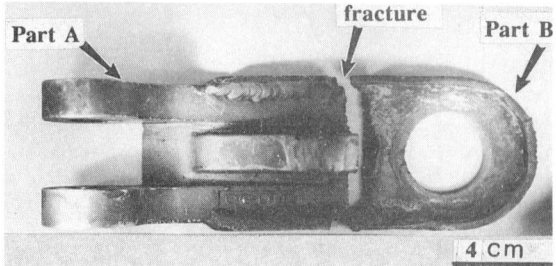

Figure 5.79 Two pieces, parts A and B, of the broken hinge.

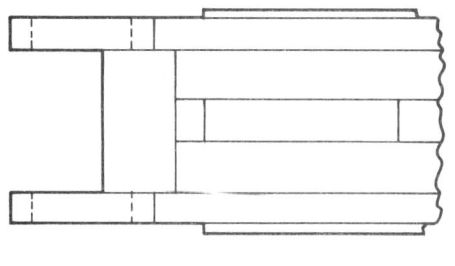

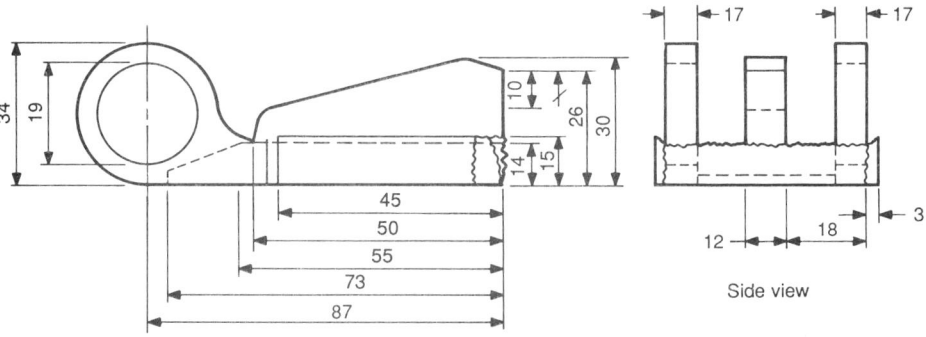

Figure 5.80 Dimensions of part A (in millimeters).

5.9.2 Experimental procedure

The first step was a visual examination of the part and photographic documentation of the two pieces. The fracture surfaces of both parts were examined carefully, both macroscopically and in a scanning electron microscope. In addition, EDS analyses were obtained from various regions of interest on the fracture surfaces.

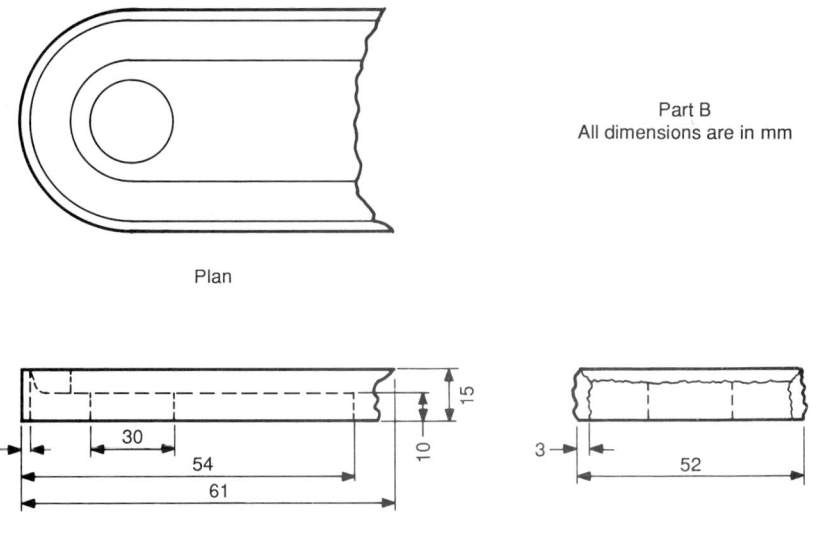

Figure 5.81 Dimensions of part B (in millimeters).

Sections were cut from part A, as shown schematically in Fig. 5.82, to yield two specimens (called samples 1 and 2) for microhardness measurements, magnetic measurements, and microstructural evaluation. Magnetic measurements were made on sample 1 in an effort to evaluate the level of delta ferrite in the microstructure.[32] Micro-

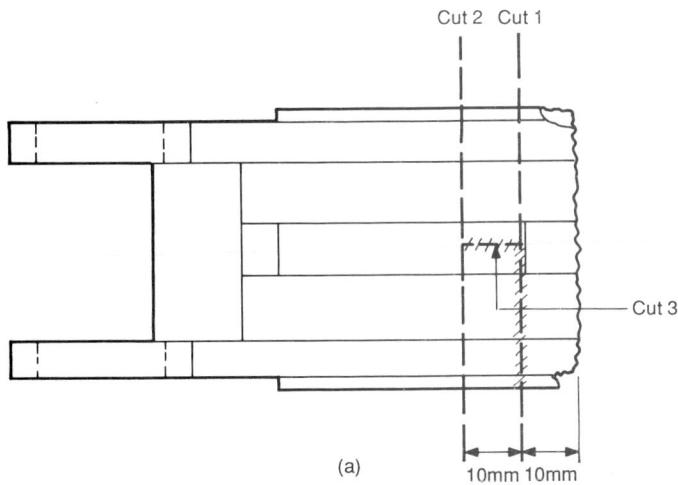

Figure 5.82 (a) Schematic representation of the sectioning scheme used to obtain specimens from part A.

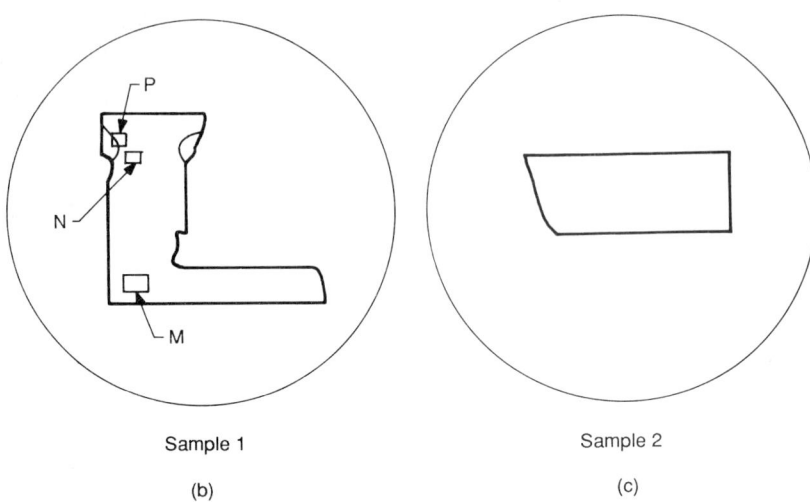

Sample 1   Sample 2
(b)   (c)

Figure 5.82 (*Continued*) (*b*), (*c*) Two specimens.

hardness measurements were also obtained from sample 1. Microstructural information was obtained by polishing and etching (Kalling's etch) samples 1 and 2 using standard metallographic procedures.

5.9.3 Results

Visual inspection. Most of the surface of part A appeared to have been painted green. There were three distinct regions in this part: a base plate, a rib on the base plate, and a strap welded on the base. These are identified in Fig. 5.83. One of the filet welds used to join the strap to the base plate is marked in Fig. 5.83a. Two stampings were in evidence on the base of part A—one on the top surface (Fig. 5.83a) and one on the undersurface (Fig. 5.83b). Some shiney surfaces (rubbed appearance) were in evidence, and these are also marked on Fig. 5.83.

Part B was much darker and dirtier in appearance than part A. The painting on this part appeared very patchy, flaky, and lighter in color than on part A. The overall appearance of part B gave the impression that the part had been exposed to high temperature. Figure 5.84 shows the top and underside of this part.

A "bearing surface" is evident around the hole in Fig. 5.84a. Figure 5.85 shows two end-on views of part B. These clearly indicate bending (plastic deformation) adjacent to the fracture location (see Sec. 4.4).

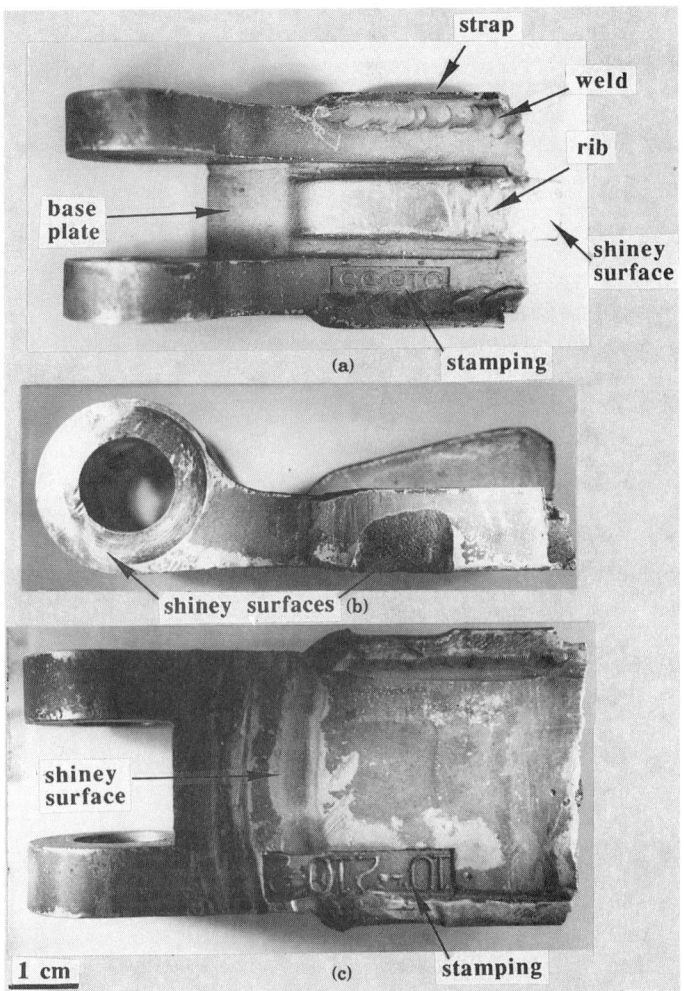

Figure 5.83 Part A. (a) Plan view. (b) Side view. (c) Underside. Stampings on part and some shiney surfaces are noted.

SEM fractography and EDS analysis. Figure 5.86 shows low-magnification (overall) fractographs of the fracture surfaces of parts A and B in the as-received condition. As can be seen in Fig. 5.86a, the fracture surface of part A was rusted in small patches and one-half was covered by a black film. The fracture surface of part B (Fig. 5.86b) was all but obscured by a black (oxide?) film. Cleaning the fracture surface of A was accomplished by repeated use of acetate replicating tape followed by ultrasonically cleaning in a solution of 10 percent oxalic acid. Al-

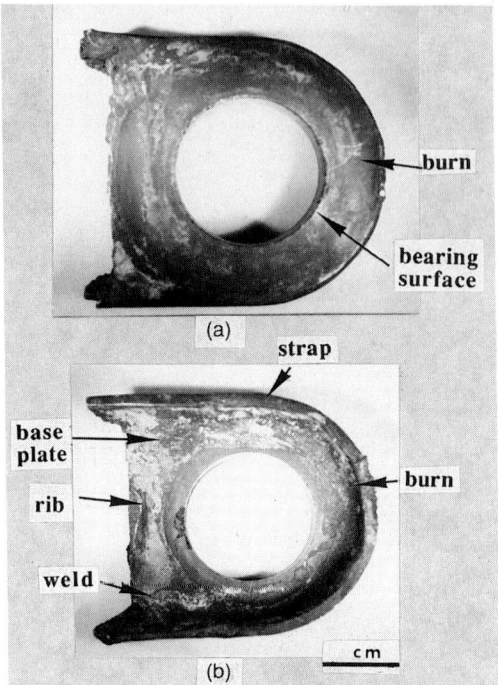

Figure 5.84 Part *B*. (*a*) Top view. (*b*) Underside.

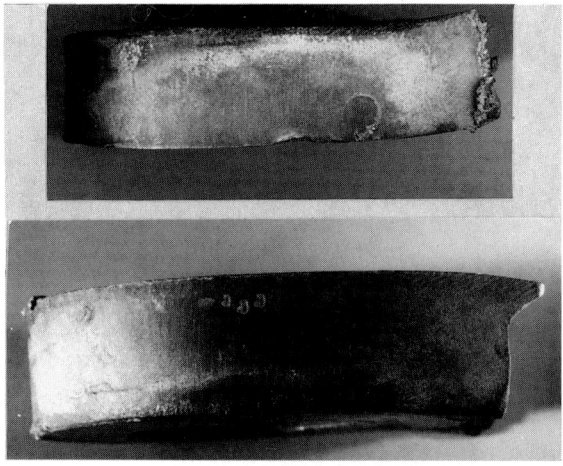

Figure 5.85 Two end views of part *B* showing bending.

Case Studies   341

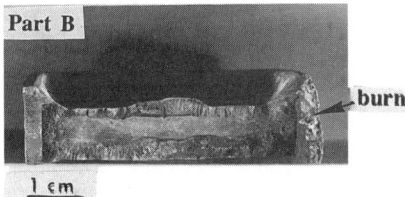

Figure 5.86 Low-magnification fractographs. (a) Part A. (b) Part B.

most all the rust was removed by this, but the black film persisted (see Sec. 1.4.1). This obviated the determination of the fracture mechanism by detailed fractography.

Figure 5.87 shows a composite of the fracture surface of part A. Two distinct regions can be seen—a bright rim with distinct radial marks and a central dull rectangular region. The radial marks point back toward the central rectangle, which has a machined appearance (see

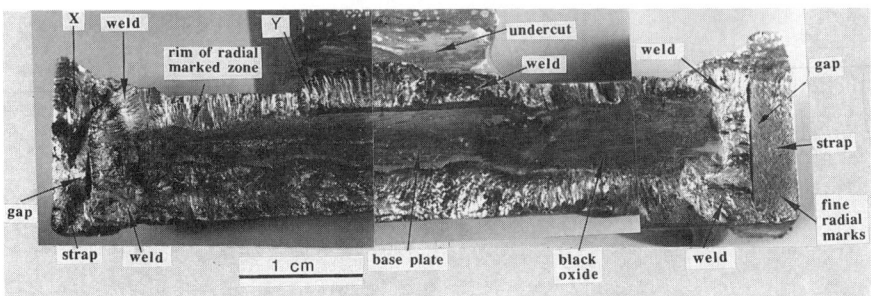

Figure 5.87 Low-magnification fractograph of part A showing two distinct regions, one bright and the other dull in appearance.

Sec. 4.2). Under a stereo microscope the bright rim consists of very sharp and mirrorlike facets. The strap is seen to be welded on the base plate by two filet welds, one at the top and one at the underside. The latter weld has been ground down. The fracture surfaces of the strap are a bit receded from that of the base plate. The top weld provides a sloped junction between the fracture surfaces of the strap and the base plate. Also the top weld is discontinuous with the strap for a majority of the weld interface with the strap. At the location marked X in Fig. 5.87, a small round discontinuity was observed at the weld. The surface of this appeared solidified as opposed to fractured. The surface was dark and around it was a crack in the weld metal which extended to the top surface of the weld. Next to this there was an eye-shaped entrapped slag with a very shiney black appearance. This in turn was surrounded by a rim of bright metal. A partial view of this is shown in Fig. 5.88. The weld between the rib and the base plate of part A shows an undercut (Fig. 5.87). As illustrated in Fig. 5.89, at location Y marked in Fig. 5.87, a fine crack and radial marks pointing to its origin are seen in the scanning electron micrograph.

EDS analyses were also carried out in the scanning electron microscope on polished and etched metallographic specimens (next section), on both the base plate and the strap. The results are shown in Table 5.6. From these it is surmised that the strap is 347 stainless steel, while the base plate is CF8C (compositions are given in Table 5.7) since some Nb was found in a previous analysis. Table 5.8 shows the results of EDS analyses on the matrix and what was suspected to be

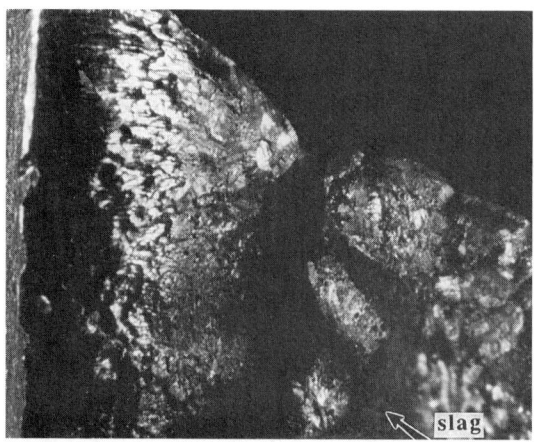

Figure 5.88 Optical photograph of slag found near location X marked in Fig. 5.87.

Case Studies 343

Figure 5.89 Optical photograph of crack found in location Y marked in Fig. 5.87.

TABLE 5.6 EDS Results from Base Plate and Strap

Base Plate			Strap		
keV	Counts	Element	keV	Counts	Element
1.75	51	Si	2.31	87	S
2.34	86	S	5.40	775	Cr
5.40	736	Cr	5.93	203	Cr
5.95	179	Cr	6.40	1745	Fe
6.39	1732	Fe	7.05	251	Fe
7.06	275	Fe	7.50	203	Ni
7.47	192	Ni	8.26	51	Ni
8.25	65	Ni			

TABLE 5.7 Chemical Composition Specification for CF8C Casting and 347 Stainless Steel

CF8C casting*		347 Stainless steel†	
Element	wt.%	Element	wt.%
Cr	18.0–21.0	Cr	17.0–19.0
Ni	8.0–11.0	Ni	9.0–13.0
C	0.08 max	C	0.08 max
Mn	1.50 max	Mn	2.00 max
Si	2.00 max	Si	1.00 max
P	0.04 max	P	0.045 max
S	0.04 max	S	0.03 max
Nb	> 8 × C but < 1.00	Nb + Ta	> 10 × min. C

*From ASTM A743-79.[33]
†From *Metals Handbook*.[25]

TABLE 5.8 EDS Results from Matrix and Delta Ferrite

Matrix			Ferrite		
keV	Counts	Element	keV	Counts	Element
1.69	121	Si	5.42	375	Cr
5.43	288	Cr	5.94	127	Cr
5.94	101	Cr	6.42	629	Fe
6.40	717	Fe	7.03	142	Fe
7.05	156	Fe	7.47	65	Ni
7.48	106	Ni			

delta ferrite in the microstructure. Indeed the results are consistent with the suspected phase being delta ferrite.

Microstructural information. Figure 5.82 schematically shows samples 1 and 2 and the locations from which they were taken. In the unetched condition, sample 1 shows a large concentration of round inclusions in the base plate and the structure is dendritic. The base plate thus is a casting. In contrast, the strap shows fewer and stringer-type inclusions, which indicates that it is made from rolled stock. One feature of note is that the strap lies inclined to the base plate, as illustrated schematically in Fig. 5.90.

Figure 5.91 is an optical micrograph taken from region M in Fig. 5.82b and is from the base plate. The background shows a cored dendritic structure and delta ferrite in the interdendritic areas. Figure 5.92 shows a micrograph from region N taken in a scanning electron microscope; the delta ferrite is very evident. A similar micrograph from region P, shown in Fig. 5.93, reveals a crack which has propagated through delta ferrite. Note that the weld metal has less delta ferrite than the base plate.

Magnetic measurements. The ferrite content of the materials used in the base plate, weld, and strap was measured using a magnetic gage.[32] Results of readings taken from sample 1 are shown on Fig. 5.94 (locations indicated). At these low ferrite levels the ferrite num-

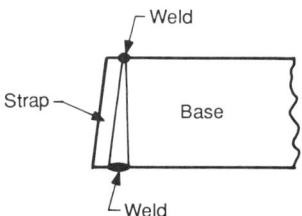

Figure 5.90 Schematic representation of relative inclination between strap and base plate.

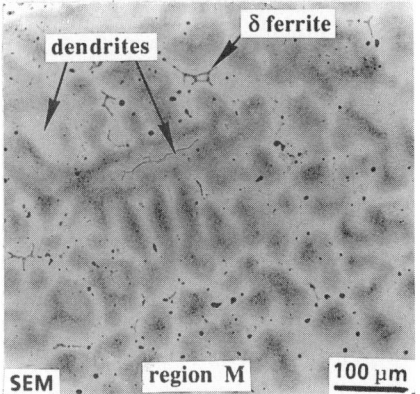

Figure 5.91 Micrograph from region *M* in Fig. 5.82*b*. Note the cored dendritic structure and delta ferrite in the interdendritic areas.

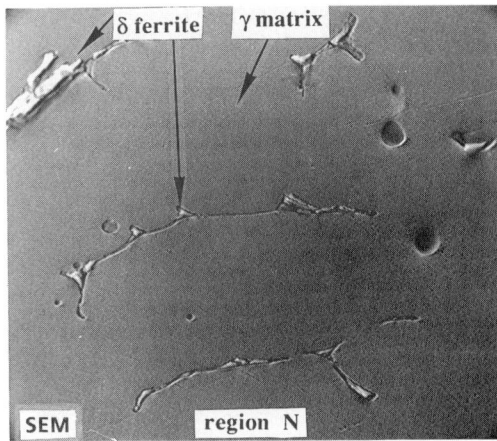

Figure 5.92 Micrograph from region *N* in Fig. 5.82*b* showing delta ferrite.

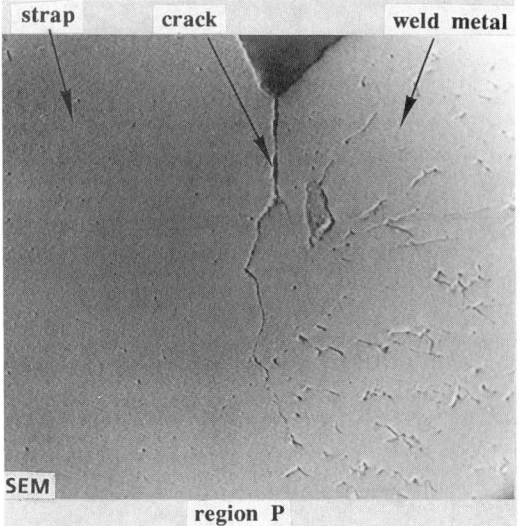

Figure 5.93 Micrograph from region *P* in Fig. 5.82*b* showing a crack through the delta ferrite.

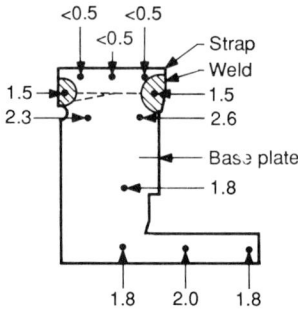

Figure 5.94 Ferrite number of the material at various locations on sample 1 as measured by a magnetic gage.

ber is approximately equal to the volume percent ferrite. From these results the range of ferrite contents is deduced to be

Strap: No ferrite

Weld metal: ~ 1.5%

Heat-affected zone in base plate: ~ 2.5%

Base plate: ~ 2.0%

Microhardness measurements. The results of microhardness measurements made on several locations on sample 1 are given in Fig. 5.95.

5.9.4 Discussion

It appears that the part was manufactured by welding two pieces of cast CF8C stainless steel. A strap of 347 stainless steel was then welded on for additional strength. The welding seems to have been done shoddily. This is evidenced by the facts that the strap did not lie flat on the base plate, a crack was found at the weld metal–strap interface, and there is a gap between the weld and strap and base plate (that is, the root of the weld did not penetrate to the point of contact between the strap and the base plate as it should have).

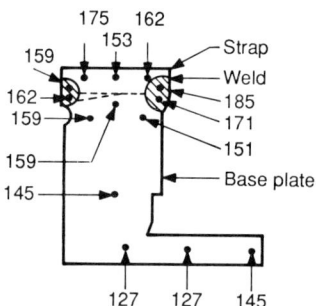

Figure 5.95 Microhardness values at various locations on sample 1.

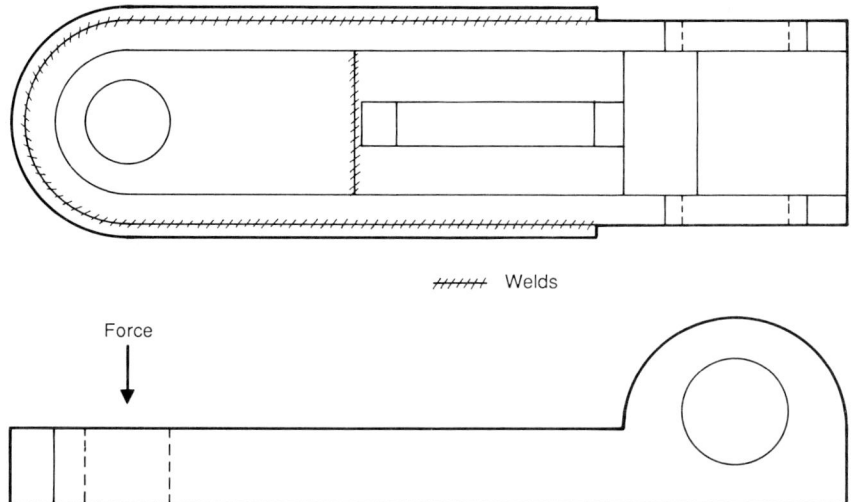

Figure 5.96 Effect of bending loads on the failed part.

Considering the part under the action of a bending load, as shown in Fig. 5.96, it can be seen that part A is stiffened by the rib while part B is not. This is a weld region and, as described, the weld also had a discontinuity which can act as an acute stress concentrator. The part seems to have failed under the action of an impact load in excess of its design limits. That the loading was of a bending type is evidenced by the deformation seen in Fig. 5.85 (see Sec. 4.4). The low ferrite content of the weld indicates that the crack seen at the discontinuity is probably a hot crack. Austenitic stainless steels need at least 5 percent delta ferrite to prevent hot cracking,[34] the prevalent theory being that ferrite scavenges tramp elements (such as S, B) which are thought to be responsible for hot cracking.[34]

It is very possible that the component was simply not designed for the type of impact it was subjected to when it failed. Further the strength level of the base plate is at the lower end of the required strength level according to ASTM A743-79.[32]

5.9.5 Conclusions

It appears that the mode of failure was bending overload, probably of the impact type. No clear mechanism of failure was discerned. Except for a slightly higher delta ferrite content in the weld metal (which is desirable), no other material faults were found. The possibility of the improper weld contributing to fracture cannot be ruled out.

References

1. C. R. Brooks, *Heat Treatment, Structure and Properties of Non-Ferrous Alloys,* American Society for Metals, Metals Park, Ohio, 1982.
2. *Metals Handbook,* 8th ed., vol. 4: *Forming,* American Society for Metals, Metals Park, Ohio, 1969, p. 423.
3. R. H. Uhlig, *Corrosion and Corrosion Control,* 2d ed., Wiley, New York, 1971.
4. M. G. Fontana and N. D. Green, *Corrosion Engineering,* 2d ed., McGraw-Hill, New York, 1978.
5. *Metals Handbook,* 9th ed., vol. 13: *Corrosion,* ASM International, Metals Park, Ohio, 1987, p. 131.
6. E. P. Polushkin and H. J. Shuldener, "Corrosion of Yellow Brass Pipes in Domestic Hot-Water Systems—A Metallographic Study," *Trans. AIME,* vol. 161, 1945, p. 214.
7. H. Leidheiser, *The Corrosion of Copper, Tin and Their Alloys,* Wiley, New York, 1971.
8. *Metals Handbook,* 9th ed., vol. 6: *Welding, Brazing and Soldering,* American Society for Metals, Metals Park, Ohio, 1983.
9. *ASHRAE Handbook—Equipment,* American Society for Heating, Refrigeration, and Air-Conditioning Engineers, Atlanta, Ga., 1983, p. 21.7.
10. *Metals Handbook,* 9th ed., vol. 2: *Properties and Selection: Nonferrous Alloys and Pure Metals,* American Society for Metals, Metals Park, Ohio, 1979.
11. *Metals Handbook,* 8th ed., vol. 8: *Metallography, Structure and Phase Diagrams,* American Society for Metals, Metals Park, Ohio, 1973.
12. S. Shivkumar, S. Ricci, Jr., C. Keller, and D. Apelian, "Effect of Solution Treatment Parameters on Tensile Properties of Cast Aluminum Alloys," *J. Heat Treat.,* vol. 8, p. 63, 1990.
13. H. E. Boyer and T. L. Gall (eds.), *Metals Handbook—Desk Edition,* American Society for Metals, Metals Park, Ohio, 1985.
14. R. A. Wallis, *Axial Flow Fans and Ducts,* Wiley, New York, 1983.
15. J. E. Hatch (ed.), *Aluminum—Properties and Physical Metallurgy,* American Society for Metals, Metals Park, Ohio, 1984.
16. *Motor Auto Repair,* 51st ed., Motor, New York, 1987, p. 14-6.
17. R. Kiessling, *Nonmetallic Inclusions in Steel: Part III,* The Metals Society, London, U.K., 1978.
18. ASTM E 140-88, "Standard Hardness Conversion Tables for Metals," American Society for Testing and Materials, Philadelphia, Pa., 1988.
19. Shale and Faschig, "Operating Characteristics of a High-Temperature Electrostatic Precipitator," Rep. 7276, U.S. Bureau of Mines, Washington, D.C., 1969.
20. ASTM E 140-86, vol. 03.01, American Society for Testing and Materials, Philadelphia, Pa., 1987.
21. H. L. Craig, T. W. Crooker, and D. M. Hoeppner (eds.), *Corrosion-Fatigue Technology,* American Society for Testing and Materials, Philadelphia, Pa., 1978, pp. 126, 168.
22. *Mechanical and Physical Properties of Austenitic Chromium-Nickel Stainless Steels at Ambient Temperatures,* International Nickel Co., New York, 1963.
23. *Metals Handbook,* 8th ed., vol. 1: *Properties and Selection of Metals,* American Society for Metals, Metals Park, Ohio, 1961.
24. J. W. Spretnak, M. G. Fontana, and H. E. Brooks, "Notched and Unnotched Tensile and Fatigue Properties of Ten Engineering Alloys at 25 and −196°C," *Trans. ASM,* vol. 43, p. 547, 1951.
25. *Metals Handbook,* 9th ed., vol. 3: *Properties and Selection: Stainless Steels, Tool Materials and Special Purpose Metals,* American Society for Metals, Metals Park, Ohio, 1980.
26. R. H. Henke, as quoted in K. G. Brickner and J. P. Defilippi, "Mechanical Properties of Stainless Steels at Cryogenic Temperature and at Room temperature," *Handbook of Stainless Steels,* McGraw-Hill, New York, 1977, p. 20-11 ff.
27. C. R. Brooks, *The Heat Treatment of Ferrous Alloys,* Hemisphere Publ. Corp./McGraw-Hill, New York, 1979.

28. *Republic Alloy Steels,* Cleveland, Ohio, 1968, p. 337.
29. G. V. Van der Voort, "Embrittlement of Steels," in *Metals Handbook,* 10th ed., vol. 1: *Properties and Selection: Irons, Steels, and High-Performance Alloys,* ASM International, Metals Park, Ohio, 1990, p. 689 ff.
30. *Metal Finishing Guidebook,* 43d ed., Metals and Plastics Publ., Hackensack, N.J., 1975, pp. 156–157.
31. S. F. Floreen, "Hydrogen Cracking in Speciality Steels," in R. A. Oriani, J. P. Hirth, and M. Smialowski (eds.), *Hydrogen Degradation of Ferrous Alloys,* Noyes Publ., Park Ridge, N.J., 1985, p. 799 ff.
32. ASTM A799/A-91, "Standard Practice for Steel Castings, Austenitic Alloy, Estimating Ferrite Content Thereof," vol. 01.02, American Society for Testing and Materials, Philadelphia, Pa., 1991.
33. ASTM A743-79, "Standard Specification for Corrosion Resistant Iron-Chromium, Iron-Chromium-Nickel, and Nickel-Base Alloy Castings for General Application," American Society for Testing and Materials, Philadelphia, Pa., 1979.
34. E. Folkhard, *Welding Metallurgy of Stainless Steels,* Springer, New York, 1988, p. 144 ff.

Bibliography

Brooks, C. R., *Heat Treatment, Structure, and Properties of Non-Ferrous Alloys,* American Society for Metals, Metals Park, Ohio, 1982.

Krauss, G., *Steels: Heat Treatment and Processing Principles,* ASM International, Metals Park, Ohio, 1990.

Appendix A

Temperature Conversions*

*Reprinted with permission from *1977 Metal Progress Databook,* American Society for Metals, Metals Park, Ohio, 1977.

Temperature Conversions

The general arrangement of this conversion table was devised by Sauveur and Boylston. The middle columns of numbers (in **boldface** type) contain the temperature readings (°F or °C) to be converted. When converting from degrees Fahrenheit to degrees Celsius, read the Celsius equivalent in the column headed "C". When converting from Celsius to Fahrenheit, read the Fahrenheit equivalent in the column headed "F".

F	** **	C	F	** **	C	F	** **	C
....	**−458**	−272.22	....	**−308**	−188.89	−252.4	**−158**	−105.56
....	**−456**	−271.11	....	**−306**	−187.78	−248.8	**−156**	−104.44
....	**−454**	−270.00	....	**−304**	−186.67	−245.2	**−154**	−103.33
....	**−452**	−268.89	....	**−302**	−185.56	−241.6	**−152**	−102.22
....	**−450**	−267.78	....	**−300**	−184.44	−238.0	**−150**	−101.11
....	**−448**	−266.67	....	**−298**	−183.33	−234.4	**−148**	−100.00
....	**−446**	−265.56	....	**−296**	−182.22	−230.8	**−146**	−98.89
....	**−444**	−264.44	....	**−294**	−181.11	−227.2	**−144**	−97.78
....	**−442**	−263.33	....	**−292**	−180.00	−223.6	**−142**	−96.67
....	**−440**	−262.22	....	**−290**	−178.89	−220.0	**−140**	−95.56
....	**−438**	−261.11	....	**−288**	−177.78	−216.4	**−138**	−94.44
....	**−436**	−260.00	....	**−286**	−176.67	−212.8	**−136**	−93.33
....	**−434**	−258.89	....	**−284**	−175.56	−209.2	**−134**	−92.22
....	**−432**	−257.78	....	**−282**	−174.44	−205.6	**−132**	−91.11
....	**−430**	−256.67	....	**−280**	−173.33	−202.0	**−130**	−90.00
....	**−428**	−255.56	....	**−278**	−172.22	−198.4	**−128**	−88.89
....	**−426**	−254.44	....	**−276**	−171.11	−194.8	**−126**	−87.78
....	**−424**	−253.33	....	**−274**	−170.00	−191.2	**−124**	−86.67
−457.6	**−422**	−252.22	....	**−272**	−168.89	−187.6	**−122**	−85.56
−454.0	**−420**	−251.11	....	**−270**	−167.78	−184.0	**−120**	−84.44
−450.4	**−418**	−250.00	....	**−268**	−166.67	−180.4	**−118**	−83.33
−446.8	**−416**	−248.89	....	**−266**	−165.56	−176.8	**−116**	−82.22
−443.2	**−414**	−247.78	....	**−264**	−164.44	−173.2	**−114**	−81.11
−439.6	**−412**	−246.67	....	**−262**	−163.33	−169.6	**−112**	−80.00
−436.0	**−410**	−245.56	....	**−260**	−162.22	−166.0	**−110**	−78.89
−432.4	**−408**	−244.44	....	**−258**	−161.11	−162.4	**−108**	−77.78
−428.8	**−406**	−243.33	....	**−256**	−160.00	−158.8	**−106**	−76.67
−425.2	**−404**	−242.22	....	**−254**	−158.89	−155.2	**−104**	−75.56
−421.6	**−402**	−241.11	....	**−252**	−157.78	−151.6	**−102**	−74.44
−418.0	**−400**	−240.00	....	**−250**	−156.67	−148.0	**−100**	−73.33

F	** **	C	F	** **	C	F	** **	C
+17.6	**−8**	−22.22	+35.6	**+2**	−16.67	+53.6	**+12**	−11.11
+21.2	**−6**	−21.11	+39.2	**+4**	−15.56	+57.2	**+14**	−10.00
+24.8	**−4**	−20.00	+42.8	**+6**	−14.44	+60.8	**+16**	−8.89
+28.4	**−2**	−18.89	+46.4	**+8**	−13.33	+64.4	**+18**	−7.78
+32.0	**±0**	−17.78	+50.0	**+10**	−12.22	+68.0	**+20**	−6.67
+71.6	**+22**	−5.56	+89.6	**+32**	−0.00	+107.6	**42**	5.56
+75.2	**+24**	−4.44	+93.2	**+34**	+1.11	111.2	**44**	6.67
+78.8	**+26**	−3.33	+96.8	**+36**	+2.22	114.8	**46**	7.78
+82.4	**+28**	−2.22	+100.4	**+38**	+3.33	118.4	**48**	8.89
+86.0	**+30**	−1.11	+104.0	**+40**	+4.44	122.0	**50**	10.00

F	** **	C	F	** **	C	F	** **	C
287.6	**142**	61.11	305.6	**152**	66.67	323.6	**162**	72.22
291.2	**144**	62.22	309.2	**154**	67.78	327.2	**164**	73.33
294.8	**146**	63.33	312.8	**156**	68.89	330.8	**166**	74.44
298.4	**148**	64.44	316.4	**158**	70.00	334.4	**168**	75.56
302.0	**150**	65.56	320.0	**160**	71.11	338.0	**170**	76.67
341.6	**172**	77.78	359.6	**182**	83.33	377.6	**192**	88.89
345.2	**174**	78.89	363.2	**184**	84.44	381.2	**194**	90.00
348.8	**176**	80.00	366.8	**186**	85.56	384.8	**196**	91.11
352.4	**178**	81.11	370.4	**188**	86.67	388.4	**198**	92.22
356.0	**180**	82.22	374.0	**190**	87.78	392.0	**200**	93.33

....	−398	−238.89	−414.8	−248	−155.56	−144.4	−98	−72.22	125.6	52	11.11	395.6	202	94.44
....	−396	−237.78	−410.8	−246	−154.44	−140.8	−96	−71.11	129.2	54	12.22	399.2	204	95.56
....	−394	−236.67	−407.2	−244	−153.33	−137.2	−94	−70.00	132.8	56	13.33	402.8	206	96.67
....	−392	−235.56	−403.6	−242	−152.22	−133.6	−92	−68.89	136.4	58	14.44	406.4	208	97.78
....	−390	−234.44	−400.0	−240	−151.11	−130.0	−90	−67.78	140.0	60	15.56	410.0	210	98.89
....	−388	−233.33	−396.4	−238	−150.00	−126.4	−88	−66.67	143.6	62	16.67	413.6	212	100.00
....	−386	−232.22	−392.8	−236	−148.89	−122.8	−86	−65.56	147.2	64	17.78	417.2	214	101.11
....	−384	−231.11	−389.2	−234	−147.78	−119.2	−84	−64.44	150.8	66	18.89	420.8	216	102.22
....	−382	−230.00	−385.6	−232	−146.67	−115.6	−82	−63.33	154.4	68	20.00	424.4	218	103.33
....	−380	−228.89	−382.0	−230	−145.56	−112.0	−80	−62.22	158.0	70	21.11	428.0	220	104.44
....	−378	−227.78	−378.4	−228	−144.44	−108.4	−78	−61.11	161.6	72	22.22	431.6	222	105.56
....	−376	−226.67	−374.8	−226	−143.33	−104.8	−76	−60.00	165.2	74	23.33	435.2	224	106.67
....	−374	−225.56	−371.2	−224	−142.22	−101.2	−74	−58.89	168.8	76	24.44	438.8	226	107.78
....	−372	−224.44	−367.6	−222	−141.11	−97.6	−72	−57.78	172.4	78	25.56	442.4	228	108.89
....	−370	−223.33	−364.0	−220	−140.00	−94.0	−70	−56.67	176.0	80	26.67	446.0	230	110.00
....	−368	−222.22	−360.4	−218	−138.89	−90.4	−68	−55.56	179.6	82	27.78	449.6	232	111.11
....	−366	−221.11	−356.8	−216	−137.78	−86.8	−66	−54.44	183.2	84	28.89	453.2	234	112.22
....	−364	−220.00	−353.2	−214	−136.67	−83.2	−64	−53.33	186.8	86	30.00	456.8	236	113.33
....	−362	−218.89	−349.6	−212	−135.56	−79.6	−62	−52.22	190.4	88	31.11	460.4	238	114.44
....	−360	−217.78	−346.0	−210	−134.44	−76.0	−60	−51.11	194.0	90	32.22	464.0	240	115.56
....	−358	−216.67	−342.4	−208	−133.33	−72.4	−58	−50.00	197.6	92	33.33	467.6	242	116.67
....	−356	−215.56	−338.8	−206	−132.22	−68.8	−56	−48.89	201.2	94	34.44	471.2	244	117.78
....	−354	−214.44	−335.2	−204	−131.11	−65.2	−54	−47.78	204.8	96	35.56	474.8	246	118.89
....	−352	−213.33	−331.6	−202	−130.00	−61.6	−52	−46.67	208.4	98	36.67	478.4	248	120.00
....	−350	−212.22	−328.0	−200	−128.89	−58.0	−50	−45.56	212.0	100	37.78	482.0	250	121.11
....	−348	−211.11	−324.4	−198	−127.78	−54.4	−48	−44.44	215.6	102	38.89	485.6	252	122.22
....	−346	−210.00	−320.8	−196	−126.67	−50.8	−46	−43.33	219.2	104	40.00	489.2	254	123.33
....	−344	−208.89	−317.2	−194	−125.56	−47.2	−44	−42.22	222.8	106	41.11	492.8	256	124.44
....	−342	−207.78	−313.6	−192	−124.44	−43.6	−42	−41.11	226.4	108	42.22	496.4	258	125.56
....	−340	−206.67	−310.0	−190	−123.33	−40.0	−40	−40.00	230.0	110	43.33	500.0	260	126.67
....	−338	−205.56	−306.4	−188	−122.22	−36.4	−38	−38.89	233.6	112	44.44	503.6	262	127.78
....	−336	−204.44	−302.8	−186	−121.11	−32.8	−36	−37.78	237.2	114	45.56	507.2	264	128.89
....	−334	−203.33	−299.2	−184	−120.00	−29.2	−34	−36.67	240.8	116	46.67	510.8	266	130.00
....	−332	−202.22	−295.6	−182	−118.89	−25.6	−32	−35.56	244.4	118	47.78	514.4	268	131.11
....	−330	−201.11	−292.0	−180	−117.78	−22.0	−30	−34.44	248.0	120	48.89	518.0	270	132.22
....	−328	−200.00	−288.4	−178	−116.67	−18.4	−28	−33.33	251.6	122	50.00	521.6	272	133.33
....	−326	−198.89	−284.8	−176	−115.56	−14.8	−26	−32.22	255.2	124	51.11	525.2	274	134.44
....	−324	−197.78	−281.2	−174	−114.44	−11.2	−24	−31.11	258.8	126	52.22	528.8	276	135.56
....	−322	−196.67	−277.6	−172	−113.33	−7.6	−22	−30.00	262.4	128	53.33	532.4	278	136.67
....	−320	−195.56	−274.0	−170	−112.22	−4.0	−20	−28.89	266.0	130	54.44	536.0	280	137.78
....	−318	−194.44	−270.4	−168	−111.11	−0.4	−18	−27.78	269.6	132	55.56	539.6	282	138.89
....	−316	−193.33	−266.8	−166	−110.00	+3.2	−16	−26.67	273.2	134	56.67	543.2	284	140.00
....	−314	−192.22	−263.2	−164	−108.89	+6.8	−14	−25.56	276.8	136	57.78	546.8	286	141.11
....	−312	−191.11	−259.6	−162	−107.78	+10.4	−12	−24.44	280.4	138	58.89	550.4	288	142.22
....	−310	−190.00	−256.0	−160	−106.67	+14.0	−10	−23.33	284.0	140	60.00	554.0	290	143.33

F	C	F	C	F	C	F	C	F	C	F	C
557.6	292	144.44	870.8	466	241.11	1832.0	1000	537.78	3398.0	1870	1021.1	4964.0	2740	1504.4
561.2	294	145.56	874.4	468	242.22	1850.0	1010	543.33	3416.0	1880	1026.7	4982.0	2750	1510.0
564.8	296	146.67	878.0	470	243.33	1868.0	1020	548.89	3434.0	1890	1032.2	5000.0	2760	1515.6
568.4	298	147.78	881.6	472	244.44	1886.0	1030	554.44	3452.0	1900	1037.8	5018.0	2770	1521.1
572.0	300	148.89	885.2	474	245.56	1904.0	1040	560.00	3470.0	1910	1043.3	5036.0	2780	1526.7
575.6	302	150.00	888.8	476	246.67	1922.0	1050	565.56	3488.0	1920	1048.9	5054.0	2790	1532.2
579.2	304	151.11	892.4	478	247.78	1940.0	1060	571.11	3506.0	1930	1054.4	5072.0	2800	1537.8
582.8	306	152.22	896.0	480	248.89	1958.0	1070	576.67	3524.0	1940	1060.0	5090.0	2810	1543.3
586.4	308	153.33	899.6	482	250.00	1976.0	1080	582.22	3542.0	1950	1065.6	5108.0	2820	1548.9
590.0	310	154.44	903.2	484	251.11	1994.0	1090	587.78	3560.0	1960	1071.1	5126.0	2830	1554.4
593.6	312	155.56	906.8	486	252.22	2012.0	1100	593.33	3578.0	1970	1076.7	5144.0	2840	1560.0
597.2	314	156.67	910.4	488	253.33	2030.0	1110	598.89	3596.0	1980	1082.2	5162.0	2850	1565.6
600.8	316	157.78	914.0	490	254.44	2048.0	1120	604.44	3614.0	1990	1087.8	5180.0	2860	1571.1
604.4	318	158.89	917.6	492	255.56	2066.0	1130	610.00	3632.0	2000	1093.3	5198.0	2870	1576.7
608.0	320	160.00	921.2	494	256.67	2084.0	1140	615.56	3650.0	2010	1098.9	5216.0	2880	1582.2
611.6	322	161.11	924.8	496	257.78	2102.0	1150	621.11	3668.0	2020	1104.4	5234.0	2890	1587.8
615.2	324	162.22	928.4	498	258.89	2120.0	1160	626.67	3686.0	2030	1110.0	5252.0	2900	1593.3
618.8	326	163.33	932.0	500	260.00	2138.0	1170	632.22	3704.0	2040	1115.6	5270.0	2910	1598.9
622.4	328	164.44	935.6	502	261.11	2156.0	1180	637.78	3722.0	2050	1121.1	5288.0	2920	1604.4
626.0	330	165.56	939.2	504	262.22	2174.0	1190	643.33	3740.0	2060	1126.7	5306.0	2930	1610.0
629.6	332	166.67	942.8	506	263.33	2192.0	1200	648.89	3758.0	2070	1132.2	5324.0	2940	1615.6
633.2	334	167.78	946.4	508	264.44	2210.0	1210	654.44	3776.0	2080	1137.8	5342.0	2950	1621.1
636.8	336	168.89	950.0	510	265.56	2228.0	1220	660.00	3794.0	2090	1143.3	5360.0	2960	1626.7
640.4	338	170.00	953.6	512	266.67	2246.0	1230	665.56	3812.0	2100	1148.9	5378.0	2970	1632.2
644.0	340	171.11	957.2	514	267.78	2264.0	1240	671.11	3830.0	2110	1154.4	5396.0	2980	1637.8
647.6	342	172.22	960.8	516	268.89	2282.0	1250	676.67	3848.0	2120	1160.0	5414.0	2990	1643.3
651.2	344	173.33	964.4	518	270.00	2300.0	1260	682.22	3866.0	2130	1165.6	5432.0	3000	1648.9
654.8	346	174.44	968.0	520	271.11	2318.0	1270	687.78	3884.0	2140	1171.1	5450.0	3010	1654.4
658.4	348	175.56	971.6	522	272.22	2336.0	1280	693.33	3902.0	2150	1176.7	5468.0	3020	1660.0
662.0	350	176.67	975.2	524	273.33	2354.0	1290	698.89	3920.0	2160	1182.2	5486.0	3030	1665.6
665.6	352	177.78	978.8	526	274.44	2372.0	1300	704.44	3938.0	2170	1187.8	5504.0	3040	1671.1
669.2	354	178.89	982.4	528	275.56	2390.0	1310	710.00	3956.0	2180	1193.3	5522.0	3050	1676.7
672.8	356	180.00	986.0	530	276.67	2408.0	1320	715.56	3974.0	2190	1198.9	5540.0	3060	1682.2
676.4	358	181.11	989.6	532	277.78	2426.0	1330	721.11	3992.0	2200	1204.4	5558.0	3070	1687.8
680.0	360	182.22	993.2	534	278.89	2444.0	1340	726.67	4010.0	2210	1210.0	5576.0	3080	1693.3
683.6	362	183.33	996.8	536	280.00	2462.0	1350	732.22	4028.0	2220	1215.6	5594.0	3090	1698.9
687.2	364	184.44	1000.4	538	281.11	2480.0	1360	737.78	4046.0	2230	1221.1	5612.0	3100	1704.4
690.8	366	185.56	1004.0	540	282.22	2498.0	1370	743.33	4064.0	2240	1226.7	5702.0	3150	1732.2
694.4	368	186.67	1007.6	542	283.33	2516.0	1380	748.89	4082.0	2250	1232.2	5792.0	3200	1760.0
698.0	370	187.78	1011.2	544	284.44	2534.0	1390	754.44	4100.0	2260	1237.8	5882.0	3250	1787.7
701.6	372	188.89	1014.8	546	285.56	2552.0	1400	760.00	4118.0	2270	1243.3	5972.0	3300	1815.5
705.2	374	190.00	1018.4	548	286.67	2570.0	1410	765.56	4136.0	2280	1248.9	6062.0	3350	1843.3
708.8	376	191.11	1022.0	550	287.78	2588.0	1420	771.11	4154.0	2290	1254.4	6152.0	3400	1871.1
712.4	378	192.22	1040.0	560	293.33	2606.0	1430	776.67	4172.0	2300	1260.0	6242.0	3450	1898.8
716.0	380	193.33	1058.0	570	298.89	2624.0	1440	782.22	4190.0	2310	1265.6			

719.6	382	194.44	1076.0	580	304.44	2642.0	1450	787.78	4208.0	2320	1271.1
723.2	384	195.56	1094.0	590	310.00	2678.0	1460	793.33	4226.0	2330	1276.7
726.8	386	196.67	1112.0	600	315.56	2678.0	1470	798.89	4244.0	2340	1282.2
730.4	388	197.78	1130.0	610	321.11	2696.0	1480	804.44	4262.0	2350	1287.8
734.0	390	198.89	1148.0	620	326.67	2714.0	1490	810.00	4280.0	2360	1293.3
737.6	392	200.00	1166.0	630	332.22	2732.0	1500	815.56	4298.0	2370	1298.9
741.2	394	201.11	1184.0	640	337.78	2750.0	1510	821.11	4316.0	2380	1304.4
744.8	396	202.22	1202.0	650	343.33	2768.0	1520	826.67	4334.0	2390	1310.0
748.4	398	203.33	1220.0	660	348.89	2786.0	1530	832.22	4352.0	2400	1315.6
752.0	400	204.44	1238.0	670	354.44	2804.0	1540	837.78	4370.0	2410	1321.1
755.6	402	205.56	1256.0	680	360.00	2822.0	1550	843.33	4388.0	2420	1326.7
759.2	404	206.67	1274.0	690	365.56	2840.0	1560	848.89	4406.0	2430	1332.2
762.8	406	207.78	1292.0	700	371.11	2858.0	1570	854.44	4424.0	2440	1337.8
766.4	408	208.89	1310.0	710	376.67	2876.0	1580	860.00	4442.0	2450	1343.3
770.0	410	210.00	1328.0	720	382.22	2894.0	1590	865.56	4460.0	2460	1348.9
773.6	412	211.11	1346.0	730	387.78	2912.0	1600	871.11	4478.0	2470	1354.4
777.2	414	212.22	1364.0	740	393.33	2930.0	1610	876.67	4496.0	2480	1360.0
780.8	416	213.33	1382.0	750	398.89	2948.0	1620	882.22	4514.0	2490	1365.6
784.4	418	214.44	1400.0	760	404.44	2966.0	1630	887.78	4532.0	2500	1371.1
788.0	420	215.56	1418.0	770	410.00	2984.0	1640	893.33	4550.0	2510	1376.7
791.6	422	216.67	1436.0	780	415.56	3002.0	1650	898.89	4568.0	2520	1382.2
795.2	424	217.78	1454.0	790	421.11	3020.0	1660	904.44	4586.0	2530	1387.8
798.8	426	218.89	1472.0	800	426.67	3038.0	1670	910.00	4604.0	2540	1393.3
802.4	428	220.00	1490.0	810	432.22	3056.0	1680	915.56	4622.0	2550	1398.9
806.0	430	221.11	1508.0	820	437.78	3074.0	1690	921.11	4640.0	2560	1404.4
809.6	432	222.22	1526.0	830	443.33	3092.0	1700	926.67	4658.0	2570	1410.0
813.2	434	223.33	1544.0	840	448.89	3110.0	1710	932.22	4676.0	2580	1415.6
816.8	436	224.44	1562.0	850	454.44	3128.0	1720	937.78	4694.0	2590	1421.1
820.4	438	225.56	1580.0	860	460.00	3146.0	1730	943.33	4712.0	2600	1426.7
824.0	440	226.67	1598.0	870	465.56	3164.0	1740	948.89	4730.0	2610	1432.2
827.6	442	227.78	1616.0	880	471.11	3182.0	1750	954.44	4748.0	2620	1437.8
831.2	444	228.89	1634.0	890	476.67	3200.0	1760	960.00	4766.0	2630	1443.3
834.8	446	230.00	1652.0	900	482.22	3218.0	1770	965.56	4784.0	2640	1448.9
838.4	448	231.11	1670.0	910	487.78	3236.0	1780	971.11	4802.0	2650	1454.4
842.0	450	232.22	1688.0	920	493.33	3254.0	1790	976.67	4820.0	2660	1460.0
845.6	452	233.33	1706.0	930	498.89	3272.0	1800	982.22	4838.0	2670	1465.6
849.2	454	234.44	1724.0	940	504.44	3290.0	1810	987.78	4856.0	2680	1471.1
852.8	456	235.56	1742.0	950	510.00	3308.0	1820	993.33	4874.0	2690	1476.7
856.4	458	236.67	1760.0	960	515.56	3326.0	1830	998.89	4892.0	2700	1482.2
860.0	460	237.78	1778.0	970	521.11	3344.0	1840	1004.4	4910.0	2710	1487.8
863.6	462	238.89	1796.0	980	526.67	3362.0	1850	1010.0	4928.0	2720	1493.3
867.2	464	240.00	1814.0	990	532.22	3380.0	1860	1015.6	4946.0	2730	1498.9

6422.0	3550	1954.4		
6512.0	3600	1982.2		
6602.0	3650	2010.0		
6692.0	3700	2037.7		
6782.0	3750	2065.5		
6872.0	3800	2093.3		
6962.0	3850	2121.1		
7052.0	3900	2148.8		
7142.0	3950	2176.6		
7232.0	4000	2204.4		
7322.0	4050	2232.2		
7412.0	4100	2260.0		
7502.0	4150	2287.7		
7592.0	4200	2315.5		
7682.0	4250	2343.3		
7772.0	4300	2371.1		
7862.0	4350	2398.8		
7952.0	4400	2426.6		
8042.0	4450	2454.4		
8132.0	4500	2482.2		
8222.0	4550	2510.0		
8312.0	4600	2537.7		
8402.0	4650	2565.5		
8492.0	4700	2593.3		
8582.0	4750	2621.1		
8672.0	4800	2648.8		
8762.0	4850	2676.6		
8852.0	4900	2704.4		
8942.0	4950	2732.2		
9032.0	5000	2760.0		
9122.0	5050	2787.7		
9212.0	5100	2815.5		
9302.0	5150	2843.3		
9392.0	5200	2871.1		
9482.0	5250	2898.8		
9572.0	5300	2926.6		
9662.0	5350	2954.4		
9752.0	5400	2982.2		
9842.0	5450	3010.0		
9932.0	5500	3037.7		
10022.0	5550	3065.5		
10112.0	5600	3093.3		

Appendix B

Metric Conversion Factors*

To convert from	To	Multiply by	To convert from	To	Multiply by
angstrom	m	1.0000×10^{-10}(a)	hp(e)	W	7.4570×10^2
atm	Pa	1.0133×10^5	hp(f)	W	7.4600×10^2
Btu(b)	J	1.054×10^3	in.	m	2.5400×10^{-2}
Btu(b)/ft$^2\cdot$h	W/m^2	3.1525	in.^{2}	m^2	6.4516×10^{-4}
Btu(b)/ft$^2\cdot$h$\cdot$°F	W/m$^2\cdot$K	5.6745	in.^{3}	m^3	1.6387×10^{-5}
Btu(b)$\cdot$ft/h$\cdot$ft$^2\cdot$°F	W/m$\cdot$K	1.7296	in. of Hg(g)	Pa	3.3864×10^3
Btu(b)/ft$^2\cdot$s	W/m^2	1.135×10^4	in. of water(c)	Pa	2.4908×10^2
Btu(b)$\cdot$in./ft$^2\cdot$h$\cdot$°F	W/m$\cdot$K	1.4413×10^{-1}	K	°C	$t_C = t_K - 273.15$
Btu(b)$\cdot$in./s$\cdot$ft$^2\cdot$°F	W/m$\cdot$K	5.1887×10^2	kgf	N	9.80665(a)
Btu(b)/lbm$\cdot$°F	J/kg$\cdot$K	4.1840×10^3	kgf/mm^2	Pa	9.80665×10^6(a)
cal(b)	J	4.1840 (a)	ksi	MPa	6.8948
cal(b)/cm$\cdot$s$\cdot$°C	W/m$\cdot$K	4.1840×10^2(a)	ksi	Pa	6.8948×10^6
cal(b)/g	J/kg	4.1840×10^3(a)	ksi$\sqrt{\text{in.}}$	MPa$\sqrt{m}$	1.089
cal(b)/g$\cdot$°C	J/kg$\cdot$K	4.1840×10^3(a)	lb(h)	kg	4.5359×10^{-1}
circ mil	m^2	5.0671×10^{-10}	lb/in.^{3}	kg/m^3	2.7680×10^4
°C	K	$t_K = t_C + 273.15$	lbf	N	4.4482
degree	rad	1.7453×10^{-2}	lbf$\cdot$in.	N$\cdot$m	1.1298×10^{-1}
dyne/cm^2	Pa	1.0000×10^{-1}(a)	lbf$\cdot$ft	N$\cdot$m	1.3558
°F	°C	$t_C = (t_F - 32)/1.8$	MPa$\sqrt{m}$	MNm$^{-3/2}$	1.0000(a)
°F	K	$t_K = (t_F + 459.67)/1.8$	µin.	m	2.5400×10^{-8}(a)
ft	m	3.0480×10^{-1}	mil	m	2.5400×10^{-5}(a)
ft^2	m^2	9.2903×10^{-2}	N/m^2	Pa	1.0000(a)
ft^3	m^3	2.8317×10^{-2}	oersted	A/m	79.578
ft of water(c)	Pa	2.9890×10^3	oz/ft^2	kg/m^2	3.0515×10^{-1}
ft^2/h (thermal diffusivity)	m^2/s	2.58064×10^{-5}(a)	psi	Pa	6.8948×10^3
ft$\cdot$lbf	J	1.3558	°R	K	$t_K = t_R/1.8$
ft$\cdot$lbf/s	W	1.3558	ton(j)	kg	9.0718×10^2
ft/s	m/s	3.0480×10^{-1}	ton(k)	kg	1.0160×10^3
gauss	T	1.0000×10^{-4}(a)	ton/in.^{2}	Pa	1.3786×10^4
gallon(d)	m^3	3.7854×10^{-3}	tonne	kg	1.0000×10^3(a)
g/cm^3	kg/m^3	1.0000×10^3(a)	torr	Pa	1.3332×10^2
g/cm^3	Mg/m^3	1.0000(a)	Ω/circ mil$\cdot$ft	Ω$\cdot$m	1.6624×10^{-9}

(a) Exactly. (b) Thermochemical. (c) At 4 °C (39.2 °F). (d) U.S. liquid. (e) Mechanical (1 hp = 550 ft$\cdot$lbf/s). (f) Electrical. (g) At 0 °C (32 °F). (g) Avoirdupois. (j) Short; equal to 2000 lbm. (k) Long; 2240 lbm.

*Reprinted with permission from H. E. Boyer and T. L. Gall (eds.), *Metals Handbook—Desk Edition*, American Society for Metals, Metals Park, Ohio, 1985.

Appendix

C

Converting Common Units from the English to the Metric (SI) System*

---

*Reprinted with permission from *1977 Metal Progress Databook,* American Society for Metals, Metals Park, Ohio, 1977.

Appendix C

The International System of Units (SI for short) is a modernized version of the metric system. It is built upon seven base units and two supplementary units. Derived units are related to base and supplementary units by formulas in the right-hand column. Symbols for units with specific names are given in parentheses. The information in this Data Sheet, adapted from the revised "Metric Practice Guide," Standard E380 ASTM, includes a selected list of factors for converting U. S. customary units to SI units.

Metric Units and Conversion Factors

Quantity	Unit	Formula
Base Units		
length	metre (m)	
mass	kilogram (kg)	
time	second (s)	
electric current	ampere (A)	
thermodynamic temperature	kelvin (K)	
amount of substance	mole (mol)	
luminous intensity	candela (cd)	
Supplementary Units		
plane angle	radian (rad)	
solid angle	steradian (sr)	
Derived Units		
acceleration	metre per second squared	m/s^2
activity (of a radioactive source)	disintegration per second	(disintegration)/s
angular acceleration	radian per second squared	rad/s^2
angular velocity	radian per second	rad/s
area	square metre	m^2
density	kilogram per cubic metre	kg/m^3
electric capacitance	farad (F)	$A \cdot s/V$
electric conductance	siemens (S)	A/V
electric field strength	volt per metre	V/m
electric inductance	henry (H)	$V \cdot s/A$
electric potential difference	volt (V)	W/A
electric resistance	ohm (Ω)	V/A
electromotive force	volt (V)	W/A
energy	joule (J)	$N \cdot m$
entropy	joule per kelvin	J/K
force	newton (N)	$kg \cdot m/s^2$
frequency	hertz (Hz)	(cycle)/s
illuminance	lux (lx)	lm/m^2
luminance	candela per square metre	cd/m^2
luminous flux	lumen (lm)	$cd \cdot sr$
magnetic field strength	ampere per metre	A/m
magnetic flux	weber (Wb)	$V \cdot s$
magnetic flux density	tesla (T)	Wb/m^2
magnetomotive force	ampere (A)	—
power	watt (W)	J/s
pressure	pascal (Pa)	N/m^2
quantity of electricity	coulomb (C)	$A \cdot s$
quantity of heat	joule (J)	$N \cdot m$
radiant intensity	watt per steradian	W/sr
specific heat	joule per kilogram-kelvin	$J/kg \cdot K$
stress	pascal (Pa)	N/m^2
thermal conductivity	watt per metre-kelvin	$W/m \cdot K$
velocity	metre per second	m/s
viscosity, dynamic	pascal-second	$Pa \cdot s$
viscosity, kinematic	square metre per second	m^2/s
voltage	volt (V)	W/A
volume	cubic metre	m^3
wavenumber	reciprocal metre	(wave)/m
work	joule (J)	$N \cdot m$

Metric Conversion Factors

To convert from	To	Multiply by
atmosphere (760 mm Hg)	Pa	$1.013\ 25 \times 10^5$
Btu (International Table)	J	$1.055\ 056 \times 10^3$
Btu (International Table)/hour	W	$2.930\ 711 \times 10^{-1}$
calorie (International Table)	J	$4.186\ 800^*$
centipoise	$Pa \cdot s$	$1.000\ 000^* \times 10^{-3}$
centistoke	m^2/s	$1.000\ 000^* \times 10^{-4}$
circular mil	m^2	$5.067\ 075 \times 10^{-10}$
degree Fahrenheit	°C	$tC = (tF - 32)/1.8$
foot	m	$3.048\ 000^* \times 10^{-1}$
foot^2	m^2	$9.290\ 304^* \times 10^{-2}$
foot^3	m^3	$2.831\ 685 \times 10^{-2}$
foot-pound-force	J	$1.355\ 818$
foot-pound-force/minute	W	$2.259\ 697 \times 10^{-2}$
foot/second^2	m/s^2	$3.048\ 000^* \times 10^{-1}$
gallon (U.S. liquid)	m^3	$3.785\ 412 \times 10^{-3}$
horsepower (electric)	W	$7.460\ 000^* \times 10^2$
inch	m	$2.540\ 000^* \times 10^{-2}$
inch^2	m^2	$6.451\ 600^* \times 10^{-4}$
inch^3	m^3	$1.638\ 706 \times 10^{-5}$
inch of mercury (60 F)	Pa	$3.376\ 85 \times 10^3$
inch of water (60 F)	Pa	$2.488\ 4 \times 10^2$
kilogram-force/centimetre^2	Pa	$9.806\ 650^* \times 10^4$
kip (1000 lbf)	N	$4.448\ 222 \times 10^3$
kip/inch^2 (ksi)	Pa	$6.894\ 757 \times 10^6$
ounce (U.S. fluid)	m^3	$2.957\ 353 \times 10^{-5}$
ounce-force (avoirdupois)	N	$2.780\ 139 \times 10^{-1}$
ounce-mass (avoirdupois)	kg	$2.834\ 952 \times 10^{-2}$
ounce-mass/ft^2	kg/m^2	$3.051\ 52 \times 10^{-1}$
ounce-mass/yard^2	kg/m^2	$3.390\ 575 \times 10^{-2}$
pint (U.S. liquid)	m^3	$4.731\ 765 \times 10^{-4}$
pound-force (lbf avoirdupois)	N	$4.448\ 222$
pound-mass (lbm avoirdupois)	kg	$4.535\ 924 \times 10^{-1}$
pound-force/inch^2 (psi)	Pa	$6.894\ 757 \times 10^3$
pound-mass/inch^3	kg/m^3	$2.767\ 990 \times 10^4$
pound-mass/foot^3	kg/m^3	$1.601\ 846 \times 10$
quart (U.S. liquid)	m^3	$9.463\ 529 \times 10^{-4}$
ton (short, 2000 lbm)	kg	$9.071\ 847 \times 10^2$
torr (mm-Hg)	Pa	$1.333\ 22 \times 10^2$
watt-hour	J	$3.600\ 000^* \times 10^3$
yard	m	$9.144\ 000^* \times 10^{-1}$
yard^2	m^2	$8.361\ 274 \times 10^{-1}$
yard^3	m^3	$7.645\ 549 \times 10^{-1}$

*Exact

Multiplication Factors	Prefix	SI Symbol
$1\ 000\ 000\ 000\ 000 = 10^{12}$	tera	T
$1\ 000\ 000\ 000 = 10^9$	giga	G
$1\ 000\ 000 = 10^6$	mega	M
$1\ 000 = 10^3$	kilo	k
$100 = 10^2$	hecto*	h
$10 = 10^1$	deka*	da
$0.1 = 10^{-1}$	deci*	d
$0.01 = 10^{-2}$	centi*	c
$0.001 = 10^{-3}$	milli	m
$0.000\ 001 = 10^{-6}$	micro	μ
$0.000\ 000\ 001 = 10^{-9}$	nano	n
$0.000\ 000\ 000\ 001 = 10^{-12}$	pico	p
$0.000\ 000\ 000\ 000\ 001 = 10^{-15}$	femto	f
$0.000\ 000\ 000\ 000\ 000\ 001 = 10^{-18}$	atto	a

*To be avoided where possible

Appendix D

Rockwell C and B Hardness Numbers for Steel*

---

*Reprinted with permission from H. E. Boyer and T. L. Gall (eds.), *Metals Handbook—Desk Edition*, American Society for Metals, Metals Park, Ohio, 1985.

Rockwell C-Scale Hardness Numbers

Rockwell C-scale hardness No.	Vickers hardness No.	Brinell hardness No., 3000-kg load, 10-mm ball Standard ball	Brinell hardness No., 3000-kg load, 10-mm ball Tungsten carbide ball	Rockwell A scale, 60-kg load, Brale indenter	Rockwell B scale, 100-kg load, 1/16-in. diam ball	Rockwell D scale, 100-kg load, Brale indenter	Rockwell superficial 15N scale, 15-kg load	Rockwell superficial 30N scale, 30-kg load	Rockwell superficial 45N scale, 45-kg load	Knoop hardness No., 500-g load and greater	Shore Scleroscope hardness No.	Tensile strength (approx), 1000 psi	Rockwell C-scale hardness No.
68	940	...	...	85.6	...	76.9	93.2	84.4	75.4	920	97	...	68
67	900	...	...	85.0	...	76.1	92.9	83.6	74.2	895	95	...	67
66	865	...	...	84.5	...	75.4	92.5	82.8	73.3	870	92	...	66
65	832	...	(739)	83.9	...	74.5	92.2	81.9	72.0	846	91	...	65
64	800	...	(722)	83.4	...	73.8	91.8	81.1	71.0	822	88	...	64
63	772	...	(705)	82.8	...	73.0	91.4	80.1	69.9	799	87	...	63
62	746	...	(688)	82.3	...	72.2	91.1	79.3	68.8	776	85	...	62
61	720	...	(670)	81.8	...	71.5	90.7	78.4	67.7	754	83	...	61
60	697	...	(654)	81.2	...	70.7	90.2	77.5	66.6	732	81	...	60
59	674	...	(634)	80.7	...	69.9	89.8	76.6	65.5	710	80	351	59
58	653	...	615	80.1	...	69.2	89.3	75.7	64.3	690	78	338	58
57	633	...	595	79.6	...	68.5	88.9	74.8	63.2	670	76	325	57
56	613	...	577	79.0	...	67.7	88.3	73.9	62.0	650	75	313	56
55	595	...	560	78.5	...	66.9	87.9	73.0	60.9	630	74	301	55
54	577	...	543	78.0	...	66.1	87.4	72.0	59.8	612	72	292	54
53	560	...	525	77.4	...	65.4	86.9	71.2	58.6	594	71	283	53
52	544	(500)	512	76.8	...	64.6	86.4	70.2	57.4	576	69	273	52
51	528	(487)	496	76.3	...	63.8	85.9	69.4	56.1	558	68	264	51
50	513	(475)	481	75.9	...	63.1	85.5	68.5	55.0	542	67	255	50
49	498	(464)	469	75.2	...	62.1	85.0	67.6	53.8	526	66	246	49
48	484	(451)	455	74.7	...	61.4	84.5	66.7	52.5	510	64	238	48
47	471	442	443	74.1	...	60.8	83.9	65.8	51.4	495	63	229	47
46	458	432	432	73.6	...	60.0	83.5	64.8	50.3	480	62	221	46
45	446	421	421	73.1	...	59.2	83.0	64.0	49.0	466	60	215	45
44	434	409	409	72.5	...	58.5	82.5	63.1	47.8	452	58	208	44
43	423	400	400	72.0	...	57.7	82.0	62.2	46.7	438	57	201	43
42	412	390	390	71.5	...	56.9	81.5	61.3	45.5	426	56	194	42
41	402	381	381	70.9	...	56.2	80.9	60.4	44.3	414	55	188	41
40	392	371	371	70.4	...	55.4	80.4	59.5	43.1	402	54	182	40
39	382	362	362	69.9	...	54.6	79.9	58.6	41.9	391	52	177	39
38	372	353	353	69.4	...	53.8	79.4	57.7	40.8	380	51	171	38
37	363	344	344	68.9	...	53.1	78.8	56.8	39.6	370	50	166	37
36	354	336	336	68.4	(109.0)	52.3	78.3	55.9	38.4	360	49	161	36
35	345	327	327	67.9	(108.5)	51.5	77.7	55.0	37.2	351	48	157	35
34	336	319	319	67.4	(108.0)	50.8	77.2	54.2	36.1	342	47	153	34
33	327	311	311	66.8	(107.5)	50.0	76.6	53.3	34.9	334	46	149	33
32	318	301	301	66.3	(107.0)	49.2	76.1	52.1	33.7	326	44	145	32
31	310	294	294	65.8	(106.0)	48.4	75.6	51.3	32.5	318	43	141	31
30	302	286	286	65.3	(105.5)	47.7	75.0	50.4	31.3	311	42	138	30
29	294	279	279	64.7	(104.5)	47.0	74.5	49.5	30.1	304	41	135	29
28	286	271	271	64.3	(104.0)	46.1	73.9	48.6	28.9	297	40	131	28
27	279	264	264	63.8	(103.0)	45.2	73.3	47.7	27.8	290	39	128	27
26	272	258	258	63.3	(102.5)	44.6	72.8	46.8	26.7	284	38	125	26
25	266	253	253	62.8	(101.5)	43.8	72.2	45.9	25.5	278	38	122	25
24	260	247	247	62.4	(101.0)	43.1	71.6	45.0	24.3	272	37	119	24
23	254	243	243	62.0	100.0	42.1	71.0	44.0	23.1	266	36	117	23
22	248	237	237	61.5	99.0	41.6	70.5	43.2	22.0	261	35	114	22
21	243	231	231	61.0	98.5	40.9	69.9	42.3	20.7	256	35	112	21

Hardness Numbers for Steel

Rockwell B-scale hardness No.	Vickers hardness No.	Brinell hardness No., 10-mm-diam ball		Rockwell hardness No.			Rockwell superficial hardness No., 1/16-in.-diam ball			Knoop hardness No., 500-g load and greater	Shore Sclero-scope hardness No.	Tensile strength (approx), 1000 psi	Rockwell B-scale hardness No.
		500-kg load	3000-kg load	A scale, 60-kg load, Brale indenter	C scale, 150-kg load, Brale indenter	F scale, 60-kg load, 1/16-in.-diam ball	15T scale, 15-kg load	30T scale, 30-kg load	45T scale, 45-kg load				

Rockwell B-Scale Hardness Numbers

B	HV	HB 500	HB 3000	A	C	F	15T	30T	45T	HK	HS	TS	B
98	228	189	228	60.2	(19.9)	...	92.5	81.8	70.9	241	34	107	98
97	222	184	222	59.5	(18.6)	...	92.1	81.1	69.9	236	33	104	97
96	216	179	216	58.9	(17.2)	...	91.8	80.4	68.9	231	32	102	96
95	210	175	210	58.3	(15.7)	...	91.5	79.8	67.9	226	...	99	95
94	205	171	205	57.6	(14.3)	...	91.2	79.1	66.9	221	31	97	94
93	200	167	200	57.0	(13.0)	...	90.8	78.4	65.9	216	30	94	93
92	195	163	195	56.4	(11.7)	...	90.5	77.8	64.8	211	...	92	92
91	190	160	190	55.8	(10.4)	...	90.2	77.1	63.8	206	29	90	91
90	185	157	185	55.2	(9.2)	...	89.9	76.4	62.8	201	28	88	90
89	180	154	180	54.6	(8.0)	...	89.5	75.8	61.8	196	27	86	89
88	176	151	176	54.0	(6.9)	...	89.2	75.1	60.8	192	...	84	88
87	172	148	172	53.4	(5.8)	...	88.9	74.4	59.8	188	26	82	87
86	169	145	169	52.8	(4.7)	...	88.6	73.8	58.8	184	26	81	86
85	165	142	165	52.3	(3.6)	...	88.2	73.1	57.8	180	25	79	85
84	162	140	162	51.7	(2.5)	...	87.9	72.4	56.8	176	...	78	84
83	159	137	159	51.1	(1.4)	...	87.6	71.8	55.8	173	24	76	83
82	156	135	156	50.6	(0.3)	...	87.3	71.1	54.8	170	24	75	82
81	153	133	153	50.0	...	...	86.9	70.4	53.8	167	...	73	81
80	150	130	150	49.5	...	...	86.6	69.7	52.8	164	23	72	80
79	147	128	147	48.9	...	...	86.3	69.1	51.8	161	...	70	79
78	144	126	144	48.4	...	...	86.0	68.4	50.8	158	22	69	78
77	141	124	141	47.9	...	...	85.6	67.7	49.8	155	22	68	77
76	139	122	139	47.3	...	...	85.3	67.1	48.8	152	...	67	76
75	137	120	137	46.8	...	99.6	85.0	66.4	47.8	150	21	66	75
74	135	118	135	46.3	...	99.1	84.7	65.7	46.8	148	21	65	74
73	132	116	132	45.8	...	98.5	84.3	65.1	45.8	145	...	64	73
72	130	114	130	45.3	...	98.0	84.0	64.4	44.8	143	20	63	72
71	127	112	127	44.8	...	97.4	83.7	63.7	43.8	141	20	62	71
70	125	110	125	44.3	...	96.8	83.4	63.1	42.8	139	...	61	70
69	123	109	123	43.8	...	96.2	83.0	62.4	41.8	137	19	60	69
68	121	107	121	43.3	...	95.6	82.7	61.7	40.8	135	19	59	68
67	119	106	119	42.8	...	95.1	82.4	61.0	39.8	133	19	58	67
66	117	104	117	42.3	...	94.5	82.1	60.4	38.7	131	...	57	66
65	116	102	116	41.8	...	93.9	81.8	59.7	37.7	129	18	56	65
64	114	101	114	41.4	...	93.4	81.4	59.0	36.7	127	18	...	64
63	112	99	112	40.9	...	92.8	81.1	58.4	35.7	125	18	...	63
62	110	98	110	40.4	...	92.2	80.8	57.7	34.7	124	...	...	62
61	108	96	108	40.0	...	91.7	80.5	57.0	33.7	122	17	...	61
60	107	95	107	39.5	...	91.1	80.1	56.4	32.7	120	...	...	60
59	106	94	106	39.0	...	90.5	79.8	55.7	31.7	118	...	...	59
58	104	92	104	38.6	...	90.0	79.5	55.0	30.7	117	...	...	58
57	103	91	103	38.1	...	89.4	79.2	54.4	29.7	115	...	...	57
56	101	90	101	37.7	...	88.8	78.8	53.7	28.7	114	...	...	56
55	100	89	100	37.2	...	88.2	78.5	53.0	27.7	112	...	...	55

(a) For carbon and alloy steels in the annealed, normalized, and quenched-and-tempered conditions; less accurate for cold worked condition and for austenitic steels. The values in **boldface type** correspond to the values in the joint SAE-ASM-ASTM hardness conversions as printed in ASTM E140, Table 2. The values in parentheses are beyond normal range and are given for information only.

Appendix E

The Relations Between ASTM Grain Size and Average Grain "Diameter"*

*Adapted from *1966 Book of ASTM Standards,* part 31, American Society for Testing and Materials, Philadelphia, Pa., 1966; © ASTM, reprinted with permission.

Appendix E

ASTM micro-grain size number	Calculated "diameter" of average grain		ASTM micro-grain size number	Calculated "diameter" of average grain	
	mm	in. $\times 10^{-3}$		mm	in. $\times 10^{-3}$
00	0.508	20.0	7.5	0.027	1.05
0	0.359	14.1	-	0.025	0.984
0.5	0.302	11.9	8.0	0.0224	0.884
1.0	0.254	10.0	-	0.0200	0.787
-	0.250	9.84	8.5	0.0189	0.743
1.5	0.214	8.41	9.0	0.0159	0.625
-	0.200	7.87	-	0.0150	0.591
-	0.180	7.09	9.5	0.0134	0.526
2.0	0.179	7.07	10.0	0.0112	0.442
2.5	0.151	5.95	-	0.0100	0.394
-	0.150	5.91	10.5	0.00944	0.372
3.0	0.127	5.00	-	0.00900	0.354
-	0.120	4.72	-	0.00800	0.315
3.5	0.107	4.20	11.0	0.00794	0.313
-	0.099	3.90	-	0.00700	0.276
4.0	0.090	3.54	11.5	0.00667	0.263
4.5	0.076	2.97	-	0.00600	0.236
-	0.070	2.76	12.0	0.00561	0.221
5.0	0.064	2.50	-	0.00500	0.197
-	0.060	2.36	12.5	0.00472	0.186
5.5	0.053	2.10	-	0.00400	0.158
-	0.050	1.97	13.0	0.00397	0.156
6.0	0.045	1.77	13.5	0.00334	0.131
-	0.040	1.58	-	0.00300	0.118
6.5	0.038	1.49	14.0	0.00281	0.111
-	0.035	1.38	-	0.00250	0.098
7.0	0.032	1.25			
-	0.030	1.18			

Appendix F

Comments on Magnification Markers

It is customary to indicate the magnification of a photograph of a microstructure (micrograph), but when the photograph size is changed (such as enlarged), the original magnification must be corrected. A method of avoiding this is to place on the micrograph a bar or marker that has the correct dimension for the magnification of the photograph. Then if the magnification of the photograph is changed, the marker also changes dimensions to maintain the correct dimension at the new magnification.

For example, consider a micrograph at 100×. Then a marker 1 cm long would correspond to a dimension of 1/100 = 0.01 cm. It is common to express this in micrometers. Since a micrometer is equal to 10^{-4} cm, then the 0.01-cm dimension could also be labeled as 100 μm (0.01 × 10,000). If the micrograph is enlarged 2.5 times, then the marker becomes 2.5 × 1 = 2.5 cm long, and it is still labeled as 100 μm.

In general,

$$M = \frac{x}{y \times 10^{-4}}$$

where M is the magnification, x is the marker length in centimeters, and y is the marker value in micrometers on the micrograph.

At very high magnification, the marker value may be less than 1. For example, if the magnification is 40,000 ×, then 1 cm would correspond to 0.25 μm. In such cases, the marker corresponding to a certain value in angstroms (Å) may be used. 1 Å is equal to 10^{-8} cm, so in this case the marker dimension y in angstroms would be obtained from the equation as $40,000 = 1/(y \times 10^8)$. Then $y = 2500$ Å.

For the SI system the nanometer (nm) is used. 1 nm equals 0.1 Å, so for the preceding case, the marker would be labeled 25,000 nm.

Such markers are very useful because the actual sizes of microstructural features can be determined. For example, if the marker is labeled 2 μm and is 1 cm long, and a particle is present which is about 0.5 cm long in the micrograph, then its *actual* dimension in the sample is about 1 μm.

Glossary*

abrasion The process of grinding or wearing away through the use of abrasives; a roughening or scratching of a surface due to *abrasive wear*.

abrasive wear The removal of material from a surface when hard particles slide or roll across the surface under pressure. The particles may be loose or may be part of another surface in contact with the surface being abraded. Compare with *adhesive wear*.

adhesive wear The removal or displacement of material from a surface by the welding together and subsequent shearing of minute areas of two surfaces that slide across each other under pressure. Compare with *abrasive wear*.

alligatoring The longitudinal splitting of flat slabs in a plane parallel to the rolled surface. Also called fishmouthing.

alligator skin See *orange peel*.

ambient Surrounding; usually used in relation to temperature, as "ambient temperature" surrounding a certain part or assembly.

annealing twin A *twin* formed in a crystal during recrystallization.

anode The electrode of an electrolytic cell at which oxidation occurs. Contrast with *cathode*.

arrest lines (marks) See *beach marks*.

asperity In *tribology*, a protuberance in the small-scale topographical irregularities of a solid surface.

axial Longitudinal, or parallel to the axis or centerline of a part. Usually refers to axial compression or axial tension.

axial strain Increase (or decrease) in length resulting from a stress acting parallel to the longitudinal axis of a test specimen.

---

*Reprinted with permission from *Metals Handbook*, 9th edition, vol. 11: *Failure Analysis and Prevention*, American Society for Metals, Metals Park, Ohio, 1986.

banded structure A segregated structure consisting of alternating, nearly parallel bands of different composition, typically aligned in the direction of primary hot working.

beach marks Macroscopic (visible) progression marks on a fracture surface that indicate successive position of the advancing crack front. The classic appearance is of irregular elliptical or semielliptical rings, radiating outward from one or more origins. Beach marks (also known as clamshell marks, tide marks, or arrest marks) are typically found on service fractures where the part is loaded randomly, intermittently, or with periodic variations in mean stress or alternating stress. Not to be confused with *striations*, which are microscopic and form differently.

breaking stress See *rupture stress*.

Brinell hardness number, HB A number related to the applied load and to the surface area of the permanent impression made by a ball indenter, computed from

$$HB = \frac{2P}{\pi D(D - \sqrt{D^2 - d^2})}$$

where P is applied load, kgf; D is diameter of ball, mm; and d is mean diameter of impression, mm.

Brinell hardness test A test for determining the hardness of a material by forcing a hard steel or carbide ball of specified diameter into it under a specified load. The result is expressed as the *Brinell hardness number*.

brinelling Damage to a solid bearing surface characterized by one or more plastically formed indentations brought about by overload. This term is often applied in the case of rolling-element bearings. See also *false brinelling*.

brittle Permitting little or no plastic (permanent) deformation prior to fracture.

brittle crack propagation A very sudden propagation of a crack with the absorption of no energy except that stored elastically in the body. Microscopic examination may reveal some deformation not noticeable to the unaided eye. Contrast with *ductile crack propagation*.

brittle erosion behavior Erosion behavior having characteristic properties (e.g., little or no plastic flow, the formation of cracks) that can be associated with *brittle fracture* of the exposed surface. The maximum volume removal occurs at an angle near 90°, in contrast to approximately 25° for *ductile erosion behavior*.

brittle fracture Separation of a solid accompanied by little or no macroscopic plastic deformation. Typically, brittle fracture occurs by rapid crack propagation with less expenditure of energy than for *ductile fracture*.

brittleness The tendency of a material to fracture without first undergoing significant *plastic deformation*. Contrast with *ductility*.

buckle (1) An indented valley in the surface of a sand casting due to expansion of the molding sand. (2) A local waviness in a metal bar or sheet, usually transverse to the direction of rolling.

buckling A compression phenomenon that occurs when, after some critical level of load, a bulge, bend, bow, kink, or other wavy condition is produced in a beam, column, plate, bar, or sheet product form.

bulk modulus See *bulk modulus of elasticity*.

bulk modulus of elasticity, K The measure of resistance to change in volume; the ratio of hydrostatic stress to the corresponding unit change in volume. This elastic constant can be expressed by

$$K = \frac{\sigma_m}{\Delta} = \frac{-p}{\Delta} = \frac{1}{\beta}$$

where K is bulk modulus of elasticity, σ_m is hydrostatic or mean stress tensor, p is hydrostatic pressure, and β is compressibility. Also known as bulk modulus, compression modulus, hydrostatic modulus, and volumetric modulus of elasticity.

carbon flotation Segregation in which free graphite has separated from the molten iron. This defect tends to occur at the upper surfaces of the cope of the castings.

casting shrinkage See *liquid shrinkage, shrinkage cavity, solidification shrinkage,* and *solid shrinkage*.

catastrophic wear Rapidly occurring or accelerating surface damage, deterioration, or change of shape caused by wear to such a degree that the service life of a part is appreciably shortened or its function is destroyed.

caustic cracking A form of *stress-corrosion cracking* most frequently encountered in carbon steels or iron-chromium-nickel alloys that are exposed to concentrated hydroxide solutions at temperatures of 200 to 250°C (400 to 480°F). Also known as caustic embrittlement.

caustic embrittlement See *caustic cracking*.

cavitation The formation and rapid collapse within a liquid of cavities or bubbles that contain vapor or gas or both. Cavitation caused by severe turbulent flow often leads to *cavitation damage*.

cavitation damage The degradation of a solid body resulting from its exposure to cavitation. This may include loss of material, surface deformation, or changes in properties or appearance.

cavitation erosion See *cavitation damage*.

centerline shrinkage Shrinkage or porosity occurring along the central plane or axis of a cast metal section.

chafing fatigue Fatigue initiated in a surface damaged by rubbing against another body. See also *fretting*.

Charpy test An impact test in which a V-notched, keyhole-notched, or U-notched specimen, supported at both ends, is struck behind the notch by a striker mounted at the lower end of a bar that can swing as a pendulum. The energy that is absorbed in fracture is calculated from the height to which the striker would have risen had there been no specimen and the height to which it actually rises after fracture of the specimen. Contrast with *Izod test.*

chevron pattern A fractographic pattern of radial marks (shear ledges) that looks like nested letters V; sometimes called a herringbone pattern. Chevron patterns are typically found on brittle fracture surfaces in parts whose widths are considerably greater than their thicknesses. The points of the chevrons can be traced back to the fracture origin.

chill A white iron structure that is produced by rapid solidification.

chord modulus A slope of the chord drawn between any two specific points on a stress-strain curve. See also *modulus of elasticity.*

clamshell marks See *beach marks.*

cleavage (1) Fracture of a crystal by crack propagation across a crystallographic plane of low index. (2) The tendency to cleave or split along definite crystallographic planes.

cleavage crack A crack that extends along a plane of easy *cleavage* in a crystalline material.

cleavage fracture A fracture, usually of a polycrystalline metal, in which most of the grains have failed by *cleavage,* resulting in bright reflecting facets. It is one type of *crystalline fracture* and is associated with low-energy brittle fracture. Contrast with *shear fracture.*

cleavage plane A characteristic crystallographic plane or set of planes in a crystal on which *cleavage fracture* occurs easily.

cold shot A small globule of metal that solidified prematurely and is embedded in but not entirely fused with the surface of the casting.

cold shut A discontinuity on or immediately beneath the surface of a casting, caused by the meeting of two streams of liquid metal that failed to merge. A cold shut may have the appearance of a crack or seam with smooth, rounded edges.

columnar structure A coarse structure of parallel, elongated grains formed by unidirectional growth that is most often observed in castings. This results from diffusional growth accompanied by a solid-state transformation.

composite material A heterogeneous, solid structural material consisting of two or more distinct components that are mechanically or metallurgically bonded together, such as a wire or filament of a high-melting substance embedded in a metal or nonmetal matrix.

compression modulus See *bulk modulus of elasticity.*

compressive Pertaining to forces on a body or part of a body that tend to crush, or compress, the body.

compressive strength The maximum *compressive stress* a material is capable of developing. With a brittle material that fails in compression by fracturing, the compressive strength has a definite value. In the case of ductile, malleable, or semiviscous materials (which do not fail in compression by a shattering fracture), the value obtained for compressive strength is an arbitrary value dependent on the degree of distortion that is regarded as effective failure of the material.

compressive stress A stress that causes an elastic body to deform (shorten) in the direction of the applied load. Contrast with *tensile stress*.

contact fatigue Cracking and subsequent pitting of a surface subjected to alternating hertzian stresses such as those produced under rolling contact or combined rolling and sliding. The phenomenon of contact fatigue is encountered most often in rolling-element bearings or in gears, where the surface stresses are high due to the concentrated loads and are repeated many times during normal operation.

corrosion The chemical or electrochemical reaction between a material, usually a metal, and its environment that produces a deterioration of the material and its properties. See also *corrosion fatigue, crevice corrosion, denickelification, dezincification, erosion-corrosion, exfoliation, filiform corrosion, fretting corrosion, galvanic corrosion, general corrosion, graphitic corrosion, impingement attack, interdendritic corrosion, intergranular corrosion, internal oxidation, oxidation, parting, pitting, poultice corrosion, rust, selective leaching, stray-current corrosion, stress-corrosion cracking,* and *sulfide stress cracking*.

corrosion fatigue Cracking produced by the combined action of repeated or fluctuating stress and a corrosive environment at lower stress levels or fewer cycles than would be required in the absence of a corrosive environment.

corrosive wear *Wear* in which chemical or electrochemical reaction with the environment is significant.

crack extension, Δa An increase in crack size. See also *crack length, effective crack size, original crack size,* and *physical crack size*.

crack length (depth), a In *fatigue* and *stress-corrosion cracking,* the *physical crack size* used to determine the crack growth rate and the *stress-intensity factor*. For a compact-type specimen, crack length is measured from the line connecting the bearing points of load application. For a center-crack tension specimen, crack length is measured from the perpendicular bisector of the central crack. See also *crack size*.

crack mouth opening displacement (CMOD) See *crack opening displacement*.

crack opening displacement (COD) On a K_{Ic} specimen, the opening displacement of the notch surfaces at the notch and in the direction perpendicular to the plane of the notch and the crack. The displacement at the tip is called the crack tip opening displacement (CTOD); at the mouth, it is called the crack mouth opening displacement (CMOD).

crack plane orientation An identification of the plane and direction of a fracture in relation to product geometry. This identification is designated by a hyphenated code, the first letter(s) representing the direction normal to the crack plane and the second letter(s) designating the expected direction of crack propagation.

crack size, a A lineal measure of a principal planar dimension of a crack. This measure is commonly used in the calculation of quantities descriptive of the stress and displacement fields. In practice, the value of crack size is obtained from procedures for measurement of *physical crack size, original crack size,* or *effective crack size,* as appropriate to the situation under consideration. See also *crack length (depth).*

crack tip opening displacement (CTOD) See *crack opening displacement.*

crack tip plane strain A stress-strain field near a crack tip that approaches *plane strain* to the degree required by an empirical criterion.

creep The time-dependent strain occurring under stress. The *creep strain* occurring at a diminishing rate is called primary or transient creep; that occurring at a minimum and almost constant rate, secondary or steady-rate creep; that occurring at an accelerating rate, tertiary creep.

creep rate The slope of the creep-time curve at a given time determined from a cartesian plot.

creep-rupture strength The stress that will cause fracture in a creep test at a given time in a specified constant environment. Also known as stress-rupture strength.

creep strain The time-dependent total strain (extension plus initial gage length) produced by applied stress during a creep test.

creep strength The stress that will cause a given *creep strain* in a creep test at a given time in a specified constant environment.

creep stress The constant load divided by the original cross-sectional area of the specimen.

crevice corrosion Localized *corrosion* of a metal surface at, or immediately adjacent to, an area that is shielded from full exposure to the environment because of close proximity between the metal and the surface of another material.

cross direction See *transverse direction.*

crush An indentation in a casting surface due to displacement of sand into the mold cavity when the mold is closed.

crystalline fracture A pattern of brightly reflecting crystal facets on the fracture surface of a polycrystalline metal, resulting from *cleavage fracture* of many individual crystals. Contrast with *fibrous fracture* and *silky fracture*; see also *granular fracture.*

cup fracture (cup-and-cone fracture) A mixed-mode fracture, often seen in tension test specimens of a ductile material, where the central portion undergoes plane-strain fracture and the surrounding region undergoes plane-stress fracture. One of the mating fracture surfaces looks like a miniature cup; it has a central depressed flat-face region surrounded by a shear lip. The other fracture surface looks like a miniature truncated cone.

cupping The condition sometimes occurring in heavily cold-worked rods and wires, in which the outside fibers are still intact and the central zone has failed in a series of cup-and-cone fractures.

cut A raised rough surface on a casting due to erosion by the metal stream of part of the sand mold or core.

cycle, N In fatigue, one complete sequence of values of applied load that is repeated periodically.

cyclic load (1) Repetitive loading, as with regularly recurring stresses on a part, that sometimes leads to fatigue fracture. (2) Loads that change value by following a regular repeating sequence of change.

cyclic stressing See *cyclic load*.

decarburization Loss of carbon from the surface layer of a carbon-containing alloy due to reaction with one or more chemical substances in a medium that contacts the surface.

deformation A change in the form of a body due to stress, thermal change, change in moisture, or other causes. Measured in units of length.

deformation bands Bands produced within individual grains during cold working which differ variably in orientation from the matrix.

deformation curve See *stress-strain diagram*.

dendrite A crystal with a treelike branching pattern. Dendrites are most evident in cast metals slowly cooled through the solidification range.

denickelification *Corrosion* in which nickel is selectively leached from nickel-containing alloys. Most commonly observed in copper-nickel alloys after extended service in fresh water. See also *selective leaching*.

depletion Selective removal of one component of an alloy, usually from the surface or preferentially from grain-boundary regions. See also *selective leaching*.

deposit attack See *poultice corrosion*.

deposit corrosion See *poultice corrosion*.

dezincification *Corrosion* in which zinc is selectively leached from zinc-containing alloys. Most commonly found in copper-zinc alloys containing less than 85% Cu after extended service in water containing dissolved oxygen. See also *selective leaching*.

diamond pyramid hardness test See *Vickers hardness test*.

dimpled rupture fracture A fractographic term describing *ductile fracture* that occurs through the formation and coalescence of microvoids (dimples) along the fracture path. The fracture surface of such a ductile fracture appears dimpled when observed at high magnification and usually is most clearly resolved when viewed in a scanning electron microscope.

distortion Any deviation from an original size, shape, or contour that occurs because of the application of *stress* or the release of *residual stress*.

ductile crack propagation Slow crack propagation that is accompanied by noticeable *plastic deformation* and requires energy to be supplied from outside the body. Contrast with *brittle crack propagation*.

ductile erosion behavior Erosion behavior having characteristic properties (i.e., considerable *plastic deformation*) that can be associated with *ductile fracture* of the exposed solid surface. A characteristic ripple pattern forms on the exposed surface at low values of angle of attack. Contrast with *brittle erosion behavior*.

ductile fracture Fracture characterized by tearing of metal accompanied by appreciable gross *plastic deformation* and expenditure of considerable energy. Contrast with *brittle fracture*.

ductility The ability of a material to deform plastically before fracturing. Measured by *elongation* or *reduction of area* in a tension test, by height of *cupping* in a cupping test, or by the radius or angle of bend in a bend test. Contrast with *brittleness*; see also *plastic deformation*.

dynamic Moving, or having high velocity. Frequently used with high strain rate (>0.1 s^{-1}) testing of metal specimens. Contrast with *static*.

effective crack size, a_e The *physical crack size* augmented for the effects of crack tip plastic deformation. Sometimes the effective crack size is calculated from a measured value of a physical crack size plus a calculated value of a plastic-zone adjustment. A preferred method for the calculation of effective crack size compares compliance from the secant of a load-deflection trace with the elastic compliance from a calibration for the type of specimen.

elastic constants The factors of proportionality that relate elastic displacement of a material to applied forces. See also *bulk modulus of elasticity, modulus of elasticity, Poisson's ratio,* and *shear modulus*.

elastic deformation A change in dimensions directly proportional to and in phase with an increase or decrease in applied force.

elasticity The property of a material by virtue of which deformation caused by *stress* disappears upon removal of the stress. A perfectly elastic body completely recovers its original shape and dimensions after release of stress.

elastic limit The maximum *stress* a material is capable of sustaining without any permanent *strain* (deformation) remaining upon complete release of the stress. See also *proportional limit*.

elastic strain See *elastic deformation*.

elongation A term used in mechanical testing to describe the amount of extension of a test piece when stressed. See also *elongation, percent* and *stress*.

elongation, percent The extension of a uniform section of a specimen expressed as percentage of the original gage length:

$$\text{Elongation, \%} = \frac{L_f - L_o}{L_o} \times 100$$

where L_o is original gage length and L_f is final gage length. See also *elongation*.

embrittlement The severe loss of ductility and/or toughness of a material, usually a metal or alloy.

endurance limit The maximum *stress* below which a material can presumably endure an infinite number of *stress cycles*. If the stress is not completely reversed, the value of the mean stress, the minimum stress, or the *stress ratio* also should be stated. Compare with *fatigue limit*.

erosion Destruction of materials by the abrasive action of moving fluids, usually accelerated by the presence of solid particles carried with the fluid. See also *erosion-corrosion*.

erosion-corrosion A conjoint action involving *corrosion* and *erosion* in the presence of a moving corrosive fluid, leading to the accelerated loss of material.

exfoliation *Corrosion* that proceeds laterally from the sites of initiation along planes parallel to the surface, generally at grain boundaries, forming corrosion products that force metal away from the body of the material, giving rise to a layered appearance.

failure A general term used to imply that a part in service (1) has become completely inoperable, (2) is still operable but is incapable of satisfactorily performing its intended function, or (3) has deteriorated seriously, to the point that it has become unreliable or unsafe for continued use.

false brinelling Damage to a solid bearing surface characterized by indentations not caused by *plastic deformation*, resulting from overload but thought to be due to other causes such as *fretting corrosion*. See also *brinelling*.

fatigue The phenomenon leading to *fracture* under repeated or fluctuating stresses having a maximum value less than the ultimate tensile strength of the material. See also *fatigue failure, high-cycle fatigue, low-cycle fatigue,* and *ultimate strength*.

fatigue crack growth rate, da/dN The rate of crack extension caused by constant-amplitude fatigue loading, expressed in terms of crack extension per cycle of load application.

fatigue failure Failure that occurs when a specimen undergoing fatigue completely fractures into two parts or has softened or been otherwise significantly reduced in stiffness by thermal heating or cracking. Fatigue failure

fatigue life The number of *stress cycles* that can be sustained prior to failure under a stated test condition.

fatigue limit The maximum *stress* that presumably leads to fatigue fracture in a specified number of *stress cycles*. If the stress is not completely reversed, the value of the mean stress, the minimum stress, or the *stress ratio* also should be stated. Compare with *endurance limit*.

fatigue notch factor, K_f The ratio of the *fatigue strength* of an unnotched specimen to the fatigue strength of a notched specimen of the same material and condition; both strengths are determined at the same number of *stress cycles*.

fatigue notch sensitivity, q An estimate of the effect of a notch or hole of a given size and shape on the fatigue properties of a material; measured by $q = (K_f - 1)/(K_t - 1)$, where K_f is the *fatigue notch factor* and K_t is the *stress-concentration factor*. A material is said to be fully notch-sensitive if q approaches a value of 1.0; it is not notch-sensitive if the ratio approaches 0.

fatigue ratio The *fatigue limit* under completely reversed flexural stress divided by the *tensile strength* for the same alloy and condition.

fatigue strength The maximum *stress* that can be sustained for a specified number of *stress cycles* without failure, the stress being completely reversed within each cycle unless otherwise stated.

fatigue striation See *striation*.

fatigue wear *Wear* of a solid surface caused by *fracture* arising from material fatigue.

fiber (1) The characteristic of wrought metal that indicates directional properties. It is revealed by etching of a longitudinal section or is manifested by the fibrous or woody appearance of a fracture. It is caused chiefly by extension of the constituents of the metal, both metallic and nonmetallic, in the direction of working. (2) The pattern of *preferred orientation* of metal crystals after a given deformation process, usually wiredrawing. See also *texture*.

fiber-reinforced composite A material consisting of two or more discrete physical phases, in which a fibrous phase is dispersed in a continuous matrix phase. The fibrous phase may be macro-, micro-, or submicroscopic, but it must retain its physical identity so that it could conceivably be removed from the matrix intact.

fiber stress Local *stress* through a small area (a point or line) on a section where the stress is not uniform, as in a beam under a bending load.

fibrous fracture A gray and amorphous *fracture* that results when a metal is sufficiently ductile for the crystals to elongate before fracture occurs. When a fibrous fracture is obtained in an impact test, it may be regarded as definite evidence of the toughness of the metal. See also *crystalline fracture* and *silky fracture*.

fibrous structure (1) In forgings, a structure revealed as laminations, not necessarily detrimental, on an etched section or as a ropy appearance on a *fracture*. (2) In wrought iron, a structure consisting of slag fibers embedded in ferrite. (3) In rolled steel plate stock, a uniform, lamination-free, fine-grained structure on a fractured surface.

filiform corrosion *Corrosion* that occurs under some coatings in the form of randomly distributed threadlike filaments.

fisheye A discontinuity found on the fracture surface of a weld in steel that consists of a small pore or inclusion surrounded by an approximately round, bright area.

fishmouthing See *alligatoring*.

flake A short, discontinuous internal crack in ferrous metals attributed to stresses produced by localized transformation and hydrogen-solubility effects during cooling after hot working. In fracture surfaces, flakes appear as bright, silvery areas with a coarse texture. In deep acid-etched transverse sections, they appear as discontinuities that are usually in the midway to center location of the section. Also termed hairline cracks and shatter cracks.

flow Movement (slipping or sliding) of essentially parallel planes within an element of a material in parallel directions; occurs under the action of *shear stress*. Continuous action in this manner, at constant volume and without disintegration of the material, is termed *yield, creep,* or *plastic deformation*.

flow lines Texture showing the direction of metal flow during hot or cold working. Flow lines often can be revealed by etching the surface or a section of a metal part.

fluting A type of *pitting* in which cavities occur in a regular pattern, forming grooves or flutes. Fluting is caused by *fretting* or by electric arcing.

fold A defect in metal, usually on or near the surface, caused by continued fabrication of overlapping surfaces.

fractography Descriptive explanation of a fracture process, especially in metals, with specific reference to photographs of the fracture surface. Macrofractography involves low magnification ($<25\times$); microfractography high magnification ($>25\times$).

fracture The irregular surface produced when a piece of metal is broken. See also *crystalline fracture, fibrous fracture, granular fracture, intergranular fracture, silky fracture,* and *transgranular fracture*.

fracture mechanics See *linear-elastic fracture mechanics*.

fracture stress See *rupture stress*.

fracture test Test in which a specimen is broken and its fracture surface is examined with the unaided eye or with a low-power microscope to determine such factors as composition, grain size, case depth, or discontinuities.

fracture toughness A generic term for measures of resistance to the extension of a crack. The term is sometimes restricted to results of fracture mechanics tests, which are directly applicable in fracture control. However, the term commonly includes results from simple tests of notched or precracked specimens not based on fracture mechanics analysis. Results from tests of the latter type are often useful for fracture control, based on either service experience or empirical correlations with fracture mechanics tests. See also *stress-intensity factor*.

fretting *Wear* that occurs between tight-fitting surfaces subjected to oscillation at very small amplitude. This type of wear can be a combination of *oxidative wear* and *abrasive wear*. See also *fretting corrosion*.

fretting corrosion The deterioration at the interface between contacting surfaces as the result of *corrosion* and slight oscillatory slip between the two surfaces.

fretting fatigue Fatigue fracture that initiates at a surface area where *fretting* has occurred.

galling A condition whereby excessive friction between high spots results in localized welding with subsequent *spalling* and a further roughening of the rubbing surfaces of one or both of two mating parts.

galvanic corrosion Accelerated *corrosion* of a metal because of an electrical contact with a more noble metal or nonmetallic conductor in a corrosive electrolyte.

gas hole A hole in a casting or weld formed by gas escaping from molten metal as it solidifies. Gas holes may occur individually or in clusters, or may be distributed throughout the solidified metal.

gas porosity Fine holes or pores within a metal that are caused by entrapped gas or by evolution of dissolved gas during solidification.

general corrosion A form of deterioration that is distributed more or less uniformly over a surface. See also *corrosion*.

glide See *slip*.

grain An individual crystal in a polycrystalline metal or alloy, including twinned regions or *subgrains* if present.

grain boundary An interface separating two grains at which the orientation of the lattice changes from that of one grain to that of the other. When the orientation change is very small, the boundary is sometimes referred to as a *subboundary structure*.

grain-boundary corrosion Same as *intergranular corrosion*; see also *corrosion* and *interdendritic corrosion*.

grain flow Fiberlike lines on polished and etched sections of forgings caused by orientation of the constituents of the metal in the direction of working during forging. Grain flow produced by proper die design can improve required *mechanical properties* of forgings.

granular fracture A type of irregular surface produced when metal is broken that is characterized by a rough, grainlike appearance, rather than a smooth or fibrous one. It can be subclassified as *transgranular* or *intergranular*. This type of fracture is frequently called *crystalline fracture*; however, the inference that the metal broke because it "crystallized" is not justified, because all metals are crystalline in the solid state. See also *fibrous fracture* and *silky fracture*.

graphitic corrosion Deterioration of gray cast iron in which the metallic constituents are selectively leached or converted to corrosion products, leaving the graphite intact; it occurs in relatively mild aqueous solutions and in buried pipe and fittings. The term "graphitization" is commonly used to identify this form of *corrosion,* but is not recommended because of its use in metallurgy for the decomposition of carbide to graphite.

hairline crack See *flake*.

hardness A measure of the resistance of a material to surface indentation or abrasion; may be thought of as a function of the *stress* required to produce some specified type of surface deformation. There is no absolute scale for hardness; therefore, to express hardness quantitatively, each type of test has its own scale of arbitrarily defined hardness. Indentation hardness can be measured by *Brinell, Knoop, Rockwell, Scleroscope,* and *Vickers hardness tests*.

Hartmann lines See *Lüders lines*.

heat-affected zone That portion of the base metal that was not melted during brazing, cutting, or welding, but whose *microstructure* and *mechanical properties* were altered by the heat.

herringbone pattern See *chevron pattern*.

high-cycle fatigue *Fatigue* that occurs at relatively large numbers of cycles. The arbitrary, but commonly accepted, dividing line between high-cycle fatigue and *low-cycle fatigue* is considered to be about 10^4 to 10^5 cycles. In practice, this distinction is made by determining whether the dominant component of the *strain* imposed during cyclic loading is elastic (high cycle) or plastic (low cycle), which in turn depends on the properties of the metal and on the magnitude of the nominal *stress*.

Hooke's law A material in which *stress* is linearly proportional to *strain* is said to obey Hooke's law. This law is valid only up to the *proportional limit*, or the end of the straight-line portion of the *stress-strain diagram*. See also *modulus of elasticity*.

hot crack See *solidification shrinkage crack*.

hot tear A crack or fracture formed before completion of solidification because of hindered contraction. A hot tear is frequently open to the surface of the casting and thus exposed to the atmosphere. This may result in *oxidation, decarburization,* or other metal-atmosphere reactions at the tear surface.

hydrogen blistering The formation of blisters on or below a metal surface from excessive internal hydrogen pressure. Hydrogen may be formed during cleaning, plating, corrosion, etc.

hydrogen damage A general term for the embrittlement, cracking, blistering, and hydride formation that can occur when hydrogen is present in some metals.

hydrogen embrittlement A condition of low ductility or hydrogen-induced cracking in metals resulting from the absorption of hydrogen. See also *hydrogen-induced delayed cracking.*

hydrogen-induced delayed cracking A term sometimes used to identify a form of *hydrogen embrittlement,* in which a metal appears to fracture spontaneously under a steady stress less than the *yield stress.* There is usually a delay between the application of stress (or exposure of the stressed metal to hydrogen) and the onset of cracking. Also referred to as static fatigue.

hydrostatic modulus See *bulk modulus of elasticity.*

impact energy The amount of energy required to fracture a material, usually measured by means of an *Izod test* or *Charpy test.* The type of specimen and test conditions affect the values and therefore should be specified.

impact load An especially severe shock load such as that caused by instantaneous arrest of a falling mass, by shock meeting of two parts (in a mechanical hammer, for example), or by explosive impact, in which there can be an exceptionally rapid buildup of stress.

impact strength See *impact energy.*

impingement attack *Corrosion* associated with turbulent flow of liquid. May be accelerated by entrained gas bubbles. See also *erosion-corrosion.*

inclusion A particle of foreign material in a metallic matrix. The particle is usually a compound (such as an oxide, sulfide, or silicate), but may be of any substance that is foreign to (and essentially insoluble in) the matrix. Inclusions are usually considered undesirable, although in some cases—such as in free-machining metals—manganese sulfides, phosphorus, selenium, or tellurium may be deliberately introduced to improve machinability.

intercrystalline See *intergranular.*

intercrystalline corrosion See *intergranular corrosion.*

intercrystalline cracking See *intergranular cracking.*

interdendritic corrosion Corrosive attack that progresses preferentially along interdendritic paths. This type of attack results from local differences in composition commonly encountered in alloy castings. See also *corrosion.*

interface The boundary between two contacting parts or regions of parts.

intergranular Between crystals or grains. Also termed intercrystalline. Contrast with *transgranular*.

intergranular corrosion *Corrosion* occurring preferentially at grain boundaries, usually with slight or negligible attack on the adjacent grains. See also *interdendritic corrosion*.

intergranular cracking Cracking or fracturing that occurs between the grains or crystals in a polycrystalline aggregate. Contrast with *transgranular cracking*.

intergranular fracture Brittle fracture of a metal in which the fracture is between the grains, or crystals, that form the metal. Contrast with *transgranular fracture*.

intergranular stress-corrosion cracking *Stress-corrosion cracking* in which the cracking occurs along grain boundaries.

internal oxidation (1) The formation of isolated particles of *corrosion* products beneath the metal surface. This occurs as the result of preferential oxidation of certain alloy constituents by inward diffusion of oxygen, nitrogen, sulfur, etc. Also called subsurface corrosion. (2) Preferential in situ oxidation of certain components of phases within the bulk of a solid alloy accomplished by diffusion of oxygen into the body. This is commonly used to prepare electrical contact materials.

intracrystalline See *transgranular*.

intracrystalline cracking See *transgranular cracking*.

Izod test A type of impact test in which a V-notched specimen, mounted vertically, is subjected to a sudden blow delivered by the weight at the end of a pendulum arm. The energy required to break off the free end is a measure of the impact strength or toughness of the material. Contrast with *Charpy test*.

J integral A mathematical expression; a line or surface integral that encloses the crack front from one crack surface to the other, used to characterize the *fracture toughness* of a material having appreciable plasticity before fracture. The J integral eliminates the need to describe the behavior of the material near the crack tip by considering the local stress-strain field around the crack front; J_{Ic} is the critical value of the J integral required to initiate growth of a preexisting crack.

Knoop hardness number, HK A number related to the applied load and to the projected area of the permanent impression made by a rhombic-based pyramidal diamond indenter having included edge angles of 172° 30' and 130° 0' computed from the equation

$$HK = \frac{P}{0.07028d^2}$$

where P is applied load, kgf; and d is long diagonal of the impression, mm. In reporting Knoop hardness numbers, the test load is stated.

Knoop hardness test An indentation hardness test using calibrated machines to force a rhombic-based pyramidal diamond indenter having specified edge angles, under specified conditions, into the surface of the material under test and to measure the long diagonal after removal of the load.

lamination (1) A type of discontinuity with separation or weakness generally aligned parallel to the worked surface of a metal. (2) In electrical components such as motors, a blanked piece of electrical sheet that is stacked up with several other identical pieces to make a stator or rotor.

lap (1) A surface imperfection on worked metal caused by folding over a fin overfill or similar surface condition, then impressing this into the surface by subsequent working without welding it. (2) A flat surface that holds an abrasive for polishing operations.

leaching See *selective leaching*.

linear-elastic fracture mechanics A method of fracture analysis that can determine the *stress* (or load) required to induce fracture instability in a structure containing a cracklike flaw of known size and shape. See also *stress-intensity factor*.

liquid metal embrittlement The decrease in *ductility* of a metal caused by contact with a liquid metal.

liquid shrinkage The reduction in volume of liquid metal as it cools to the liquidus.

longitudinal direction That direction parallel to the direction of maximum elongation in a worked material. See also *normal direction* and *transverse direction*.

low-cycle fatigue *Fatigue* that occurs at relatively small numbers of cycles ($<10^4$ cycles). Low-cycle fatigue may be accompanied by some plastic, or permanent, deformation. Compare with *high-cycle fatigue*.

Lüders lines Elongated surface markings or depressions, often visible to the unaided eye, that form along the length of a tension specimen at an angle of approximately 45° to the loading axis. Caused by localized *plastic deformation,* they result from discontinuous (inhomogeneous) yielding. Also known as Lüders bands, Hartmann lines, Piobert lines, or stretcher strains.

macroscopic Visible at magnifications at or below 25×.

macroshrinkage Isolated, clustered, or interconnected voids in a casting that are detectable macroscopically. Such voids are usually associated with abrupt changes in section size and are caused by feeding that is insufficient to compensate for *solidification shrinkage*.

macrostructure The structure of metals as revealed by macroscopic examination of a specimen. The examination may be carried out using an as-polished or a polished and etched specimen.

magnification The ratio of the length of a line in the image plane (for example, ground glass or a photographic plate) to the length of the same line in the object. Magnifications are usually expressed in linear terms and in units called diameters.

malleability The characteristic of metals that permits *plastic deformation* in compression without fracture. See also *ductility*.

matrix The continuous or principal phase in which another constituent is dispersed.

maximum strength See *ultimate strength*.

mechanical (cold) crack A crack or fracture in a casting resulting from rough handling or from thermal shock, such as may occur at shakeout or during heat treatment.

mechanical properties The properties of a material that reveal its elastic and inelastic (plastic) behavior when force is applied, thereby indicating its suitability for mechanical (load-bearing) applications. Examples are *elongation, fatigue limit, hardness, modulus of elasticity, tensile strength,* and *yield strength*. Compare with *physical properties*.

mechanical twin A *twin* formed in a crystal by simple shear under external loading.

metal penetration An imperfection on the surface of a casting caused by the penetration of molten metal into voids between refractory particles of the mold.

microcrack A crack of microscopic proportions. Also termed microfissure.

microfissure See *microcrack*.

microporosity Extremely fine *porosity* in castings.

microscopic Visible only at magnifications above $25\times$.

microshrinkage A casting imperfection consisting of interdendritic voids. Microshrinkage results from contraction during solidification where the opportunity to supply filler material is inadequate to compensate for shrinkage. Alloys with wide ranges in solidification temperature are particularly susceptible.

microstructure The structure of metals and alloys as revealed after polishing and etching a specimen, at magnifications greater than $25\times$.

misrun A casting not fully formed because of solidification of metal before the mold is filled.

mode One of the three classes of crack (surface) displacements adjacent to the crack tip. These displacement modes are associated with stress-strain fields around the crack tip and are designated I, II, and III. See also *crack tip plane strain* and *crack opening displacement*.

modulus of elasticity, E (1) The measure of rigidity or stiffness of a metal;

the ratio of stress, below the *proportional limit,* to the corresponding strain. In terms of the *stress-strain diagram,* the modulus of elasticity is the slope of the stress-strain curve in the range of linear proportionality of stress to strain. Also known as Young's modulus. (2) For materials that do not conform to *Hooke's law* throughout the elastic range, the slope of either the tangent to the stress-strain curve at the origin or at low stress, the secant drawn from the origin to any specified point on the stress-strain curve, or the chord connecting any two specific points on the stress-strain curve is usually taken to be the modulus of elasticity. In these cases, the modulus is referred to as the *tangent modulus, secant modulus,* or *chord modulus,* respectively.

modulus of rigidity See *shear modulus.*

modulus of rupture Nominal stress at fracture in a bend test or torsion test. In bending, modulus of rupture is the bending moment at fracture divided by the section modulus. In torsion, modulus of rupture is the torque at fracture divided by the polar section modulus. See also *modulus of rupture in bending* and *modulus of rupture in torsion.*

modulus of rupture in bending, S_b The value of maximum tensile or compressive stress (whichever causes failure) in the extreme fiber on a beam loaded to failure in bending, computed from

$$S_b = \frac{Mc}{I}$$

where M is maximum bending moment, computed from the maximum load and the original moment arm; c is initial distance from the neutral axis to the extreme fiber where failure occurs; and I is initial moment of inertia of the cross section about the neutral axis. See also *modulus of rupture.*

modulus of rupture in torsion, S_s The value of maximum shear stress in the extreme fiber of a member of circular cross section loaded to failure in torsion, computed from

$$S_s = \frac{Tr}{J}$$

where T is maximum twisting moment, r is original outer radius, and J is polar moment of inertia of the original cross section. See also *modulus of rupture.*

necking (1) Reduction of the cross-sectional area of metal in a localized area by stretching. (2) Reduction in the diameter of a portion of the length of a cylindrical shell or tube.

neutron embrittlement Embrittlement resulting from bombardment with neutrons, usually encountered in metals that have been exposed to a neutron flux in the core of a reactor. In steels, neutron embrittlement is evidenced by a rise in the ductile-to-brittle transition temperature. See also *radiation damage.*

nominal strength See *ultimate strength.*

normal direction That direction perpendicular to the plane of working in a worked material. See also *longitudinal direction* and *transverse direction*.

notch See *stress concentration*.

notch acuity Relates to the severity of the *stress concentration* produced by a given notch in a particular structure. If the depth of the notch is very small compared with the width (or diameter) of the narrowest cross section, acuity may be expressed as the ratio of the notch depth to the notch root radius. Otherwise acuity is defined as the ratio of one-half the width (or diameter) of the narrowest cross section to the notch root radius.

notch brittleness Susceptibility of a material to *brittle fracture* at points of *stress concentration*. For example, in a notch tension test, the material is said to be notch brittle if the notch strength is less than the tensile strength of an unnotched specimen. Otherwise it is said to be notch ductile.

notch depth The distance from the surface of a test specimen to the bottom of the notch. In a cylindrical test specimen, the percentage of the original cross-sectional area removed by machining an annular groove.

notch rupture strength The ratio of applied load to original area of the minimum cross section in a stress-rupture test of a notched specimen.

notch sensitivity A measure of the reduction in strength of a metal caused by the presence of *stress concentration*. Values can be obtained from static, impact, or fatigue tests.

notch strength The maximum load on a notched tension-test specimen divided by the minimum cross-sectional area (the area at the root of the notch). Also called notch tensile strength.

notch tensile strength See *notch strength*.

orange peel A surface roughening in the form of a pebble-grained pattern where a metal of unusually coarse grain is stressed beyond its elastic limit. Also known as pebbles and alligator skin.

original crack size, a_o The *physical crack size* at the start of testing.

oxidation (1) A reaction in which there is an increase in valence resulting from a loss of electrons. Contrast with *reduction*. (2) A *corrosion* reaction in which the corroded metal forms an oxide; usually applied to reaction with a gas containing elemental oxygen, such as air.

oxidative wear A type of *wear* resulting from the sliding action between two metallic components that generates oxide films on the metal surfaces. These oxide films prevent the formation of a metallic bond between the sliding surfaces, resulting in fine wear debris and low wear rates.

parting The selective *corrosion* of one or more components of a solid-solution alloy.

pebbles See *orange peel*.

physical crack size, a_p The distance from a reference plane to the observed crack front. This distance may represent an average of several measurements along the crack front. The reference plane depends on the specimen form, and it is normally taken to be either the boundary or a plane containing either the load line or the centerline of a specimen or plate.

physical properties Properties of a metal or alloy that are relatively insensitive to structure and can be measured without the application of force; for example, density, electrical conductivity, coefficient of thermal expansion, magnetic permeability, and lattice parameter. Does not include chemical reactivity. Compare with *mechanical properties*.

pinhole A small, rounded hole just below the surface of a casting, sometimes visible only after machining. Such holes, often localized, have bright interior surfaces.

Piobert lines See *Lüders lines*.

pitting (1) *Corrosion* of a metal surface, confined to a point or small area, that takes the form of cavities. (2) In *tribology,* a type of wear characterized by the presence of surface cavities formed by processes such as fatigue, local adhesion, or cavitation.

plane strain The stress condition in *linear elastic fracture mechanics* in which there is zero strain in a direction normal to both the axis of applied *tensile stress* and the direction of crack growth (i.e., parallel to the crack front); most nearly achieved in loading thick plates along a direction parallel to the plate surface. Under plane-strain conditions, the plane of fracture instability is normal to the axis of the principal tensile stress.

plane-strain fracture toughness, K_{Ic} The crack extension resistance under conditions of *crack tip plane strain*. See also *stress-intensity factor*.

plane stress The stress condition in *linear elastic fracture mechanics* in which the stress in the thickness direction is zero; most nearly achieved in loading very thin sheet along a direction parallel to the surface of the sheet. Under plane-stress conditions, the plane of fracture instability is inclined 45° to the axis of the principal *tensile stress*.

plane-stress fracture toughness, K_c The value of the crack-extension resistance at the instability condition determined from the tangency between the *R curve* and the critical crack-extension force curve of the specimen. See also *stress-intensity factor*.

plastic deformation The permanent (inelastic) distortion of metals under applied stresses that strain the material beyond its *elastic limit*.

plowing In *tribology,* the formation of grooves by *plastic deformation* of the softer of two surfaces in relative motion.

Poisson's ratio The absolute value of the ratio of the transverse strain to the corresponding axial strain, in a body subjected to uniaxial stress; usually applied to elastic conditions.

polycrystalline Comprising an aggregate of more than one crystal and usually a large number of crystals.

pore (1) A small void in the body of a metal. (2) A minute cavity in a powder metallurgy compact, sometimes intentional. (3) A minute perforation in an electroplated coating.

porosity Fine holes or pores within a metal.

poultice corrosion A term used in the automotive industry to describe the *corrosion* of vehicle body parts due to the collection of road salts and debris on ledges and in pockets that are kept moist by weather and washing. Also called deposit attack or deposit corrosion.

preferred orientation A condition of a polycrystalline aggregate in which the crystal orientations are not random, but rather exhibit a tendency for alignment with a specific direction in the bulk material, commonly related to the direction of working. See also *fiber* and *texture*.

primary creep The first, or initial, stage of *creep*, or time-dependent deformation.

principal stress (normal) The maximum or minimum value of the normal stress at a point in a plane considered with respect to all possible orientations of the considered plane. On such principal planes the shear stress is zero. There are three principal stresses on three mutually perpendicular planes. The state of stress at a point may be (1) uniaxial, a state of stress in which two of the three principal stresses are zero; (2) biaxial, a state of stress in which only one of the three principal stresses is zero; or (3) triaxial, a state of stress in which none of the principal stresses is zero. Multiaxial stress refers to either biaxial or triaxial stress.

proportional limit The maximum *stress* at which *strain* remains directly proportional to stress; the upper end of the straight-line portion of the stress-strain or load-elongation curve. See also *elastic limit*.

quasicleavage fracture A fracture mode that combines the characteristics of *cleavage fracture* and *dimpled rupture fracture*. An intermediate type of fracture found in certain high-strength metals.

quenching crack A crack formed as a result of thermal stresses produced by rapid cooling from a high temperature.

radial marks Lines on a fracture surface that radiate from the fracture origin and are visible to the unaided eye or at low magnification. Radial marks result from the intersection and connection of brittle fractures propagating at different levels. Also known as shear ledges. See also *chevron pattern*.

radiation damage A general term for the alteration of properties of a material arising from exposure to ionizing radiation (penetrating radiation), such as x rays, gamma rays, neutrons, heavy-particle radiation, or fission fragments in nuclear fuel material. See also *neutron embrittlement*.

ratchet marks Lines on a fatigue fracture surface that result from the intersection and connection of fatigue fractures propagating from multiple origins. Ratchet marks are parallel to the overall direction of crack propagation and are visible to the unaided eye or at low magnification.

rattail A shallow, indented, and irregular line on a casting surface due to sand expansion.

R curve A plot of crack-extension resistance as a function of stable crack extension, which is the difference between either the *physical crack size* or the *effective crack size* and the *original crack size*. R curves normally depend on specimen thickness and, for some materials, on temperature and strain rate.

reduction (1) In cupping and deep drawing, a measure of the percentage decrease from blank diameter to cup diameter, or of diameter reduction in redrawing. (2) In forging, rolling, and drawing, either the ratio of the original to final cross-sectional area or the percentage decrease in cross-sectional area. (3) A reaction in which there is a decrease in valence resulting from a gain in electrons. Contrast with *oxidation*.

reduction in area See *reduction of area*.

reduction of area The difference between the original cross-sectional area of a tension specimen and the smallest area at or after fracture as specified for the material being tested. Also known as reduction in area.

residual stress Stress present in a body that is free of external forces or thermal gradients.

river pattern A characteristic pattern of cleavage steps running parallel to the local direction of crack propagation on the fracture surfaces of grains that have separated by *cleavage*.

rock candy fracture A fracture that exhibits separated-grain facets; most often used to describe an *intergranular fracture* in a large-grained metal.

Rockwell hardness number, HR A number derived from the net increase in the depth of impression as the load on an indenter is increased from a fixed minor load to a major load and then returned to the minor load. Rockwell hardness numbers are always quoted with a scale symbol representing the penetrator, load, and dial used.

Rockwell hardness test An indentation hardness test using a calibrated machine that utilizes the depth of indentation, under constant load, as a measure of hardness. Either a 120° diamond cone with a slightly rounded point or a 1.6- or 3.2-mm ($\frac{1}{16}$- or $\frac{1}{8}$-in)-diameter steel ball is used as the indenter.

Rockwell superficial hardness number See *Rockwell hardness number* and *Rockwell superficial hardness test*.

Rockwell superficial hardness test Same as *Rockwell hardness test,* except that smaller minor and major loads are used.

rupture stress The stress at failure. Also known as breaking stress or fracture stress.

rust A *corrosion* product consisting primarily of hydrated iron oxide. A term properly applied only to ferrous alloys.

sand hole A pit in the surface of a sand casting resulting from a deposit of loose sand on the surface of the mold.

scab A raised and rough area on the surface of a casting due to sand being dislodged from the surface of the mold.

Scleroscope hardness number, HSc or HSd A number related to the height of rebound of a diamond-tipped hammer dropped on the material being tested. It is measured on a scale determined by dividing into 100 units the average rebound of the hammer from a quenched (to maximum hardness) and untempered AISI W-5 tool steel test block.

Scleroscope hardness test A dynamic indentation hardness test using a calibrated instrument that drops a diamond-tipped hammer from a fixed height onto the surface of the material being tested. The height of rebound of the hammer is a measure of the hardness of the material.

scoring In *tribology,* a severe form of *wear* characterized by the formation of extensive grooves and scratches in the direction of sliding.

scratching In *tribology,* the mechanical removal and/or displacement of material from a surface by the action of abrasive particles or protuberances sliding across the surfaces. See also *plowing.*

scuffing A form of *adhesive wear* that produces superficial scratches or a high polish on the rubbing surfaces. It is observed most often on inadequately lubricated parts.

seam An unfused fold or lap that appears as a crack on a metal surface.

season cracking Cracking resulting from the combined effects of corrosion and internal stress. A term usually applied to *stress-corrosion cracking of brass.*

secant modulus The slope of the secant drawn from the origin to any specified point on a stress-strain curve. See also *modulus of elasticity.*

secondary creep See *creep.*

segregation Nonuniform distribution of alloying elements, impurities, or phases.

selective leaching *Corrosion* in which one element is preferentially removed from an alloy, leaving a residue (often porous) of the elements that are more resistant to the particular environment. See also *decarburization, denickelification, dezincification,* and *graphitic corrosion.*

sensitization In austenitic stainless steels, the precipitation of chromium carbides, usually at grain boundaries, on exposure to temperatures of about 550 to 850°C (1000 to 1550°F), leaving the grain boundaries depleted of chromium and therefore susceptible to preferential attack by a corroding (oxidizing) medium.

shatter crack See *flake.*

shear bands Bands in which *deformation* has been concentrated inhomogeneously in sheets that extend across regional groups of grains. Only one system is usually present in each regional group of grains, different systems being present in adjoining groups. The bands are noncrystallographic and form on planes of maximum *shear stress* (55° to the compression direction). They carry most of the deformation at large strains.

shear fracture A *ductile fracture* in which a crystal (or a polycrystalline mass) has separated by sliding or tearing under the action of shear stresses. See also *shear stress.*

shear ledges See *radial marks.*

shear lip A narrow, slanting ridge along the edge of a fracture surface. The term sometimes also denotes a narrow, often crescent-shaped, fibrous region at the edge of a fracture that is otherwise of the cleavage type, even though this fibrous region is in the same plane as the rest of the fracture surface.

shear modulus, G The ratio of *shear stress* to the corresponding *shear strain* for shear stresses below the *proportional limit* of the material. Values of shear modulus are usually determined by torsion testing. Also known as modulus of rigidity.

shear strain The tangent of the angular change, due to force, between two lines originally perpendicular to each other through a point in a body.

shear strength The maximum *shear stress* that a material is capable of sustaining. Shear strength is calculated from the maximum load during a shear or torsion test and is based on the original dimensions of the cross section of the specimen.

shear stress (1) A *stress* that exists when parallel planes in metal crystals slide across each other. (2) The stress component tangential to the plane on which the forces act. Also known as tangential stress.

shock load The sudden application of an external force that results in a very rapid buildup of *stress*—for example, piston loading in internal combustion engines.

shrinkage See *casting shrinkage.*

shrinkage cavity A void left in cast metals as a result of *solidification shrinkage.* Shrinkage cavities occur in the last metal to solidify after casting.

silky fracture A metal fracture in which the broken metal surface has a fine texture, usually dull in appearance. Characteristic of tough and strong metals. Contrast with *crystalline fracture* and *granular fracture*.

slant fracture A type of fracture appearance, typical of *plane-stress* fractures, in which the plane of metal separation is inclined at an angle (usually about 45°) to the axis of the applied stress.

slip *Plastic deformation* by the irreversible shear displacement (translation) of one part of a crystal relative to another in a definite crystallographic direction and usually on a specific crystallographic plane. Sometimes called glide. See also *flow*.

S-N curve A plot of stress (S) against the number of cycles to failure (N). The stress can be the maximum stress (S_{max}) or the alternating stress amplitude (S_a). The stress values are usually nominal stress, i.e., there is no adjustment for stress concentration. The diagram indicates the S-N relationship for a specified value of the mean stress (S_m) or the stress ratio (A or R) and a specified probability of survival. For N a log scale is almost always used. For S a linear scale is used most often, but a log scale is sometimes used. Also known as S-N diagram.

S-N diagram See *S-N curve*.

solidification shrinkage The reduction in volume of metal from beginning to end of solidification.

solidification shrinkage crack A crack that forms, usually at elevated temperature, because of the internal (shrinkage) stresses that develop during solidification of a metal casting. Also termed hot crack.

solid shrinkage The reduction in volume of metal from the solidus to room temperature.

spalling The cracking and flaking of particles out of a surface.

static Stationary or very slow. Frequently used in connection with routine tension testing of metal specimens. Contrast with *dynamic*.

static fatigue See *hydrogen-induced delayed cracking*.

steady-rate creep See *creep*.

strain The unit of change in the size or shape of a body due to force.

strain hardening An increase in hardness and strength caused by *plastic deformation* at temperatures below the recrystallization range. Also known as work hardening.

stray-current corrosion *Corrosion* caused by electric current from a source external to the intended electric circuit, for example, extraneous current in the earth.

stress The intensity of the internally distributed forces or components of forces that resist a change in the volume or shape of a material that is or has been subjected to external forces. Stress is expressed in force per unit area and is calculated on the basis of the original dimensions of the cross section of the specimen. Stress can be either direct (tension or compression) or shear. Usually expressed in pounds per square inch (lb/in^2) or megapascals (MPa).

stress amplitude One-half the algebraic difference between the maximum and minimum *stress* in one cycle of a repetitively varying stress.

stress concentration A change in contour or a discontinuity that causes local increases in *stress* in materials under load. Typical are sharp-cornered grooves or notches, threads, fillets, holes, etc. Also called stress raiser.

stress-concentration factor, K_t A multiplying factor for applied *stress* that allows for the presence of a structural discontinuity such as a notch or hole; K_t equals the ratio of the greatest stress in the region of the discontinuity to the nominal stress for the entire section. Also known as theoretical stress-concentration factor.

stress-corrosion cracking, SCC A cracking process that requires the simultaneous action of a corrodent and sustained *tensile stress*. This excludes corrosion-reduced sections that fail by fast fracture. It also excludes intergranular or transgranular corrosion, which can disintegrate an alloy without applied or residual stress. See also *corrosion*.

stress cycle The smallest segment of the stress-time function that is repeated periodically.

stress-intensity factor, K A scaling factor used in *linear-elastic fracture mechanics* to describe the intensification of applied *stress* at the tip of a crack of known size and shape. At the onset of rapid crack propagation in any structure containing a crack, the factor is called the critical stress-intensity factor, or the *fracture toughness*. Various subscripts are used to denote different loading conditions or fracture toughnesses:

K_c Plane-stress fracture toughness. The value of stress intensity at which crack propagation becomes rapid in sections thinner than those in which plane-strain conditions prevail.

K_I Stress-intensity factor for a loading condition that displaces the crack faces in a direction normal to the crack plane. Also known as the opening mode of deformation.

K_{Ic} Plane-strain fracture toughness. The minimum value of K_c for any given material and condition, which is attained when rapid crack propagation in the opening mode is governed by plane-strain conditions.

K_{Id} Dynamic fracture toughness. The fracture toughness determined under dynamic loading conditions; it is used as an approximation of K_{Ic} for very tough materials.

K_{Iscc} Threshold stress intensity for stress-corrosion cracking when loading conditions meet plane-strain requirements.

K_Q Provisional value for plane-strain fracture toughness.

K_{th} Threshold stress intensity for stress-corrosion cracking. A value of stress intensity characteristic of a specific combination of material, material condition, and corrosive environment above which stress-corrosion crack propagation occurs and below which the material is immune from stress-corrosion cracking.

ΔK   The range of the stress-intensity factor during a fatigue cycle.

stress raiser   See *stress concentration*.

stress ratio, A or R   The algebraic ratio of two specified *stress* values in a *stress cycle*. Two commonly used stress ratios are (1) the ratio of the alternating stress amplitude to the mean stress, $A = S_a/S_m$, and (2) the ratio of the minimum stress to the maximum stress, $R = S_{min}/S_{max}$.

stress-rupture strength   See *creep-rupture strength*.

stress-strain curve   See *stress-strain diagram*.

stress-strain diagram   A graph in which corresponding values of *stress* and *strain* are plotted against each other. Values of stress are usually plotted vertically (ordinate or y axis) and values of strain horizontally (abscissa or x axis). Also known as deformation curve and stress-strain curve.

stretcher strains   See *Lüders lines*.

striation   A fatigue fracture feature often observed in electron micrographs that indicates the position of the crack front after each succeeding cycle of *stress*. The distance between striations indicates the advance of the crack front across that crystal during one *stress cycle*, and a line normal to the striation indicates the direction of local crack propagation. Not to be confused with *beach marks*, which are much larger (macroscopic) and form differently.

stringer   In wrought materials, an elongated configuration of microconstituents or foreign material aligned in the direction of working. The term is commonly associated with elongated oxide or sulfide inclusions in steel.

subboundary structure (subgrain structure)   A network of low-angle boundaries, usually with misorientations less than 1° within the main grains of a microstructure.

subgrain   A portion of a crystal or *grain*, with an orientation slightly different from the orientation of neighboring portions of the same crystal.

subsurface corrosion   See *internal oxidation*.

sulfidation   The reaction of a metal or alloy with a sulfur-containing species to produce a sulfur compound that forms on or beneath the surface of the metal or alloy.

sulfide stress cracking, SSC   Brittle failure by cracking under the combined action of *tensile stress* and *corrosion* in the presence of water and hydrogen sulfide.

tangential stress   See *shear stress*.

tangent modulus   The slope of the stress-strain curve at any specified *stress* or *strain*. See also *modulus of elasticity*.

temper brittleness Brittleness that results when certain steels are held within, or are cooled slowly through, a certain range of temperatures below the transformation range. The brittleness is manifested as an upward shift in ductile-to-brittle transition temperature, but only rarely produces a low value of *reduction of area* in a smooth-bar tension test of the embrittled material.

tensile strength In *tension testing,* the ratio of the maximum load to the original cross-sectional area. See also *ultimate strength*; compare with *yield strength*.

tensile stress A *stress* that causes two parts of an elastic body, on either side of a typical stress plane, to pull apart. Contrast with *compressive stress.*

tensile testing See *tension testing.*

tension The force or load that produces *elongation.*

tension testing A method of determining the behavior of materials subjected to uniaxial loading, which tends to stretch the metal. A longitudinal specimen of known length and diameter is gripped at both ends and stretched at a slow, controlled rate until rupture occurs. Also known as tensile testing.

tertiary creep See *creep.*

texture In a polycrystalline aggregate, the state of distribution of crystal orientations. In the usual sense, it is synonymous with *preferred orientation,* in which the distribution is not random. See also *fiber.*

theoretical stress-concentration factor See *stress-concentration factor.*

thermal fatigue Fracture resulting from the presence of temperature gradients that vary with time in such a manner as to produce cyclic stresses in a structure.

thermal shock The development of a steep temperature gradient and accompanying high stresses within a structure.

thermal stresses Stresses in metal resulting from nonuniform temperature distribution.

tide marks See *beach marks.*

torsion A twisting action applied to a shaftlike or cylindrical member. The twisting may be either reversed (back and forth) or unidirectional (one way).

torsional stress The *shear stress* on a transverse cross section resulting from a twisting action.

transcrystalline See *transgranular.*

transcrystalline cracking See *transgranular cracking.*

transgranular Through or across crystals or grains. Also called intracrystalline or transcrystalline.

transgranular cracking Cracking or fracturing that occurs through or across a crystal or grain. Also called transcrystalline cracking. Contrast with *intergranular cracking*.

transgranular fracture Fracture through or across the crystals or grains of a metal. Also called transcrystalline fracture or intracrystalline fracture. Contrast with *intergranular fracture*.

transient creep See *creep* and *primary creep*.

transverse direction Literally, "across," usually signifying a direction or plane perpendicular to the direction of working. In rolled plate or sheet, the direction across the width is often called long transverse; the direction through the thickness, short transverse.

tribology The science concerned with the design, friction, lubrication, and wear of contacting surfaces that move relative to each other.

tuberculation The formation of localized *corrosion* products that appear on a surface as knoblike prominences (tubercules).

twin Two portions of a crystal with a definite orientation relationship; one may be regarded as the parent, the other as the twin. The orientation of the twin is a mirror image of the orientation of the parent across a twinning plane or an orientation that can be derived by rotating the twin portion about a twinning axis. See also *annealing twin* and *mechanical twin*.

twin bands Bands across a crystal grain, observed on a polished and etched section, where crystallographic orientations have a mirror-image relationship to the orientation of the matrix grain across a composition plane that is usually parallel to the sides of the band.

ultimate strength The maximum *stress* (tensile, compressive, or shear) a material can sustain without fracture, determined by dividing maximum load by the original cross-sectional area of the specimen. Also known as nominal strength or maximum strength.

Vickers hardness number, HV A number related to the applied load and the surface area of the permanent impression made by a square-based pyramidal diamond indenter having included face angles of 136°, computed from

$$HV = 2P \sin\frac{\alpha/2}{d^2} = \frac{1.8544P}{d^2}$$

where P is applied load, kgf; d is mean diagonal of impression, mm; and α is face angle of indenter, 136°.

Vickers hardness test An indentation hardness test employing a 136° diamond pyramid indenter (Vickers) and variable loads, enabling the use of one hardness scale for all ranges of hardness—from very soft lead to tungsten carbide. Also known as diamond pyramid hardness test.

volumetric modulus of elasticity See *bulk modulus of elasticity*.

Wallner lines A distinct pattern of intersecting sets of parallel lines, usually producing a set of V-shaped lines, sometimes observed when viewing brittle fracture surfaces at high magnification in an electron microscope. Wallner lines are attributed to interaction between a shock wave and a brittle crack front propagating at high velocity. Sometimes Wallner lines are misinterpreted as fatigue striations.

wear Damage to a solid surface, generally involving progressive loss of material, due to relative motion between that surface and a contacting surface or substance.

wear rate The rate of material removal or dimensional change due to *wear* per unit of exposure parameter—for example, quantity of material removed (mass, volume, thickness) in unit distance of sliding or unit time.

whiskers (1) Metallic filamentary growths, often microscopic in size, that attain very high strengths. (2) Oxide whiskers, such as sapphire, which because of their strength and inertness at high temperatures are used as reinforcements in metal-matrix composites.

work hardening See *strain hardening*.

yield Evidence of *plastic deformation* in structural materials. See also *creep* and *flow*.

yield point The first *stress* in a material, usually less than the maximum attainable stress, at which an increase in *strain* occurs without an increase in stress. Only certain metals—those that exhibit a localized, heterogeneous type of transition from elastic to plastic deformation—produce a yield point. If there is a decrease in stress after yielding, a distinction may be made between upper and lower yield points. The load at which a sudden drop in the flow curve occurs is called the upper yield point. The constant load shown on the flow curve is the lower yield point.

yield strength The stress at which a material exhibits a specified deviation from proportionality of *stress* and *strain*. The specified deviation is usually 0.2 percent for most metals. Compare with *tensile strength*.

yield stress The stress level of highly ductile materials, such as structural steels, at which large *strains* take place without further increase in *stress*.

Young's modulus See *modulus of elasticity*.

References to the Glossary

Merriman, A. D., *A Dictionary of Metallurgy*, Pitman, 1958.

"Standard Definitions of Terms Relating to Corrosion and Corrosion Testing," G 15, *Annual Book of ASTM Standards*, vol. 03.02, ASTM, Philadelphia, Pa., 1984, pp. 133–137.

"Standard Definitions of Terms Relating to Fatigue Testing and the Statistical Analysis of Fatigue Data," E 206, *Annual Book of ASTM Standards*, vol. 03.01, ASTM, Philadelphia, Pa., 1984, pp. 340–345.

"Standard Definitions of Terms Relating to Methods of Mechanical Testing," E 6, *Annual Book of ASTM Standards,* vol. 03.01, ASTM, Philadelphia, Pa., 1984, pp. 119–129.

"Standard Terminology Relating to Erosion and Wear," G 40, *Annual Book of ASTM Standards,* vol. 03.02, ASTM, Philadelphia, Pa., 1984, pp. 239–246.

"Standard Terminology Relating to Fracture Testing," E 616, *Annual Book of ASTM Standards,* vol. 03.01, ASTM, Philadelphia, Pa., 1984, pp. 671–684.

Index

1,3-di-n-butyl-2-thiourea, 60
2-butyne-1,4-diol, 53, 60
302 austenitic stainless steel, 319
304 austenitic stainless steel, 265, 266, 319
316L stainless steel, 317, 318
347 stainless steel, 342

Ablation, 4
Abrasive cutoff wheel, 55
Acetic acid, 53, 61, 71
Acetone, 31, 46, 49, 58–62
Acids, 56, 60
Acrylic lacquer, 46
Aging, 286, 290, 291, 295
Air blast, 48, 62
AISI 1008 steel, 266
AISI 1030 steel, 247
AISI 1035 steel, 227
AISI 1046 steel, 242
AISI 1050 steel, 232, 238, 242
AISI 1085 steel, 50, 53, 54
AISI 10B21 steel, 234, 248–250, 252, 253
AISI 10B62 steel, 229
AISI 4140L steel, 234, 252, 253
AISI 4340 steel, 217–219, 223, 254–264
AISI 4817 steel, 239
AISI 6118 steel, 226
AISI 8640 steel, 235, 243
Alcohol, 31, 49, 58, 62
Alconox, 50–52, 58–60, 62, 63, 66–70
Alkaline solutions, 53
Alternating torsion, 243
Aluminum casting alloys, 285, 287, 288, 292, 293
Ammonium citrate, 53, 71
Ammonium oxalate, 53, 71
Anaglyph method, 39
Annealed condition, 319
Annealing twins, 128, 129, 275, 276, 278, 279
Anode, 51, 60
Aperture, 8, 11, 14

Arcing, 56
Arrest marks, 48, 64, 234
Artifacts, 200, 206–209
Assembly errors, 2
ASTM X-ray diffraction data file, 30
Auger electrons, 18
Austenitic stainless steel, 184, 187, 279
Axial stress, 157
Axle grease, 46

Backscattered electrons, 18, 20, 21
Basal plane, 125
Base, 60
Base metal, 44, 46, 307, 308, 310
Beach marks, 234
Beam formula, 323
Bearing surface, 338
Bellows, 271, 272, 276, 278
Bending, 45, 56, 82, 83, 141, 158, 226, 230, 240, 243, 340
Bending load, 347
Bending overload, 227, 229–232, 249, 347
Bending stress, 281, 294
Bragg's law, 25, 26, 30
Brass, 279
Brazing, 279
Brittle, 3, 81–83, 105, 151, 171, 186, 214, 219, 223, 225, 226, 228–230, 325, 330
Brittle behavior, 107
Brittle cleavage rupture, 230
Brittle crack, 170
Brittle fracture, 90, 98, 105, 107, 151, 221, 225
Bubbles, 283
Burnishing, 45
Burns, 4

Carbides, 4, 63
Carbontetrachloride, 49
Carburization, 247, 324–327
Carcinogens, 49
Case hardened, 4, 45
Case studies, 271

402 Index

Cast iron, 4
Casting, 4, 185, 286
Catastrophic fracture, 107, 108
Cathode, 51, 60
Cathodic cleaning, 48, 51–54, 58, 60
Causes of failure, 3, 5, 6
Caustic embrittlement, 3
Cavitation, 4
Cavities, 307, 312
Cellulose acetate, 31, 33, 46, 58, 59, 62
Centrifugal force, 290, 294
Centrifugal stress, 281
CF8C stainless steel, 342
Chalk, 226, 228
Charpy test, 47, 145, 152, 193, 250, 255, 256, 261, 265
Check valve, 334
Chemical damage, 43, 44
Chemical etching, 52, 53
Chemical-etch cleaning, 48
Chevron marks, 105, 106, 222, 232
Chevron pattern, 107, 222
Chlorine, 56
Chord length, 294
Chromatic aberration, 9
Chromic acid, 53, 61, 71
Chromium plating, 321
Clamshell marks, 234
Cleaning, 31, 57, 58, 62, 71
Cleavage, 5, 63–66, 120, 125–127, 129, 131–134, 136–139, 151–154, 170, 175, 188, 189, 254, 261, 266, 267, 307, 325, 327
Cleavage fracture, 128
Cleavage plane, 131–133, 135, 170, 266
Cleavage step, 132, 135
Cleave, 124
Close-packed directions, 123
Close-packed planes, 123
Cold worked condition, 275, 276, 278, 279, 317, 319
Collimate, 8
Composition, 4, 287, 317, 343
Compressed air, 45, 58, 175–177, 234–236, 238, 239, 242, 243, 250, 265, 299, 314
Conchoidal marks, 175–177, 234–236, 238, 239, 242, 243, 249, 250, 265, 299, 301
Condenser lens, 8
Constraint, 88, 90, 219
Constructive interference, 26
Coolants, 55

Corrosion, 3, 48, 53, 54, 56, 186, 278, 280, 315, 317, 319
Corrosion-assisted fatigue, 69, 187, 188
Corrosion fatigue, 3, 44, 48
Corrosion inhibitors, 53, 58
Corrosion pit, 316
Corrosion products, 42, 50, 59, 62, 203, 205, 273, 274, 279, 314, 315
Crack, 4, 108, 239
Crack blunting, 163
Crack closure, 44
Crack detection, 56
Crack front, 44, 64
Crack initiation, 235, 258
Crack initiator, 159
Crack nucleation, 184
Crack opening, 163
Crack propagation, 101, 134, 137, 138, 152, 160, 161, 165, 168, 170, 174, 187
Crack tip, 45, 55, 163
Creep, 3, 107, 115, 177
Creep cavities, 182, 183
Creep curve, 116
Creep deformation, 115
Creep fracture, 116, 183
Creep rate, 115
Creep voids, 183
Critical resolved normal stress, 126
Critical resolved shear stress, 120, 121, 124, 126
Critical stress intensity factor, 111
Cross slip, 124
Crows-feet markings, 185
Crystalline, 107
Crystallographic twins, 128
Cup & cone fracture, 97, 106, 215, 217, 223, 225
Curling, 209

Datum plane, 40
Dealloying, 278
Debris, 31, 34, 46, 48, 50, 51, 58, 63, 66, 69, 203
Decarburization, 4, 309, 324
Decohesion, 141, 142, 153, 155, 156
Defects, 159
Deformation bands, 158, 275, 276, 279
Delamination, 4
Delta Ferrite, 337, 344, 345, 347
ΔK, 113
Dendrites, 179, 180, 306, 307, 344, 345
Depth of field, 14–17, 22

Depth of focus, 36
Design errors, 2, 3
Desiccant, 42, 45
Desiccator, 45
Detergent, 50, 58, 59, 305
Dezincification, 4, 278–280
Diamond pyramid hardness (DPH), 307, 310
Diffraction, 25, 26, 30, 32
Diffractometer, 26, 27
Dimple, 53, 139, 141, 143–146, 148, 149, 151–153, 155, 156, 178, 180, 181, 191, 192, 200–202, 248, 250–256, 258–262, 265, 301, 315, 325, 327, 330, 331
Dimple rupture, 179
Distortions, 36
Dosage, 4
Double prism viewers, 38
Drag, 290, 294
Drawing lines, 225
Ductile, 53, 81, 82, 84, 105, 108, 139, 214, 215, 217, 225, 226, 228, 229, 259
Ductile behavior, 107
Ductile cleavage, 148
Ductile crack, 161
Ductile fracture, 38, 97, 105, 106, 225
Ductile hairlines, 186
Ductile overload, 214
Ductile tearing, 53, 325
Ductility, 45, 77, 78, 107
Dust, 58
Dyes, 58

Energy dispersive spectrometry (EDS), 27, 28, 58, 69, 273–275, 277, 282, 284, 297, 302, 303, 329, 330, 336, 339, 342
Effluents, 312, 317
Elastic deformation, 3, 74
Elastic energy, 108–110
Elastic limit, 74
Elastic modulus, 77, 78, 109
Elastic strain energy, 110
Electrolyte, 51
Electromagnetic deflection, 36
Electron detector, 20, 21
Electron diffraction, 30, 32
Electron shell, 23
Electrostatic precipitator, 312, 313
Elongated dimples, 143, 146
Elongated inclusions, 302, 303
Elongated porosity, 284
Elongation, 78

Embrittlement, 4
Endox, 60, 62
Endurance limit, 295
Engineering strain, 76
Engineering stress, 75
Engineering stress-engineering strain diagram, 77, 78, 101, 109
Environmental damage, 43
Environmentally assisted fracture, 184
Equiaxed dimples, 144, 146
Erosion, 4
Escape distance, 21
Etching, 64
Eucentric-tilt, 37
Eutectic melting, 179
Eutectic structure, 285, 286
EX16 steel, 52
Exothermic reaction, 304
Extension rate, 214
Extraction replica, 58
Extrusions, 158, 159
Eyepiece, 8, 13

Fabrication, 3
Failure, 2
Failure analysis, 2
Failure mode, 57
Fan, 134
Fan blade, 280, 281
Fast fracture, 226
Fatigue, 56, 159, 188, 234, 235, 296, 301, 314
Fatigue crack, 158, 162, 163, 175, 226, 229, 233, 236, 237, 246, 250, 303, 316
Fatigue crack growth, 114, 115, 161
Fatigue crack propagation rate, 115
Fatigue cracking, 3, 44, 47
Fatigue curve, 114, 319
Fatigue damage, 65, 66, 69
Fatigue fracture, 63, 66, 107, 157, 164–167, 169–174, 232, 237–241, 243–245, 247, 248, 255, 261, 265, 303, 314–317
Fatigue fracture topography, 157
Fatigue limit, 113
Fatigue loading, 112, 157, 158
Fatigue marks, 242
Fatigue precrack, 50, 51
Fatigue rupture, 175
Fatigue strength, 113, 295, 317
Fatigue striations, 50, 63–65, 163–171, 173–175, 187, 188, 233, 250, 253, 255, 256, 258, 259, 261, 265, 301, 303, 315

Fatigue test, 113
Fatty acids, 58
Feeding, 312
Ferrite, 63, 264
Ferrite number, 344
Fiber stress, 320, 323, 325
Fibrous fracture, 106
Fibrous zone, 105, 215, 217–224, 252, 253, 258, 330
Filet weld, 338
Filter, 9
Fine topography, 119
Fins, 4
Fishmouth rupture, 116
Flame cutting, 55
Flat fracture, 97, 99, 101, 107
Fluorescent X-rays, 18, 23, 28, 29, 32, 34
Flutes, 187, 189
Flux, 279, 280
Flywheel flex plate, 296, 297, 302, 303
Fractography, 6, 8, 119
Fracture initiating site, 45
Fracture mechanics, 7, 56, 107, 112
Fracture mechanism, 5, 119
Fracture micromechanics, 119
Fracture mode, 5, 186, 213, 214
Fracture plane, 160
Fracture toughness, 50, 111, 112, 326
Freon, 49
Fretting fatigue, 3

Gage length, 75
Galling, 4
Glide plane decohesion, 148
Glossary, 369–399
Gouges, 4
Grain boundaries, 134–136, 148, 149, 153, 156, 182, 183, 185, 266, 324–326, 328, 333
Grain growth, 4
Grain size, 302, 365, 366
Granular, 107
Graphite, 60
Graphitization, 4
Gravitational force, 290
Grease, 49, 58, 62, 71
Grinding, 4
Growth, 45
Gum, 58

H11 steel, 263–265
Hairline patterns, 185
Handling fractures, 42, 43

Hardness, 6
Hardness conversions, 361–363
Heat-affected zone (HAZ), 4, 307–309
Heat treatment, 3
Herringbone pattern, 105, 107, 135, 136
Hexamethylene tetramine, 60, 71
High cycle axial fatigue, 249, 251, 255, 257, 259, 263, 266
High-temperature fracture, 177, 179
Hinge, 334
History, 5
Hot crack, 347
Hot short splits, 4
Hot tearing, 181
Howell & Boyde method, 39, 41
Human eye, 13
Hydrochloric acid, 53, 56, 60, 62, 71
Hydrogen, 51, 60
Hydrogen blisters, 62
Hydrogen embrittlement, 3, 156, 185–187, 327, 328, 334

Image, 9
Impact, 4, 47, 152, 330
Impact energy, 250
Impact fracture, 133, 264
Impact samples, 102, 248, 254
Impact test, 254, 258
Incipient melting, 180
Inclusions, 4, 50, 52, 143, 146, 201, 247, 302, 303, 309, 310, 329, 344
Index of refraction, 10
Induction hardening, 232, 247
Inhibitor compounds, 46
Inspection, 3, 56
Interdendritic porosity, 284, 285
Interference, 10
Intergranular, 275
Intergranular embrittlement, 334
Intergranular failure, 65
Intergranular fracture, 66, 148, 149, 153, 155–157, 171, 179, 180, 182–185, 205, 266–268, 330–332, 334
Intrusion, 158

Kalling's etch, 338
Ketones, 49, 58
Keyways, 246, 247
K_I, 111
K_t, 85, 87, 317

Lacquer, 55
Lateral shift method, 36, 38, 39

Leaching, 278
Lens error, 9
Lenticular silver screen, 39
Lift, 290
Lift coefficient, 294
Light gathering angle, 10
Lines of force, 84
Liquid metal embrittlement (LME), 44, 48
Liquid penetrants, 56
Litigation, 7
Load-elongation curve, 75
Loading, 74, 157, 239
Loading cycles, 233
Loading type, 241, 248
Local melting, 3, 179
Local tearing, 200
Low angle boundary, 134
Low cycle axial fatigue, 202, 255, 258
Lubricant, 2

Machining, 4
Macroexamination, 7, 45
Macrofractographic features, 248
Macrofractography, 120, 213, 214
Macromechanism, 214
Macroscopic orientation, 73
Magnetic gage, 344
Magnetic measurements, 337
Magnetic particle inspection, 56
Magnification, 13, 367, 368
Maintenance, 2, 3
Maraging steel, 180, 181
Martensite, 332, 334
Material selection, 3
Mating fractures, 42
Matte texture, 106
Maximum normal stress, 80–83, 85, 87, 88, 226
Maximum shear stress, 81, 82, 93, 95
Maximum stress, 257–259, 263
Mechanical advantage, 320
Mechanical damage, 43–55
Mechanical properties, 291–293, 318
Mechanical twins, 127, 128, 131–133, 135
Mechanics, 73
Melting, 190
Metallography, 45
Metric conversions, 357–360
Microanalysis, 45
Microchemical analysis, 28
Microcracks, 153, 181
Microexamination, 7, 45

Microfractographic features, 248
Microfractography, 119
Micropores, 185, 186
Microstructure, 8
Microvoid coalescence, 63–66
Microvoids, 65, 181
Minimum normal stress, 80, 81
Minimum stress, 257–259, 263
Mirror stereoscope, 38
Misuse, 2
Mixed mechanism, 151, 152, 327
Mixed mode, 153
Moment, 294, 323
Moment of inertia, 294, 323
Monochromatic beam, 9, 30
Mud cracks, 204, 205

Naptha, 47, 49
Necked region, 75
Necking, 75, 96, 98, 102, 107, 214, 216, 217, 223
Necks, 181
Nitric acid, 53, 61, 71
Nodular iron, 47
Nominal strain, 76
Nominal stress, 75, 79, 85
Nominal stress–nominal strain curve, 101, 109
Nondestructive testing, 6, 43, 56
Normal stress, 79, 80, 91, 124, 225
Normalized, 258, 260, 263
Notch, 85, 108, 141, 145, 220, 222–224, 238, 239, 258
Notch root, 87
Notch sensitivity, 317, 319
Notch severity, 239–241, 243, 248
Numerical aperture, 10–12

Objective lens, 8, 12, 13
Oil, 49, 58, 62, 71
Oil-immersion lens, 10
Optical microscope, 8, 16, 17, 22
Optical path, 17
Organic fiber brush, 48
Organic solvents, 49, 58, 71
Orthophosphoric acid, 53, 61, 71
Overload, 188
Overload fracture, 175
Oxalic acid, 339
Oxidation, 3, 44, 46, 48, 59, 178, 190, 235
Oxidation products, 71
Oxides, 58–60, 62, 177, 339
Oyster shell marks, 234

Parallax, 37
Pasty state, 179
Pearlite, 64, 69, 203, 258, 261, 325, 329, 332, 334
Pearlitic steel, 100
Peening, 247
Penetration, 4
Perspective, 36
Phase diagram, 286, 289
Phosphoric acid, 53, 61, 71
Photogrammetry, 39, 40
Photographic plane, 40
Photomultiplier, 20
Pinhole photographs, 11
Pitting, 315–317, 319
Planar slip, 188
Planck's constant, 25
Plane strain, 92, 94–96, 102, 103
Plane stress, 91–93, 102
Plastic deformation, 3, 74, 129, 139, 141, 153, 157, 158, 170, 181, 182, 219, 227, 228, 233, 277, 282, 297, 321, 330, 338
Plate fracture, 106
Plate thickness, 222
Plating, 4, 185, 327, 328–331, 334
Pliers, 56, 322
Pocket viewers, 38
Poisson's ratio, 89
Polarized method, 39
Polyvinyl alcohol, 33
Porosity, 4, 278, 283, 285, 289
Precipitation, 4
Precipitation hardening, 288, 289, 295
Preservation techniques, 43, 45
Primary cracks, 44, 56
Primary fracture, 55
Principal normal stress, 80, 82, 83
Principal planes, 80
Principal point, 40
Principal shear stresses, 80, 82, 83
Principal stresses, 79–81, 85, 230
Prior austenite grains, 309, 324–326, 328, 333
Processing, 4d
Projector lens, 15
Properties, 6

Quality control, 3
Quantitative stereoscopy, 39
Quasicleavage, 151–154, 186, 200, 201, 254, 264
Quench cracking, 4

Radial lines, 254
Radial marks, 103–105, 215, 218, 219, 222–224, 226, 230, 232, 248–250, 252–255, 258, 261, 264, 265, 282, 283, 305, 307, 341
Radial shear marks, 219
Radial zone, 217, 221–224
Radiation damage, 4
Railroad rails, 304
Rapid crack propagation, 113
Raster, 17
Ratchet marks, 237, 238, 242
Refraction, 9, 10
Reilly inhibitor, 71, 72
Replica, 7, 31, 32–34, 42, 48, 49, 58, 59, 61, 62, 134–136, 150, 158, 168, 172, 173, 184, 191, 195–203, 206–209, 339
Residual stress, 4
Resolution, 10, 12, 13, 15, 21
Resolved normal stress, 124, 126
Resolved shear stress, 121, 125, 126
Resolving power, 10, 13
Retained austenite, 4, 328, 329, 332–334
Reticulation, 207, 208
Reversed bending, 236, 237
Reversed bending fatigue, 242
Reversed torsion, 243–247
Ripple marks, 148, 150, 234
River pattern, 132, 135, 138, 152
Rock candy fracture, 330
Rodine # 50, 72
Rolling contact fatigue, 3
Rosettes, 151, 153, 154
Rotating beam fatigue, 253
Rotating bending, 236, 238–240, 242
Rotation method, 36
Rub marks, 204, 205
Rubbing, 176, 226, 297, 301, 338
Rupture, 156
Rupture life, 115
Rust, 31, 32, 34, 49, 52, 71, 341
Rust inhibitors, 55
Rustarest, 71, 72

SAE 1050 steel, 231
Safety factor, 295
Salt steam spray, 50
Saltwater, 335
Sample preparation, 31
Sand casting, 286
Scanning electron microscopy (SEM), 17, 19, 20, 22, 34
Scintillator, 20

Scraping, 204, 206
Season cracking, 279, 280
Secondary cracks, 43, 45, 55, 56, 138, 168–170, 174, 179, 202, 230, 253, 255, 325, 327
Secondary electrons, 18, 21
Secondary fluorescence, 29, 30
Secondary rupture, 132
Sectioning, 7, 45, 53, 55
Segregation, 4, 157
Seizing, 4, 63, 274, 277, 282, 284, 313, 325, 327, 329, 339
SEM fractographs, 190, 200
Serpentine glide, 148, 150
Service conditions, 4
Service damage, 3
Shadowing, 32
Shaft, 230, 232, 238, 240
Shallow dimples, 147, 151, 152
Shear, 3, 45
Shear dimples, 143–145, 201, 248, 261
Shear fracture, 217
Shear lip, 93, 102, 104, 107, 215, 217–222, 224, 228, 248, 249, 252–254
Shear loading, 143
Shear stress, 79, 80, 120, 124, 225
Shrinkage cavities, 180
Shrinkage porosity, 181
Silky texture, 106
Simulation, 6, 7
Single crystal, 120, 122, 140
Slag, 342
Slant fracture, 93, 94, 97, 99, 101, 102, 107
Slip, 120–123, 125–127, 129, 139, 140, 148, 163
Slip bands, 121, 135, 158, 163
Slip direction, 124
Slip lines, 121, 135, 139, 158
Slip planes, 121, 124, 148, 163
Slip system, 124
Slip traces, 124, 148, 149, 204
Sodium carbonate, 51, 60
Sodium cyanide, 51, 52, 60
Sodium hydroxide, 51, 53, 60, 61, 71
Soft brushes, 58
Solidification, 312
Solution treating, 286
Soot, 326
Span length, 294
Spherical aberration, 9
Splined shaft, 226, 244–246
Spring, 229

Stable crack, 45
Stable crack growth, 54, 101
Stage I of fatigue, 158–161
Stage II of fatigue, 159, 161, 163, 174, 175
Stage III of fatigue, 175
Stainless steel, 334
Stamping, 4
Steel punch, 328
Steps in failure analysis, 5
Stereo angle, 37, 38
Stereo image, 38
Stereo imaging, 36, 190
Stereomicroscopy, 31, 36, 45, 48, 59–61
Stereo pairs, 38–40, 191–194, 200–202, 332
Stereo projection, 39
Stereo viewer, 31
Stereopsis, 38
Stop marks, 234
Strain hardening, 75, 96
Strain rate, 100, 101, 151, 179
Stress-strain diagram, 76–78
Stress analysis, 5, 290
Stress concentration, 45, 84, 108, 112, 233, 240
Stress concentration factor, 85, 86, 317
Stress concentrator, 87, 347
Stress corrosion cracking (SCC), 4, 44, 48, 69, 157, 184, 185, 187–189, 205, 279, 280
Stress intensity, 85
Stress intensity factor, 64
Stress level, 236, 237, 239–241, 243, 248, 321
Stress raisers, 3, 87
Stretching, 148, 201, 204
Striation (*see* Fatigue striations)
Striation spacing, 168
Stringers, 192, 310
Subcritical crack growth, 113
Subgrains, 136
Subjects, 8
Sulfamic acid, 71
Sulfuric acid, 51
Surface coating, 45, 46
Surface energy, 110
Swirl pattern, 226, 227, 261

Tear, 131, 132, 188, 207
Tear dimples, 143, 145
Tear loading, 143
Tear ridges, 132, 151–154

408 Index

Tearing, 3, 4, 153, 155, 204
Tearing topography surface, 153, 155, 156
Tectyl 506, 46–48
TEM fractographs, 190
Temper embrittlement, 157, 328
Temperature, 100
Temperature conversions, 351–355
Tempering, 256, 259, 261, 262, 326
Tempered martensite embrittlement (TME), 328
Tensile bar, 97
Tensile ductility, 295
Tensile fracture, 84, 95, 225, 267, 268
Tensile loading, 82, 160
Tensile overload, 214, 223, 248
Tensile sample, 75, 103, 105, 252, 253, 266
Tensile strength, 76, 77, 282, 317
Tensile stress, 158, 160, 163, 164
Tensile tearing, 144
Tensile test, 56, 74, 100, 101, 107, 108, 112, 157, 193, 252, 254, 256, 260, 263
Tension, 63, 82, 83, 249
Tension overload, 202
Thermal cycling, 3
Thermal expansion, 327
Thermal shock, 4
Thermal stress, 312
Thermite process, 304, 305
Thin sheet, 100
Tide marks, 234
Tilt boundary, 138
Tilt method, 36–38, 40
Tire tracks, 171–173, 204
Toluene, 49, 58, 62
Tongue, 133, 137, 138
Topology, 7
Torsion, 45, 82, 83, 143, 145, 261
Torsion deformation bands, 225
Torsion fatigue, 247
Torsion overload, 225–229
Torsion test, 262
Torsional dimples, 145
Toughness, 6, 113
Tramp elements, 347
Transformation stress, 312
Transgranular cleavage, 261
Transgranular fracture, 171, 185, 187, 275, 279
Transition temperature, 151
Transmission, 296

Transmission electron microscopy (TEM), 15–17, 22
Triaxial stress, 88–90, 96, 228
Trichloroethylene, 49
True strain, 78
True stress, 78
True stress-true strain curve, 78, 100
Twin, 131, 133, 134, 137, 138
Twin plane, 133
Twinning, 120, 127
Twist boundary, 138

Ultimate strength, 76
Ultimate tensile strength, 76, 323
Ultrasonic cleaning, 31, 46, 49, 50–52, 58–60, 62, 305, 339
Underbead cracking, 4
Undercuts, 4
Unstable crack growth, 63, 102, 113

Vacuum bellows, 271, 272, 276, 278
Vacuum deposition, 33
Valence, 23
Varnish, 58
Vibration, 4, 295, 304, 317
Virtual image, 13
Visual examination, 5, 45
Void coalescence, 5, 105, 139–141, 143, 147, 151, 152, 155, 175, 177, 179, 181, 248–250, 255, 261, 263–265, 301, 325
Void nucleation, 142, 143
Voids, 146, 182

Wallner lines, 203
Warping, 3
Water quenching, 286
Wavelength dispersive spectroscopy (WDS, WDXA), 25
Wear, 4, 189
Web, 307
Wedge cracks, 182, 183
Weld, 45, 137, 143, 308–312, 338, 342
Weld metal, 309, 347
Weld root, 346
Welding, 4, 304
Width-to-length ratio, 221
Width-to-thickness ratio, 99
Wire cutters, 319–322
Wires, 312
Woody fracture, 146, 147
Work hardening, 75, 96
Working, 4

X-ray analysis, 66
X-ray diffraction, 30
X-ray emission, 24
X-ray photon, 24
X-ray spectrum, 27–29
Xylene, 49, 58

Yield point, 76, 186
Yield point behavior, 77
Yield point stress, 76
Yield strength, 6, 76, 77, 107, 112, 115, 157, 323, 325
Young's modulus, 78

ABOUT THE AUTHORS

CHARLIE R. BROOKS is a professor of metallurgical engineering at the University of Tennessee, Knoxville. Professor Brooks received his B.S. in chemical engineering, and his M.S. and Ph.D. in metallurgical engineering from the University of Tennessee. He has received numerous teaching awards, including the Tennessee Tomorrow Professor award, Outstanding Teacher of the Materials Science and Engineering Department, and a Nancy and Leon Cole Superior Teaching Award of the College of Engineering. He has authored or coauthored 90 research papers and written four other books.

ASHOK CHOUDHURY has a Bachelor of Technology degree from the Indian Institute of Technology—Kharagpur and M.S. and Ph.D. in metallurgical engineering from the University of Tennessee, Knoxville. He is a development staff member at the Metals and Ceramics Division of the Oak Ridge National Laboratory in Oak Ridge, Tennessee.